16.00

D1083620

# THE LORENTZ GROUP AND
# HARMONIC ANALYSIS

# The Mathematical Physics Monograph Series

A. S. Wightman, Editor
*Princeton University*

Ralph Abraham
*Princeton University*     FOUNDATIONS OF MECHANICS

Vladimir I. Arnold
*University of Moscow*
André Avez
*University of Paris*     ERGODIC PROBLEMS OF
CLASSICAL MECHANICS

Freeman J. Dyson
*The Institute for*
*Advanced Study*     SYMMETRY GROUPS IN NUCLEAR
AND PARTICLE PHYSICS

Robert Hermann
*Argonne National*
*Laboratory*     LIE GROUPS FOR PHYSICISTS

Rudolph C. Hwa
*State University of*
*New York at Stony Brook*
Vigdor L. Teplitz
*Massachusetts Institute*     HOMOLOGY AND FEYNMAN
*of Technology*     INTEGRALS

John R. Klauder
E. C. G. Sudarshan     FUNDAMENTALS OF
*Syracuse University*     QUANTUM OPTICS

André Lichnerowicz     RELATIVISTIC HYDRODYNAMICS
*College de France*     AND MAGNETOHYDRODYNAMICS

George W. Mackey     THE MATHEMATICAL FOUNDATIONS
*Harvard University*     OF QUANTUM MECHANICS

Roger G. Newton
*Indiana University*     THE COMPLEX $j$-PLANE

David Ruelle
*Institut des Hautes*
*Etudes Scientifiques*     STATISTICAL MECHANICS

W. Rühl     THE LORENTZ GROUP
*CERN, Geneva*     AND HARMONIC ANALYSIS

R. F. Streater
*Imperial College of*
*Science and Technology*
A. S. Wightman     PCT, SPIN AND STATISTICS,
*Princeton University*     AND ALL THAT

James D. Talman
*The University of*
*Western Ontario*     SPECIAL FUNCTIONS

René Thom
*Institut des Hautes*     STABILITE STRUCTURELLE ET
*Etudes Scientifiques*     MORPHOGENESE

# THE LORENTZ GROUP AND HARMONIC ANALYSIS

**W. RÜHL**

**CERN, Geneva**

*W. A. Benjamin, Inc.*

*New York*

**1970**

**THE LORENTZ GROUP AND HARMONIC ANALYSIS**

Standard Book Number: 8053-8362-10
Library of Congress Catalog Card Number 70-91509
Manufactured in the United States of America
12345R32109

*The manuscript was put into production on June 1, 1969;
this volume was published on February 1, 1970.*

W. A. BENJAMIN, INC.
New York, New York 10016

# Preface

The present book is based on a series of lectures given by the author before an audience of physicists at the Theoretical Study Division at CERN during the winter of 1967–1968. In this course we intended to outline the concepts and methods used in the theory of representations of the inhomogeneous Lorentz group and its subgroups as they are applied in present-day elementary particle physics. The applications we had in mind in particular are the theories of Lorentz invariant equations and Regge and Toller poles. Naimark's classic textbook on representations of the Lorentz group already contains an extensive treatment of invariant equations and their main problem, the construction of covariant operators (generalized Dirac matrices). Instead of repeating Naimark's results, we preferred to look at the problem of covariant operators from another point of view, namely, we stressed their relation with trilinear invariant functionals. Such functionals can be constructed with the same elementary algorithm as used for bilinear invariant functionals in these lectures. Moreover they might be of special interest to physicists who are used to deriving their invariant equations from invariant Lagrange functions. On the other hand, a full chapter is devoted to Regge and Toller poles. This gives us the opportunity to apply our formalism of Fourier transformations of distributions on the groups $SU(1, 1)$ and $SL(2, C)$ to a physical problem. We believe, however, that the theory of invariant equations could also be pushed forward considerably by the use of such techniques. So far only a rather limited application appears in the literature [28].

This textbook is primarily intended to introduce graduate students of theoretical physics to this subject, and to serve as a source of reference for physicists working actively in this field. For this reason we have tried to present many of our formulas in a standard form, for example, we have chosen the definition and phase convention of Edmonds and of Andrews and Gunson, respectively, for the representation functions of $SU(2)$ and $SU(1, 1)$. Besides a general knowledge of calculus, a fair acquaintance with the basic definitions and techniques of the theory of distributions is presupposed, as presented, for example, in the first volume of Gelfand and Shulov's series on generalized functions. We have used only the most elementary notions of all the other subjects which are unavoidably touched upon in a course like this: abstract groups and algebras, linear topological space, Banach and Hilbert spaces, and special functions.

The material is divided into chapters and sections. At the end of each chapter (in the case of the first chapter, at the end of each section) the reader can find some "remarks" which contain references. All the references have been accumulated at the end of the book, but our list is, of course, not intended to be complete. Sometimes we have to evaluate integrals analytically. Whenever possible we refer to the integral tables of Gradshteyn and Ryzhik. A formula of this table, say formula 8.703, is referred to as *GR* 8.073 in the text.

If an author who counts himself among the physicists writes a textbook on a mathematical subject, he always feels obliged to apologize: to the mathematicians, because every page of the manuscript proves his limited knowledge of general mathematics; to the physicists, because he is not always in a position to convince them that the notions he uses can perhaps one day be thought of in physical terms. In a situation like this, it is most comfortable for an author if he is able to appeal to a great master. We therefore take this opportunity to pay reverence to A. Sommerfeld who would have celebrated his hundredth birthday this year. However, the author calls on A. Sommerfeld not only because of this occasion but also because Sommerfeld's book on "Partial Differential Equations of Physics" first propagated the *eigentliche Eigenschwingungen*, which later became known as Regge poles, and which in this book reappear in the framework of analytic functionals on the group SU(1,1).

Those readers who are acquainted with the original literature in this field will notice the extent to which the author is indebted to the work of Gelfand and his school. We can only hope that in the course of adopting his ideas, not too much of the beauty and lucidity of Gelfand's own presentation was lost. Finally, we would like to thank our audience at CERN, whose patience stimulated the preparation of this manuscript, and in particular Dr. G. Sommer for several critical comments.

W. Rühl

CERN, Geneva
September 1968

# Contents

# THE LORENTZ GROUP AND
# HARMONIC ANALYSIS

# Chapter 1

# Definitions, Notations, and Concepts of Representation Theory

In the first chapter we present a conglomerate of many different things, most of which serve for later reference. We define the inhomogeneous Lorentz group, or as we shall call it the Poincaré group, by two fundamental matrix representations. Then we introduce several abstract mathematical notions such as functions on groups, the invariant measure or Haar measure on a group, and group representations. Finally, we give a few theorems and results which remain unproved in this course. These theorems illustrate the abstract mathematical background for problems to be attacked in later chapters. Their main purpose is, however, to guarantee the uniqueness of the solutions found for these problems by explicit construction.

## 1–1  BASIC PROPERTIES OF THE POINCARÉ GROUP

In relativistic mechanics and classical electrodynamics the homogeneous Lorentz group is, in general, regarded to be represented by real orthogonal 4 × 4 matrices. Physical quantities such as the position in space-time, the energy-momentum, and the electromagnetic field transform like vectors or tensors under this group. After the discovery of the Dirac equation physicists learned of another basic representation of the homogeneous Lorentz group, namely, the group of complex unimodular 2 × 2 matrices. We introduce the homogeneous Lorentz group by means of both matrix representations, describe their relations by explicit formulas, and study some elementary algebraic properties of these matrix groups.

### a.  Definition of the Poincaré Group

We consider a linear four-dimensional space of real vectors

$$x = (x^0, x^1, x^2, x^3)$$

1

with the scalar product

$$xy = x^0 y^0 - x^1 y^1 - x^2 y^2 - x^3 y^3 \equiv g_{\mu\nu} x^\mu y^\nu \tag{1-1}$$

(Here and in the sequel we make use of the summation convention for repeated indices.) Such a space is called a Minkowski space. We call $xx = (x)^2$ the length squared of the vector $x$. A vector is denoted timelike if its length squared is positive, spacelike if it is negative, and lightlike or isotropic if the length vanishes. The manifold of all lightlike vectors forms the light cone.

Linear transformations in Minkowski space defined by

$$\begin{aligned} x \rightarrow x' &= \Lambda x + y \\ x'^\mu &= \Lambda_\nu{}^\mu x^\nu + y^\mu \end{aligned} \tag{1-2}$$

which satisfy

$$g_{\mu\nu} \Lambda_\rho{}^\mu \Lambda_\sigma{}^\nu = g_{\rho\sigma} \tag{1-3}$$

are called inhomogeneous Lorentz transformations. We denote them by the symbol $(\Lambda, y)$. The homogeneous Lorentz transformations $(\Lambda, 0)$ leave the scalar product invariant

$$x'y' = g_{\mu\nu} x'^\mu y'^\nu = g_{\mu\nu} \Lambda_\rho{}^\mu \Lambda_\sigma{}^\nu x^\rho y^\sigma = xy$$

They can be characterized by this property.

The inhomogeneous Lorentz transformations form a group, the Poincaré group P. The multiplication law of this group can be written as

$$(\Lambda_1, y_1)(\Lambda_2, y_2) = (\Lambda_1 \Lambda_2, y_1 + \Lambda_1 y_2) \tag{1-4}$$

The homogeneous Lorentz transformations constitute the Lorentz group L. The group of translations is denoted $T_4$ and we write

$$P = L \times T_4$$

By the symbol $C = A \times B$ we mean a semidirect product, that is, A is a subgroup and B is an invariant subgroup of C. the Poincaré group is a Lie group. All groups which we are going to study are Lie groups.

In the natural topology of matrices the Lorentz group L possesses four disconnected pieces. The relation

$$g_{\mu\nu} \Lambda_0{}^\mu \Lambda_0{}^\nu = 1$$

following from (1–3) implies the inequality

$$(\Lambda_0{}^0)^2 = 1 + \sum_{k=1}^{3} (\Lambda_0{}^k)^2 \geq 1$$

Consequently we have either

$$\Lambda_0{}^0 \geq 1 \quad \text{or} \quad \Lambda_0{}^0 \leq -1$$

The four pieces of L are assigned as follows:

$L_+^\uparrow$: det $\Lambda = +1$, sign $\Lambda_0{}^0 = +1$. This piece contains the identity $I$ and is itself a group. We call it the proper, orthochronous Lorentz group.

$L_-^\uparrow$: det $\Lambda = -1$, sign $\Lambda_0{}^0 = +1$. This manifold contains an element $I_s$ defined by

$$I_s x = (x^0, -x^1, -x^2, -x^3)$$

$L_-^\downarrow$: det $\Lambda = -1$, sign $\Lambda_0{}^0 = -1$. This manifold contains an element $I_t$ defined by

$$I_t x = (-x^0, x^1, x^2, x^3)$$

$L_+^\downarrow$: det $\Lambda = +1$, sign $\Lambda_0{}^0 = -1$. This manifold contains the element $I_{st} = I_s I_t$.

$I_s$ is called a space inversion and $I_t$ a time reversal operation.

From these four pieces we can build the following subgroups:

the orthochronous Lorentz group     $L^\uparrow = L_+^\uparrow \cup L_-^\uparrow$,
the proper Lorentz group     $L_+ = L_+^\uparrow \cup L_+^\downarrow$,
the orthochorous Lorentz group     $L_0 = L_+^\uparrow \cup L_-^\downarrow$.

The proper orthochronous Lorentz group is identical with the group $SO(3, 1)$; it is, moreover, isomorphic to the group $SL(2, C)/Z_2$. The latter assertion is proved and explained in Section 1–1b.

Instead of the contravariant vectors $x^\mu$ we can also use covariant vectors $x'_\mu$. Both types of vectors can be transformed into each other by means of the metric matrix $g_{\mu\nu}$.

$$x'_\mu = g_{\mu\nu} x^\nu$$

b.    *The Group SL(2, C)*

The group $SL(2, C)$ is defined to consist of all complex $2 \times 2$ matrices with determinant one,

$$a = \begin{pmatrix} a_{11} & a_{12} \\ a_{21} & a_{22} \end{pmatrix} \qquad a_{ik} \text{ complex}$$

$$\det a = a_{11} a_{22} - a_{12} a_{21} = 1$$

In the natural topology of matrices this group is simply connected. It is often very convenient to parametrize an element $a \in SL(2, C)$ by

$$a = a_0 e + \sum_{k=1}^{3} a_k \sigma_k$$

where $\sigma_k$ are Pauli matrices,

$$\sigma_1 = \begin{pmatrix} 0 & 1 \\ 1 & 0 \end{pmatrix} \qquad \sigma_2 = \begin{pmatrix} 0 & -i \\ i & 0 \end{pmatrix} \qquad \sigma_3 = \begin{pmatrix} 1 & 0 \\ 0 & -1 \end{pmatrix}$$

and $a_0, a_k$, $k = 1, 2, 3$, are complex numbers. The term $e$ is the unit matrix.

Any point of Minkowski space can be represented by a Hermitian $2 \times 2$ matrix

$$x \rightarrow \mathbf{x} = x^0 e + \sum_{k=1}^{3} x^k \sigma_k \tag{1-5}$$

We can form a linear real vector space with these Hermitian matrices as well. By means of an element $a \in SL(2, C)$ we define a linear mapping of this space onto itself by

$$\mathbf{x}' = a\mathbf{x}a^\dagger \tag{1-6}$$

The corresponding operation in Minkowski space is a linear operation and preserves the scalar product since

$$(x')^2 = \det \mathbf{x}' = \det \mathbf{x} = (x)^2$$

We write the correspondence as

$$\mathbf{x}' = a\mathbf{x}a^\dagger \rightarrow x' = \Lambda(a)x \tag{1-7}$$

The explicit expression of the orthogonal transformation $\Lambda(a)$ in terms of the parameters $a_0, a_k$ of $a$ is (the $\varepsilon$ symbols are fixed by $\varepsilon^{123} = \varepsilon^{0123} = +1$)

$$\Lambda_0{}^0 = |a_0|^2 + \sum_{k=1}^{3} |a_k|^2$$

$$\Lambda_0{}^k = a_0 \bar{a}_k + \bar{a}_0 a_k + i\varepsilon^{klm} a_l \bar{a}_m$$

$$\Lambda_0{}^k = a_0 \bar{a}_k + \bar{a}_0 a_k - i\varepsilon^{klm} a_l \bar{a}_m \tag{1-8}$$

$$\Lambda_k{}^l = \delta_k{}^l \left( |a_0|^2 - \sum_{t=1}^{3} |a_t|^2 \right) + a_k \bar{a}_l + \bar{a}_k a_l$$

$$+ i\varepsilon^{klm} (\bar{a}_0 a_m - a_0 \bar{a}_m)$$

We note in particular the useful relation

$$\Lambda_\nu{}^\nu(a) = \mathrm{Tr}\,\Lambda(a) = |\mathrm{Tr}\,a|^2 = 4|a_0|^2$$

Because SL(2, C) is connected and the mapping into L is a continuous homomorphism, the image of SL(2, C) must be a subgroup of $L_+^\uparrow$.

Now we give an element $\Lambda \in L_+^\uparrow$ and try to invert the relations (1–8). If

$$\Lambda_\nu{}^\nu = 4|a_0|^2 \neq 0$$

we obtain

$$a_0 e + \sum_{k=1}^{3} a_k \sigma_k = D^{-1}\left\{\Lambda_\nu{}^\nu e + \sum_{k=1}^{3}(\Lambda_0{}^k + \Lambda_k{}^0 + i\varepsilon_\tau^{0k\rho}\Lambda_\rho{}^\tau)\sigma_k\right\} \qquad (1\text{-}9)$$

with

$$D^2 = 4 - \mathrm{Tr}\,(\Lambda\Lambda) + (\mathrm{Tr}\,\Lambda)^2 + i\varepsilon_{\rho\tau}^{\mu\lambda}\Lambda_\mu{}^\rho\Lambda_\lambda{}^\tau \qquad (1\text{-}10)$$

The sign of the denominator $D$ is undetermined. Since the smallest subgroup of $L_+^\uparrow$ that contains all elements with $\mathrm{Tr}\Lambda \neq 0$ is $L_+^\uparrow$ itself, the image of SL(2, C) is the whole of $L_+^\uparrow$.

It is easy to find the elements $a \in$ SL(2, C) which go into $L_+^\uparrow$ in the case $\mathrm{Tr}\Lambda$ also. If

$$\sum_{k=1}^{3}(\Lambda_0{}^k)^2 \neq 0$$

$\Lambda$ describes a rotation with an angle $\pi$, as can be inspected from

$$a_0 = 0 \qquad a = \sum_{k=1}^{3} a_k \sigma_k \qquad aa = \sum_{k=1}^{3} a_k{}^2 e = -e$$

$$\Lambda(a)\Lambda(a) = \Lambda(aa) = \Lambda(-e) = I$$

The 3 ×3 matrix

$$M_k{}^l = \delta_k{}^l \Lambda_0{}^0 + \Lambda_k{}^l$$

is symmetric and possesses the eigenvalues

$$0 \qquad \Lambda_0{}^0 - 1 \qquad \Lambda_0{}^0 + 1$$

The eigenvalue zero belongs to the normalized eigenvector $\xi_0$ defined by

$$(\xi_0)^k = \left[\sum_{l=1}^{3} (\Lambda_0{}^l)^2\right]^{-1/2} \Lambda_0{}^k$$

We denote the normalized real eigenvector corresponding to the eigenvalue $\Lambda_0{}^0 + 1$ by $\xi_1$. We can then express the components $a_k$ in terms of the vectors $\xi_1$ and $\xi_1 \times \xi_0$ in the fashion

$$a_k = \pm\{[\tfrac{1}{2}(\Lambda_0{}^0 - 1)]^{1/2}(\xi_1 \times \xi_0)^k + i[\tfrac{1}{2}(\Lambda_0{}^0 + 1)]^{1/2}(\xi_1)^k\} \qquad (1\text{-}11)$$

Again $a$ is determined only up to a sign. The remaining case

$$\Lambda_v{}^v = 0 \qquad \Lambda_0{}^k = 0 \qquad \text{for all} \quad k = 1, 2, 3$$

is contained in (1–11) as the limit $\Lambda_0{}^0 = 1$,

$$a_k = \pm i(\xi_1)^k$$

The sign ambiguity of $a = a(\Lambda)$ means in particular that the unit $I$ of $L_+^\uparrow$ is the image of both central elements $e_\pm$ of SL(2, C)

$$e_\pm = \pm e$$

(group elements are denoted central if they commute with all group elements; central elements form the center of the group). Our explicit calculation establishes, therefore, the isomorphism

$$L_+^\uparrow \cong \text{SL(2, C)}/Z_2$$

where $Z_2$ denotes the center of SL(2, C) consisting of the elements $e_\pm$.

The proof that the kernel of the homomorphism SL(2, C) $\to L_+^\uparrow$ coincides with the center of SL(2, C) can be accomplished also in the following simple manner. By definition the element $a$ is in the kernel of the homomorphism if for all Hermitian matrices x

$$\mathbf{x} = a\mathbf{x}a^\dagger$$

Putting x = $e$ we get

$$a^\dagger = a^{-1}$$

which allows us to rewrite the condition on $a$ as

$$x a = a x$$

Inserting for $x$ the Pauli matrices and remembering that $\det a = 1$ leaves the only solutions $e_\pm$.

### c. Complex Rotations

It is sometimes convenient to consider the group $SL(2, C)$ as a complex form of the rotation group $SU(2)$ in the sense that the real parameters of $SU(2)$ are extended into the complex domain. Since an element of the rotation group can be characterized by a real three-vector, the rotation vector, whose length determines the rotation angle, one way of complexifying $SU(2)$ is to introduce a complex rotation vector and a complex rotation angle. It turns out that in this fashion we get all elements of $SL(2, C)$ except a small set of singular elements.

As in the case of the rotation group $SU(2)$ the complex rotation angle is connected with the eigenvalues of the matrix $a \in SL(2, C)$ or the matrix $\Lambda(a) \in L_+^\uparrow$. We may define it by

$$a_0 = \cosh \tfrac{1}{2}(\eta + i\varphi) \qquad \eta \geq 0$$

If $\eta \neq 0$ or $\varphi \not\equiv 0 \bmod 2\pi$, the eigenvalues of $a$ are different from each other and $a$ can be diagonalized

$$a = a_1 \, \delta a_1^{-1}$$

$$\delta = \begin{pmatrix} e^{(\eta + i\varphi)/2} & 0 \\ 0 & e^{-(\eta + i\varphi)/2} \end{pmatrix} \tag{1-12}$$

(we want to reserve the notation $d$ for matrices $\delta$ with $\varphi = 0$). The corresponding decomposition of $\Lambda$ is

$$\Lambda(a) = \Lambda(a_1)\Lambda(\delta)\Lambda(a_1)^{-1}$$

$$\Lambda(\delta) = \begin{pmatrix} \cosh \eta & 0 & 0 & \sinh \eta \\ 0 & \cos \varphi & \sin \varphi & 0 \\ 0 & -\sin \varphi & \cos \varphi & 0 \\ \sinh \eta & 0 & 0 & \cosh \eta \end{pmatrix} \tag{1-13}$$

The eigenvalues of the matrix $\Lambda$ are therefore $\exp(\pm\eta)$ and $\exp(\pm i\varphi)$. $\Lambda$ has in general two complex conjugate spacelike eigenvectors [for complex vectors spacelike means also $(x)^2 < 0$] and two real isotropic eigenvectors. The matrix $\Lambda(\delta)$ can be considered a normal form of $\Lambda$.

When $\eta = 0$, $\varphi \not\equiv 0 \bmod 2\pi$, we call $\Lambda$ a pure rotation. On the other hand, we call $\Lambda$ a pure Lorentz transformation when $\varphi \cong 0 \bmod 2\pi$ and $\eta \neq 0$. Matrices that can be brought to a normal form $\Lambda(\delta)$ as in $(1-13)$ form a class $\gamma(\eta, \varphi)$ of equivalent elements. Each of these classes belongs to a unique complex rotation angle if we restrict $\varphi$ to $0 \leq \varphi < 2\pi$. One class $\gamma(\eta, \varphi)$ of elements of $\Lambda$ corresponds to two classes $\gamma(\eta, \varphi)$ and $\gamma(\eta, \varphi + 2\pi)$ of elements of $SL(2, C)$.

However, if $\eta = 0$ and $\varphi \cong 0 \bmod 2\pi$, an additional possibility occurs. It may happen that $a$ cannot be diagonalized but only brought to one of the normal forms.

$$\pm \begin{pmatrix} 1 & 1 \\ 0 & 1 \end{pmatrix}$$

We denote these two exceptional classes of $SL(2, C)$ by $\gamma(exc^\pm)$ and the one corresponding class of $L_+^\uparrow$ by $\gamma(exc)$.

We make the ansatz

$$a = \exp \frac{i}{2} \sum_{k=1}^{3} \beta_k \sigma_k \qquad (1\text{-}14)$$

and define $\beta = (\beta_1, \beta_2, \beta_3)$ to be the complex rotation vector. For the subgroup $SU(2)$ of $SL(2, C)$ this vector is real. We relate the length of this rotation vector and the rotation angle by

$$\sum_{k=1}^{3} \beta_k{}^2 = -(\eta + i\varphi)^2$$

If $\eta + i\varphi \neq 0$, we find

$$a_0 = \cosh \tfrac{1}{2}(\eta + i\varphi)$$

$$a_k = i\beta_k \frac{\sinh (1/2)(\eta + i\varphi)}{\eta + i\varphi} \qquad (1\text{-}15)$$

whereas

$$\eta + i\varphi = 0$$

yields

$$a_0 = +1 \qquad a_k = \frac{i}{2}\beta_k$$

The condition

$$\sum_{k=1}^{3} \beta_k^2 = 0$$

implies therefore that $a$ belongs to $\gamma(0, 0)$ if in addition

$$\sum_{k=1}^{3} |\beta_k|^2 = 0$$

but to $\gamma(\text{exc}^+)$ if

$$\sum_{k=1}^{3} |\beta_k|^2 > 0$$

We notice that the class $\gamma(\text{exc}^-)$ does not occur. If we solve (1–15) in turn for the vector $\beta$ we obtain

$$\frac{i}{2} \beta_k = a_k \frac{\log[a_0 + (a_0^2 - 1)^{1/2}]}{(a_0^2 - 1)^{1/2}} \tag{1-16}$$

where the logarithm is defined on its principal sheet. With this definition the complex rotation vector $\beta$ is holomorphic in $a_0$ except for a cut extending from $-1$ to $-\infty$. The branch point corresponds to the class $\gamma(\text{exc}^-)$. The real interval $-1 \leqq a_0 \leqq +1$ and the imaginary $a_k$ are taken on by the group SU(2). Apart from the cut the whole group SL(2, C) can therefore be parametrized uniquely by complex rotation vectors.

From this discussion we should particularly keep in mind that in general not all elements of a group of matrices can be reached by exponentiation, that is, expressed by an ansatz

$$a = \exp i\beta$$

where $\beta$ is an element of the Lie algebra. We shall learn of a complexification of SU(2) which does not lead to singularities in Section 2–4.

### d. Subgroups of SL(2, C)

The group SL(2, C) possesses a series of important subgroups, some of which play crucial roles in our further investigations. These subgroups correspond to subgroups of $L_+^\uparrow$ as well. But because of the physical equivalence of $L_+^\uparrow$ and SL(2, C) (see our remarks in Section 1–1e) and since SL(2, C) is simpler to handle, we prefer to deal with the group SL(2, C) $\times$ $T_4$, which we shall call the inhomogeneous SL(2, C), and its subgroups.

The group SU(2) was already mentioned in Section 1–1c. It consists of those elements $u \in SL(2, C)$ which are, moreover, unitary

$$u^\dagger = u^{-1} \tag{1-17}$$

One convenient possibility of parametrizing SU(2) is

$$u = u_0 e + i \sum_{k=1}^{3} u_k \sigma_k$$

$$u_0^2 + \sum_{k=1}^{3} u_k^2 = 1$$

where $u_0$ and $u_k$, $k = 1, 2, 3$, are real numbers. Other parametrizations will be introduced later.

The group SU(1, 1) contains those elements $v \in SL(2, C)$ which satisfy

$$v^\dagger \sigma_3 = \sigma_3 v^{-1} \tag{1-18}$$

We parametrize them in the fashion

$$v = v_0 e + v_1 \sigma_1 + v_2 \sigma_2 + i v_3 \sigma_3$$

$$v_0^2 - v_1^2 - v_2^2 + v_3^2 = 1$$

The numbers $v_0$ and $v_k$, $k = 1, 2, 3$, are real.

The group SL(2, R) consists of those elements of SL(2, C) which are submitted to the constraint

$$a^\dagger \sigma_2 = \sigma_2 a^{-1}$$

They can be represented as

$$a = a_0 e + a_1 \sigma_1 + i a_2 \sigma_2 + a_3 \sigma_3$$

with real parameters $a_0$ and $a_k$, $k = 1, 2, 3$. The matrix $a$ is a real $2 \times 2$ matrix. By a rotation

$$\exp\left\{i \frac{\pi}{4} \sigma_1\right\} = 2^{-1/2}(e + i\sigma_1)$$

in the 2, 3-plane we can map SU(1, 1) on SL(2, R)

$$a = \exp\left\{-i \frac{\pi}{4} \sigma_1\right\} v \exp\left\{i \frac{\pi}{4} \sigma_1\right\} \tag{1-19}$$

$$a_0 = v_0 \qquad a_1 = v_1 \qquad a_2 = -v_3 \qquad a_3 = v_2$$

Later we shall refer to this one-to-one mapping of the two matrix groups SU(1, 1) and SL(2, R) onto each other as the standard isomorphism.

The group of triangular matrices

$$(\varphi, \mu) \equiv \begin{pmatrix} e^{-i(\varphi/2)} & 0 \\ \mu e^{-i(\varphi/2)} & e^{i(\varphi/2)} \end{pmatrix}$$

$$\mu \text{ complex} \qquad 0 \leqq \varphi < 4\pi$$

with the multiplication law

$$(\varphi_1, \mu_1)(\varphi_2, \mu_2) = (\varphi_1 + \varphi_2(\pm 4\pi), \mu_1 + e^{i\varphi_1}\mu_2)$$

is isomorphic to the group of Euclidian motions on the Riemann plane of the function $z^{1/2}$. The corresponding subgroup of $L_+^\uparrow$ is isomorphic to the group of motions in the complex $z$-plane itself. We denote this group $U(1) \times T_2$ if we mean the subgroup of $L_+^\uparrow$ and $U(1)' \times T_2$ the subgroup of SL(2, C).

### e.  Remarks

The Poincaré group P is the basic invariance group of quantum physics. The first systematic study of its properties in this context goes back to Wigner [37]. He proved that only its continuous unitary representations up to a factor need to be considered in quantum physics. He showed further that this amounts to an investigation of the continuous unitary representations of the universal covering group SL(2, C) $\times T_4$ of P. Bargmann [6] investigated representations up to a factor of the Galilean group in the same context. A detailed discussion of relativistic invariance can be found in Wightman [36]. A less extensive presentation is contained in Streater and Wightman [31]. The algebraic formula expressing $a \in SL(2, C)$ by $\Lambda \in L_+^\uparrow$ appeared first in Joos [18].

## 1–2   FUNCTIONS ON GROUPS

Functions on the real line can be submitted to the following operations. We may exert translations on them

$$f(t) \rightarrow T_{t_1} f(t) = f(t - t_1)$$

and integrate them in a translation invariant manner

$$f(t) \rightarrow I_f = \int f(t)\, dt$$

provided they are integrable. In this course we understand integrability on a noncompact manifold always in a sense which implies absolute integrability. A pair of integrable functions can be convoluted

$$f_1 \cdot f_2(t) = \int f_1(t_1) f_2(t - t_1)\, dt_1$$

This convolution product is of importance in the theory of Fourier transformations on the real line. We generalize these operations now for functions defined on Lie groups (which are always tacitly assumed to have finite dimension) and consider some linear spaces of such functions.

*a.   Translations*

We can define real or complex-valued functions on any group G. If this group is a topological group, we can consider continuous functions; if it is a Lie group, we can study differentiable or analytic functions. Sets of functions on a group can be used to construct linear vector spaces. If these spaces are appropriately defined, the following operations can be performed in them for arbitrary group elements $g_1$:

(i) a left translation

$$T^l_{g_1} f(g) = f(g_1^{-1} g)$$

(ii) a right translation

$$T^r_{g_1} f(g) = f(g g_1)$$

Since

$$T^l_{g_1} T^l_{g_2} = T^l_{g_1 g_2} \quad \text{and} \quad T^r_{g_1} T^r_{g_2} = T^r_{g_1 g_2}$$

these operations themselves constitute groups which are homomorphic images of the original group G.

*b.   Invariant Measures*

On any Lie group G two types of invariant measures (Haar measures) can be defined by requiring invariance against right translations

$$\int f(g)\, d\mu_r(g) = \int f(g g_1)\, d\mu_r(g)$$

or against left translations

$$\int f(g)\, d\mu_l(g) = \int f(g_1^{-1}g)\, d\mu_l(g)$$

for all continuous functions with compact support on G. These measures are unique up to a constant factor. For a certain class of groups which includes all compact Lie groups both measures coincide. Most but not all of the groups occurring in this course are of this type.

We give the invariant measures for the group $SL(2, C)$ and some of its subgroups. For $SL(2, C)$ we use the parameters

$$a = a_0 e + \sum_{k=1}^{3} a_k \sigma_k$$

introduced in Section 1–1b. Then we have

$$d\mu(a) = c_0{}^2\, \delta\left(a_0{}^2 - \sum_{k=1}^{3} a_k{}^2 - 1\right) \prod_{i=0}^{3} Da_i \qquad (1\text{-}20)$$

For any complex variable $z = x + iy$ we use the notations

$$Dz = dx\, dy \quad \text{and} \quad \delta(z) = \delta(x)\, \delta(y)$$

The arbitrary constant $c_0$ in (1–20) will be fixed later in a convenient fashion.

We proceed similarly in the case of the groups $SU(2)$, $SU(1, 1)$, and $SL(2, R)$. The delta functions are now meant in the one-dimensional sense.

$$SU(2): \qquad d\mu(u) = c_0\, \delta\left(\sum_{j=0}^{3} u_j{}^2 - 1\right) \prod_{i=0}^{3} du_i \qquad (1\text{-}21)$$

$$SU(1, 1): \quad d\mu(v) = c_0\, \delta(\det v - 1) \prod_{i=0}^{3} dv_i \qquad (1\text{-}22)$$

$$SL(2, R): \quad d\mu(a) = c_0\, \delta(\det a - 1) \prod_{i=0}^{3} da_i \qquad (1\text{-}23)$$

In the case of the group $SU(2)$ the manifold of parameters $u_0$ and $u_k$ is restricted by

$$u_0{}^2 + \sum_{k=1}^{3} u_k{}^2 = 1$$

and is therefore compact. The total measure of the group is finite. We find

$$\int_{SU(2)} d\mu(u) = \tfrac{1}{2}c_0 \Omega_4$$

where $\Omega_4$ is the area of the surface of the unit sphere in Euclidian four-space,

$$\Omega_4 = 2\pi^2$$

We normalize the measure on SU(2) and set

$$c_0 = \pi^{-2} \qquad\qquad\qquad (1\text{-}24)$$

We use the same value for the free constants $c_0$ in the measures of the other groups, too, although these groups are noncompact and have an infinite total measure.

We consider finally a Lie group of the semidirect product type G = H × T, whose invariant subgroup T is Abelian and a vector space and whose elements multiply as

$$g = (h, t) \qquad h \in H \qquad t \in T$$
$$(h_1, t_1)(h_2, t_2) = (h_1 h_2, t_1 + A_{h_1}(t_2))$$

$A_h$ is a linear transformation in T. The invariant measure on G is the product of the invariant measures on H and T if the modulus of the determinant of $A_h$ is 1. For the two groups of interest, SL(2, C) × $T_4$ and U(1)′ × $T_2$ we have

$$d\mu(t) = \prod_{\mu=0}^{3} dx^\mu \qquad |\det A_a| = |\det \Lambda| = 1$$

respectively,

$$d\mu(t) = D\mu \qquad |A_\varphi| = |e^{i\varphi}| = 1$$

so that the condition on $A_h$ is satisfied in both cases.

### c.   The Group Algebra

Among all possible linear spaces of complex-valued functions on a Lie group G two are of fundamental importance. The first one consists of functions which are integrable with respect to the invariant measure

$$\int |f(g)| \, d\mu(g) < \infty$$

For simplicity we assume that the invariant measure is two-sided invariant and moreover invariant against inversions

$$\int f(g) \, d\mu(g) = \int f(g^{-1}) \, d\mu(g)$$

We call this function space $\mathscr{L}^1(G)$. In this space we may introduce a norm by

$$\|f\|_1 = \int |f(g)| \, d\mu(g)$$

which makes this space a Banach space. By the convolution integral

$$(f \cdot h)(g) = \int f(g_1) h(g_1^{-1} g) \, d\mu(g_1)$$

we may define a multiplication which is continuous in both factors since

$$\|f \cdot h\|_1 \leq \|f\|_1 \|h\|_1$$

We introduce an involution operation by

$$f^\dagger(g) = \overline{f(g^{-1})}$$

with the property

$$\|f^\dagger\|_1 = \|f\|_1$$

This set of operations allows us to regard the Banach space $\mathscr{L}^1(G)$ as a Banach algebra with involution. Apart from the case of discrete groups the algebra $\mathscr{L}^1(G)$ does not contain a unit element. In fact, a unit element would have the meaning of a delta-function on the group, which lies outside $\mathscr{L}^1(G)$ for nondiscrete groups. However, we may always adjoin a unit element formally by considering formal sums.

$$\lambda e + f$$

where $\lambda$ is a complex number, $f$ is in $\mathscr{L}^1(G)$, and $e$ is the unit,

$$e \cdot f = f \cdot e = f$$

The norm and involution can be carried over by

$$\|\lambda e + f\|_1 = |\lambda| + \|f\|_1$$
$$(\lambda e + f)^\dagger = \bar{\lambda} e + f^\dagger$$

The algebra obtained in this fashion is called the group algebra $\mathscr{R}(G)$. Both algebras $\mathscr{L}^1(G)$ and $\mathscr{R}(G)$ play important roles in relating the theory of group representations to the theory of Banach algebras and their representations.

It can be shown that translations form operations in $\mathscr{L}^1(G)$ which are continuous in the group element with respect to the norm in $\mathscr{L}^1(G)$. The space $\mathscr{L}^1(G)$ carries, therefore, a continuous representation of the group G.

### d.   Square Integrable Functions

Let G again be a Lie group with two-sided invariant measure. We consider the space $\mathscr{L}^2(G)$ of functions on G which are measurable and have finite norm $\|f\| = (f, f)^{1/2}$, where $(f_1, f_2)$ is the scalar product

$$(f_1, f_2) = \int \overline{f_1(g)} f_2(g) \, d\mu(g)$$

This space is also of great importance, in particular for our later investigations. It is a Hilbert space. Translations are continuous in the group elements as in the case of $\mathscr{L}^1(G)$. Since translations preserve the norm of $\mathscr{L}^2(G)$ they are unitary operations and form a unitary, continuous representation of the group G (see Section 1–3). These representations are called the right-regular and left-regular representations, respectively. The problem of decomposing these representations into irreducible components is the basic issue of harmonic analysis.

### e.   Remarks

The proof for the existence and uniqueness of the Haar measure can be found in textbooks. Pontrjagin [27] gives it for compact topological groups and Naimark [25] for locally compact separable groups. The latter book contains also detailed material on Banach algebras, group algebras, and group representations.

## 1–3   UNITARY REPRESENTATIONS

The content of this section is grouped around the notion of unitary representations. We define irreducibility and equivalence of unitary

representations. Finally we give some general results on the decomposition of
unitary representations into their irreducible components.

## a.   *Definitions*

We recall the standard definition of a unitary representation of a Lie
group. A unitary representation is a homomorphism of the group G into the set
of unitary operators $U$ in a Hilbert space $\mathcal{H}$

$$g \to U_g \qquad U_{g_1} U_{g_2} = U_{g_1 g_2} \qquad U_e = E$$

where $E$ is unit operator in $\mathcal{H}$.

In general one assumes in addition (we shall always do so) that the
representation be continuous

$$\|u_g \xi - u_{g_0} \xi\| \to 0 \qquad \text{as} \quad g \to g_0$$

for any vector $\xi \in \mathcal{H}$. In the case of unitary representations we can introduce the
notion of irreducibility in the form of subspace irreducibility: A unitary
representation in a Hilbert space $\mathcal{H}$ is irreducible if and only if the only
invariant subspaces of $\mathcal{H}$ are $\mathcal{H}$ itself and the null space. The definition of
equivalence of two unitary representations in Hilbert spaces is as elementary.
Two representations $g \to U_g$, $g \to U'_g$ in Hilbert spaces $\mathcal{H}$, respectively $\mathcal{H}'$, are
called equivalent if an isometric mapping $U$ from $\mathcal{H}$ onto $\mathcal{H}'$ exists such that

$$\xi \in \mathcal{H} \qquad \xi' \in \mathcal{H}' \qquad \xi' = U\xi$$

implies

$$U'_g \xi' = U U_g \xi$$

for all $g$.

In the course of our investigations we shall also deal with nonunitary
representations. We define a unitary or nonunitary representation as a
homomorphic mapping $g \to T_g$ of the Lie group G on a set of bounded operators
$T_g$ in a Banach space, that is, these operators obey the group law

$$T_{g_1} T_{g_2} = T_{g_1 g_2}$$

and $T_e$ is the identity operator. Continuity of a representation can be defined as
in the case of unitary representations, it is also presupposed in general. The
notion of irreducibility of nonunitary representations in Banach spaces is,
however, less elementary. We skip it in this course and later prove irreducibility

only for the unitary representations. The notion of equivalence of two nonunitary representations is of greater practical importance. We define it in the
context where we need it.

### b.    Groups of Type One

In the remainder of this section we consider solely unitary
representations. The main problems in the study of unitary representations
of a given Lie group are the following ones:

(i) to construct all unitary irreducible representations;
(ii) to decompose a given unitary representation into its irreducible
components.

Issue (i) is solved for the groups in which we are interested in Chapters
2, 3, and 5, though we do not prove that the list of representations we give is
complete. Some general methods which enable us to construct all unitary
irreducible representations for a great class of groups and certain classes of
representations of most other Lie groups are sketched below. For groups of
type one problem (ii) possesses, in principle, a general solution. Let us start
with this problem and first explain what a group of type one is.

We consider the algebra $\mathscr{B}(\mathscr{H}_0)$ of all bounded operators in a Hilbert
space $\mathscr{H}_0$ of countable dimension (finite or infinite). We form a direct
orthogonal sum of a countable number of copies of such Hilbert spaces

$$\mathscr{H} = \sum_t {}^{\oplus} \mathscr{H}_0(t)$$

This means that we consider the linear space consisting of sequences

$$f = \{f_t\} \qquad t = 1, 2, \ldots$$

with finite norm $\|f\| = (f, f)^{1/2}$ where $(f^{(1)}, f^{(2)})$ is the scalar product

$$(f^{(1)}, f^{(2)}) = \sum_t (f_t^{(1)}, f_t^{(2)})_{\mathscr{H}_0}$$

For each operator $A_0$ in $\mathscr{B}(\mathscr{H}_0)$ we define an operator in $\mathscr{H}$ by

$$Af = \{A_0 f_t\}$$

A factor of type one can be described as any algebra of operators $A'$ in a certain
Hilbert space $\mathscr{H}'$, which is isomorphic to such an algebra of operators $A$ in the

space $\mathscr{H}$, where by isomorphism we mean that an isometric operator $U$ exists for which

$$\mathscr{H}' = U\mathscr{H} \qquad A' = UAU^{-1} \qquad \text{for all } A, A'$$

In particular one distinguishes between factors of type one with different dimensions of $\mathscr{H}_0$.

Let us remember the group algebra $\mathscr{R}(G)$ of the Lie group G. For simplicity we assume again that G possesses a two-sided invariant and inversion invariant measure. Any unitary continuous representation $g \to U_g$ of G in a Hilbert space $\mathscr{H}$ induces a representation of the group algebra $\mathscr{R}(G)$ in the same space. In fact, for any element $\lambda e + f$ of $\mathscr{R}(G)$ we can define an operator $A_{\lambda e + f}$ in $\mathscr{H}$ by

$$A_{\lambda e + f} = \lambda E + \int f(g) U_g \, d\mu(g)$$

where $E$ is the unit operator in $\mathscr{H}$. This operator $A_{\lambda e + f}$ has the properties

$$A_f A_h = A_{f \cdot h}$$

$$(A_{\lambda e + f})^\dagger = A_{(\lambda e + f)^\dagger}$$

$$\|A_{\lambda e + f}\| \leq |\lambda| + \|f\|_1 = \|\lambda e + f\|_1$$

as can be verified easily. The norm of $A_{\lambda e + f}$ is the usual operator norm in a Hilbert space. The second and the third of these properties guarantee the symmetry and the continuity of the representation of $\mathscr{R}(G)$. In order to compare such representation of $\mathscr{R}(G)$ with a factor of type one, we must extend it, since some bounded operators of the same structure as the operators $A_{\lambda e + f}$ need not yet be contained in the algebra, for example, the unitary operators $U_g$ themselves if G is a continuous group. To include all missing operators it suffices, however, to close the algebra with respect to the weak topology.

Since the connection between representations of the group and the group algebra is invertible, we may say that a unitary representation of the group G is a factorial representation of type one, if the representation of the group algebra corresponding to it after the closure is a factor of type one. A group is denoted as type one if all multiples of any of its unitary representations are of type one.

First we are interested to know which groups are of type one. We express the result in the form of two theorems.

THEOREM 1.    Semisimple connected Lie groups are of type one.

THEOREM 2.    If a Lie group is a semidirect product with an Abelian invariant subgroup such that all irreducible representations can be induced from unitary representations of little groups and if each little group is of type one, then the group is itself of type one.

The notations "little group" and "induced" will be explained in Section 1–4b. The two theorems cover all cases of interest for us.

### c.   The Central Decomposition

For groups of type one an important theorem can be proved. For its formulation and also for our later investigations we need the notion of a direct integral of Hilbert spaces which we may introduce as follows. Let S be a locally compact topological space with the measure $\rho$. We ascribe to each point $s$ of this space S a Hilbert space $\mathcal{H}_s$ of dimension $n_s$. This dimension may be finite or countably infinite. If $n_s$ is not constant we assume that it is measurable over S. On S we define vector-valued functions $\xi(s)$ such that each $\xi(s)$ is in $\mathcal{H}_s$ for any fixed $s$. We consider first the case where $n_s$ is constant (finite or infinite). Then we may identify all the spaces $\mathcal{H}_s$ with one Hilbert space $\mathcal{H}_0$. The set of functions $\xi(s)$ which are such that

(i) $(\xi(s), h)$ is measurable over S for any fixed vector $h\epsilon\mathcal{H}_0$,
(ii) their norm $\|\xi\| = (\xi, \xi)^{1/2}$ is finite, where $(\xi_1, \xi_2)$ is the scalar product

$$(\xi_1, \xi_2) = \int \overline{\xi_1(s)}\xi_2(s)\, d\rho(s)$$

forms a Hilbert space, which we denote the direct integral of the Hilbert spaces $\mathcal{H}_s$. If the dimension is not constant, we decompose S into spaces $S_n$ on which $n_s$ is equal to $n$, define the direct integral on each subspace $S_n$, and finally take the direct orthogonal sum of these integrals of Hilbert spaces. We write a direct integral of Hilbert spaces as

$$\mathcal{H} = \int^{\oplus} \mathcal{H}_s\, d\rho(s)$$

It is obvious that two equivalent measures, namely, two measures whose null sets are identical, yield the same direct integral Hilbert space.

With the help of direct integrals of Hilbert spaces we formulate the following "central decomposition" theorem.

THEOREM 3.   Any continuous, unitary representation of a Lie group of type one in a space $\mathcal{H}$ can be decomposed into a direct integral of factorial representations of type one in the following sense. The Hilbert space is a direct integral

$$\mathcal{H} = \int^{\oplus} \mathcal{H}_s\, d\rho(s)$$

There is a null set $N \subset S$

$$\int_N d\rho(s) = 0$$

such that

(1) for $s \in S - N$

$$(U_g \xi)(s) = U_g(s)\xi(s)$$

where $U_g(s)$ acts in the space $\mathcal{H}_s$ and is a factorial representation of type one;

(2) for $s, s' \in S - N$, $s \neq s'$, $U_g(s)$ and $U_g(s')$ are multiples of inequivalent representations;

(3) the decomposition is unique up to a replacement of the measure $\rho$ by an equivalent measure.

This decomposition into "disjoint" factorial representations of type one can be followed by a decomposition of each factorial representation into a direct sum of irreducible components. We may write for $s \in S - N$

$$\mathcal{H}_s = \sum_t^{\oplus} \mathcal{H}_{s,t} \qquad U_g(s) = \sum_t^{\oplus} U_g(s, t)$$

where $U_g(s, t)$ acts in $\mathcal{H}_{s,t}$ and any pair of representations $U_g(s, t)$, $U_g(s, t')$ is equivalent.

The space S can be interpreted as a topological space of the equivalence classes of the unitary, irreducible representations of the group of type one under consideration. The actual problem in a decomposition consists in finding the measure $\rho$ on this space and determining the multiplicity in the factorial representations. Theorem 3 itself does not tell us how to find this measure explicitly. The method of induced representations, which in many cases can serve for an explicit construction of the representations, can in some cases guide us also in the problem of an actual decomposition. We turn now to an investigation of this method.

### d. Remarks

The notion of factors of different types goes back to Von Neumann. Wigner uses the notation "factorial representation" in his classical paper on the representations of the Poincaré group [37]. The whole subject of Section 1–3 (and of course other topics, also) is discussed in great detail in a review article by Mackey [22]. For Theorem 2 see his §6. For Theorem 3 we refer also to this review article or to Naimark's textbook [25]. The latter book contains a bibliography of the classic articles on this subject; Mackey's bibliography starts in the year 1945.

## 1–4   INDUCED REPRESENTATIONS

Whereas Section 1–3 was devoted to the properties of unitary representations, we are now concerned with the problem of how to construct them. The idea which we investigate is to use linear spaces of functions over manifolds which are supplied by the group itself. In a natural manner the group maps homomorphically on a group of transformations of these manifolds, which in turn induces a group of operators in the function spaces. The main issue to be settled is to make these groups of operators in the function spaces unitary and irreducible representations. In the case of the inhomogeneous SL(2, C) we find in this way all unitary irreducible representations.

### a.   Multiplier Representations

For a Lie group G with closed subgroup H the set of right cosets G/H forms a differentiable manifold to which we refer as a homogeneous space. Any point $\xi$ of G/H can be represented by the class of group elements $hg(\xi)$, where $g(\xi)$ is one representative of $\xi$ and $h$ runs over H. Multiplying this class from the right with an arbitrary element $g$ of G transforms it into another class $\xi'$

$$hg(\xi)g = h'g(\xi') \qquad (1\text{-}25)$$

For simplicity throughout we use the notation

$$\xi' = \xi_g$$

for the class which results from transforming the class $\xi$ with the group element $g$. This set of transformations of the homogeneous space G/H induced by the elements of G is transitive in the sense that to any pair of points $\xi$, $\xi'$ there is a group element $g$ such that

$$\xi' = \xi_g$$

We can build linear spaces of functions on G/H and use them to construct representations of the group G just as we did in the case of the regular representation in Section 1–2d. The regular representation can in fact be regarded as the representation obtained in the special case H = $\{e\}$. The place of the invariant measure on G is taken by a quasi-invariant measure on G/H. Quasi-invariance is defined as invariance of the null sets of the measure under transformations induced by the group elements of G. This property is necessary for the existence of a Jacobian.

If $\mu$ is such measure, we can introduce a scalar product into a space of functions on G/H by

$$(f_1, f_2) = \int \overline{f_1(\xi)} f_2(\xi) \, d\mu(\xi)$$

obtaining a Hilbert space $\mathcal{H}$. For a unitary transformation in this space we make the ansatz

$$U_g f(\xi) = \alpha(\xi, g) f(\xi_g) \tag{1-26}$$

where the function $\alpha(\xi, g)$ is called the multiplier. In order to make the scalar product invariant, the multiplier must fulfill the constraint

$$|\alpha(\xi, g)|^2 = \frac{d\mu(\xi_g)}{d\mu(\xi)} \tag{1-27}$$

The right-hand side of this relation is the Jacobian whose existence was required earlier. If the multiplier satisfies the functional equation

$$\alpha(\xi, g_1 g_2) = \alpha(\xi, g_1)\alpha(\xi_{g_1}, g_2) \tag{1-28}$$

the operators $U_g$ of (1−26) form a unitary representation of G in $\mathcal{H}$. This form of a representation is sometimes denoted a multiplier representation.

We simplify the functional equation (1−28) for the multiplier. In order to get rid of the constraint (1−27) we introduce the quantity

$$\beta(\xi, g) = \left[ \frac{d\mu(\xi_g)}{d\mu(\xi)} \right]^{-1/2} \alpha(\xi, g)$$

of modulus 1, which obviously also obeys a condition such as (1−28). If $\varepsilon$ denotes that point of G/H which corresponds to the subgroup H itself, we write

$$\beta(g) = \beta(\varepsilon, g)$$

Setting

$$\xi = \varepsilon \qquad g_1 = h \qquad g_2 = g$$

(1−28) takes the simple form

$$\beta(hg) = \beta(h)\beta(g) \tag{1-29}$$

It follows from (1–29) in particular that $\beta$ is a one-dimensional unitary representation of H. If a solution of the reduced relation (1–29) has been found, the full equation (1–28) is solved uniquely by

$$\beta(\xi, g) = \frac{\beta(g(\xi)g)}{\beta(g(\xi))} \tag{1-30}$$

### b.   Induced Representations

The preceding discussion suggests the following changes in our definitions which allow us to generalize the whole concept considerably. We consider the function $f(\xi)$ as a function on the  group G which is constant on the right cosets with respect to H,

$$f(g) = f(hg(\xi)) = f(g(\xi)) \equiv f(\xi)$$

for $g \,\epsilon\, \xi$. We define new functions by

$$\varphi(g) = \beta(g)f(g)$$

They have the property

$$\varphi(hg) = \beta(h)\varphi(g) \tag{1-31}$$

We say that $\varphi(g)$ is covariant on right cosets with respect to the subgroup H and call (1–31) a covariance constraint. These covariant functions transform like (the first equation is a definition)

$$U'_{g_1}\varphi(g) = \beta(g)U_{g_1}f(g) = \left[\frac{d\mu(\xi_{g_1})}{d\mu(\xi)}\right]^{1/2}\varphi(gg_1) \tag{1-32}$$

where the class $\xi$ contains the element $g$. The invariant norm is

$$\|\varphi\| = \left\{\int |\varphi(g)|^2 \, d\mu(\xi)\right\}^{1/2} \tag{1-33}$$

Definition (1–33) makes sense because the modulus of $\beta(h)$ is 1 and $|\varphi(g)|^2$ is constant on cosets. We say that the representation obtained in this fashion has been induced from the unitary one-dimensional representation $\beta(h)$ of H.

There is in fact no need to induce from one-dimensional representations of H. Let us assume that any unitary representation $u_h$ of H in a Hilbert space $\mathcal{H}_0$ is given. We consider vector-valued functions on G with values in $\mathcal{H}_0$ which are submitted to a covariance constraint on the right cosets of H,

$$\varphi(hg) = u_h\,\varphi(g) \tag{1-34}$$

Let $\varphi(g)$ be such a function, vector valued and covariant on cosets as required by (1–34) for which the following hold:

(i) $(\varphi(g(\xi)), \phi)$ is a measurable function of $\xi$ for any fixed vector $\phi$ in $\mathscr{H}_0$ and a particular choice of representatives $g(\xi)$ in the cosets [in the cases of interest to us it is, for example, always possible to choose $g(\xi)$ from analytic submanifolds of G];

(ii) the norm

$$\|\varphi\| = \left\{\int \|\varphi(g)\|^2_{\mathscr{H}_0}\,d\mu(\xi)\right\}^{1/2}$$

is finite.

Such functions span a direct integral Hilbert space

$$\mathscr{H} = \int^\oplus \mathscr{H}_0(\xi)\,d\mu(\xi) \qquad \mathscr{H}_0(\xi) \cong \mathscr{H}_0$$

in which a unitary representation of G can be defined by

$$U_{g_1}\varphi(g) = \left[\frac{d\mu(\xi_{g_1})}{d\mu(\xi)}\right]^{1/2}\varphi(gg_1) \tag{1-35}$$

We say that this representation has been induced by the representation $u_h$ of H. One can show that the specific choice of the measure is not significant.

This method can be used to find unitary representations of many Lie groups. It is, however, of fundamental importance for the theory of representations of groups which are semidirect products $G = F \times T$ with Abelian invariant subgroup T. Let $g = (f, t)$ be an element of G. We write the multiplication rule for G as

$$g_1\,g_2 = (f_1, t_1)(f_2, t_2) = (f_1 f_2, t_1 + A_{f_1}(t_2)) \tag{1-36}$$

where we assume that T is a real vector space and $A_f$ a linear transformation in T.

Since T is Abelian, the irreducible unitary representations of T are one-dimensional

$$t \to \exp i \sum_\alpha t^\alpha p_\alpha$$

where the $t_\alpha$ are the components of the vector $t \epsilon T$ and the real numbers $p_\alpha$ characterize the representations of T. We refer to the set $p = \{p_\alpha\}$ as the character of the representation of T. The linear mapping of T onto itself caused by an element $f \epsilon F$ is in matrix form

$$t'^\alpha = A_f(t)^\alpha = \sum_\beta (A_f)_\beta{}^\alpha t^\beta$$

We can define transformations of characters (considering them dual to the vectors $t$) by

$$\sum_\alpha t'^\alpha p_\alpha = \sum_\alpha t^\alpha p'_\alpha = \sum_{\alpha, \beta} t^\alpha (A_f)_\alpha{}^\beta p_\beta$$

$$p'_\alpha = \sum_\beta (A_f)_\alpha{}^\beta p_\beta \qquad\qquad (1\text{-}37)$$

For a given reference character $p^R$ the set of all characters generated from $p^R$ by applying all elements of F to $p^R$ is denoted the orbit containing $p^R$. The stationarity condition

$$\sum_\beta (A_f)_\alpha{}^\beta p_\beta = p_\alpha$$

determines a certain subgroup $F_p$ of F, which we call the little group belonging to the character $p$. Different points on the same orbit give rise to isomorphic little groups.

We consider now the group

$$H_{p^R} = F_{p^R} \times T$$

for a fixed character $p^R$. It is obvious that the cosets $G/H_{p^R} \cong F/F_{p^R}$ can be identified with the points of that orbit which contains $p^R$. The orbits are therefore the homogeneous spaces of the group F with respect to the little groups $F_{p^R}$. Let us apply the formalism of induced representations to these homogeneous spaces.

If $f$ varies over one of the right cosets $\xi$ of $F/F_{p^R}$, the character

$$p_\alpha = p_\alpha(\xi) = \sum_\beta (A_f)_\alpha{}^\beta p_\beta{}^R$$

remains constant. This equation establishes therefore the one-to-one correspondence of the cosets $F/F_{p^R}$ with the orbit containing $p^R$. We replace the cosets $\xi$ by the characters $p$ and the measure $d\mu(\xi)$ by a measure $d\mu(p)$ on an orbit. Since in the cases of interest for us the linear transformations $A_f$ (1–37) are unimodular, the measure $d\mu(p)$ can be chosen invariant. As can be verified easily, we obtain a unitary representation of $H_{pR}$ as the direct product of the one-dimensional representation

$$\exp i \sum_\alpha t^\alpha p_\alpha^R$$

of T and any unitary representation $u_f$ of $F_{pR}$ in a Hilbert space $\mathcal{H}_0$. Next we consider functions $\varphi(g) = \varphi(f, t)$ on G. Since T is a subgroup of $H_{pR}$, the covariance constraint (1–34) requires

$$\varphi(f, t) = \exp\left\{i \sum_\alpha t^\alpha p_\alpha^R\right\} \varphi(f, 0) \tag{1-38}$$

A function $\varphi(g)$ is therefore completely determined by the function

$$\varphi(f) \equiv \varphi(f, 0)$$

We assume next that a rule exists which ascribes a unique representative $f(p) \in F$ to almost all points $p$ on a given orbit such that

$$p_\alpha = \sum_\beta (A_{f(p)})_\alpha^\beta p_\beta^R \tag{1-39}$$

We call $f(p)$ a boost. In the cases of interest for us, $f(p)$ can be made piecewise regular analytic. The covariance constraint (1–34) implies

$$\varphi(f_1 f(p)) = u_{f_1} \varphi(f(p)) \qquad f_1 \in F_{pR} \tag{1-40}$$

We may therefore introduce a third set of functions by

$$\varphi(p) \equiv \varphi(f(p))$$

We can use any of the three kinds of functions $\varphi(g)$, $\varphi(f)$, or $\varphi(p)$ to construct a carrier Hilbert space for the unitary representation of G. In the last case we require that $\varphi(p)$ be measurable, in the other cases we assume that $\varphi(g)$ or $\varphi(f)$ corresponds to a measurable function $\varphi(p)$ via the constraints (1–38) and (1–40). In addition the norm squared

$$\|\varphi\|^2 = \int \|\varphi(g)\|_{\mathcal{H}_0}^2 \, d\mu(p)$$

$$= \int \|\varphi(f)\|_{\mathcal{H}_0}^2 \, d\mu(p) \tag{1-41}$$

$$= \int \|\varphi(p)\|_{\mathcal{H}_0}^2 \, d\mu(p)$$

is to be finite. The cosets $F/F_{pR}$ transform like

$$(f(p), 0)(f, t) = (r(f, p), \sum_\beta (A_{f(p)})_\beta{}^\alpha t^\beta)(f(p_f), 0) \qquad (1\text{-}42)$$

where

$$r(f, p) = f(p) f f(p_f)^{-1} \qquad (1\text{-}43)$$

is in $F_{pR}$ and

$$(p_f)_\alpha = \sum_\beta p_\beta (A_f)_\alpha{}^\beta$$

The unitary representation can consequently be defined in any of the three equivalent fashions $[g_1 = (f_1, t_1)]$:

$$U_{g_1} \varphi(g) = \varphi(gg_1) \qquad (1\text{-}44)$$

$$U_{g_1} \varphi(f) = \exp i\left\{\sum_\alpha (A_f t_1)^\alpha p_\alpha{}^R\right\} \varphi(ff_1) \qquad (1\text{-}45)$$

$$U_{g_1} \varphi(p) = \exp i\left\{\sum_\alpha t_1{}^\alpha p_\alpha\right\} u_{r(f_1, p)} \varphi(p_{f_1}) \qquad (1\text{-}46)$$

We shall sometimes refer to the element $r(f, t)$ [Eq. (1–43)] of the little group as Wigner's rotation.

The following two theorems hold in the case that all little groups are of type one.

THEOREM 4.    A representation obtained from an irreducible unitary representation of a little group by induction is itself unitary irreducible.

THEOREM 5.    If the orbits satisfy a certain smoothness condition, any irreducible unitary representation of $G = F \times T$ can be obtained by induction from an irreducible unitary representation of a little group.

For the Poincaré group, its universal covering group $SL(2, C) \times T_4$, and the groups $U(1)' \times T_2$ and $U(1) \times T_2$, the premises are fulfilled.

c.    *Remarks*

The method of induced representations is nearly as old as group theory itself. Wigner [37] applied it to construct the representations of the Poincaré group. The mathematical theory of induced representations for continuous groups was developed by Mackey [19–21]. The methods used to construct the representations for the groups SU(1, 1) and SL(2, C) by Bargmann [5] and Gelfand and Naimark [11] (see also Naimark [24]) are also based on inductions. A short account of the theory of induced representations is given by Mackey [22].

# Chapter 2

# Unitary Representations of the Inhomogeneous SL(2, C) Group

The technique of inducing representations from known representations of subgroups, which we developed in Section 1—4, is now to be applied to the group $SL(2, C) \times T_4$. The task left is to give those objects which were introduced in general terms in Section 1—4 a concrete form for the special case of the group $SL(2, C) \times T_4$: the orbits, which are certain manifolds in momentum space with a geometry related to Lobachevskian spaces; the little groups; and the boosts. This is done in Sections 2—1 to 2—4. The remaining sections of Chapter 2 are devoted to the two classes of representations of $SL(2, C) \times T_4$ with timelike momenta and their connection with classical free fields. The classes of representations of the inhomogeneous $SL(2, C)$ for other momenta are in principle known after the representations of the corresponding little groups have been studied in Chapters 3 and 5. To write down explicit formulas such as (1—44) to (1—46) for these representations is straightforward.

The discrete transformations $I_s$, $I_t$, and $I_{st}$ (Section 1—1a) contained in the Poincaré group are completely neglected in the sequel. Only the so-called *CPT* transformation, whose role in nature is more fundamental to our present state of knowledge, will be studied in Chapter 8. That chapter contains also other interesting results on the inhomogeneous $SL(2, C)$ group for example, the decomposition of tensor products of unitary representations.

We call attention to Section 2—5, where we derive matrix elements for finite-dimensional nonunitary representations of $SL(2, C)$ using linear spaces of homogeneous polynomials. Though this method is not new, it might be of interest since our investigations of infinite-dimensional representations of $SL(2, C)$ and $SL(2, R)$ are based on a generalization of such spaces.

## 2—1  ORBITS

The unitary irreducible representations of the translation group $T_4$ are characterized by a covariant four-vector

$$p = (p_0, p_1, p_2, p_3)$$

which, instead of using the general notation "character," we call the "momentum." This momentum, however, must not yet be identified with the momentum of an elementary particle, since the latter "momentum" in contravariant notation transforms as the contravariant coordinate (1–7), which transformation behavior disagrees with (1–37). In the strict sense the character $p$ is the argument of a wave function, whereas the momentum $p$ is an (improper) eigenvalue of the infinitesimal displacement operator. This will hopefully become clear after a remark on "plane wave states" in Section 2–7.

The whole four-space of such momentum vectors can be decomposed into orbits in a fashion which satisfies the smoothness condition required in Theorem 5, Section 1–4b. These orbits can be classified as follows.

(i) Each of the two shells of the hyperboloid

$$(p)^2 = g^{\mu\nu}p_\mu p_\nu = M^2 > 0 \tag{2-1}$$

forms an orbit; we denote the orbit positive timelike if $p_0 > 0$ and negative timelike if $p_0 < 0$. To describe a timelike orbit we have to give both $M^2$ and $\text{sign}(p_0)$. The invariant measure on these orbits is (up to a constant factor, of course)

$$d\mu(p) = \varepsilon((p)^2 - M^2)\Theta(\pm p_0)\, d^4p \tag{2-2}$$

(ii) The single shell of the hyperboloid

$$(p)^2 = g^{\mu\nu}p_\mu p_\nu = -\mu^2 \qquad \mu > 0 \tag{2-3}$$

is an orbit, we call it a spacelike orbit. Each spacelike orbit is therefore given by one real parameter $\mu$ (or by a negative value of $M^2 = -\mu^2$). The invariant measure is

$$d\mu(p) = \delta((p)^2 + \mu^2)\, d^4p \tag{2-4}$$

(iii) The light cone consisting of all isotropic vectors

$$(p)^2 = 0$$

is composed of three separate orbits. These are the positive light cone with $p_0 > 0$, the negative light cone with $p_0 < 0$, and the origin in $p$-space in between. The invariant measure in the first two cases is

$$d\mu(p) = \delta((p)^2)\Theta(\pm p_0)\, d^4p \tag{2-5}$$

In the third case the measure is trivial. In all these cases we call the label $M^2$ the "mass squared."

We turn now to a brief discussion of the geometric properties of these orbits. This will facilitate the understanding of some peculiarities which show up in the definition of boosts. Positive timelike orbits are examples of real Lobachevskian spaces which themselves represent a particular kind of space with constant curvature. Since all positive timelike orbits possess the same geometry, we need only consider one copy, say the orbit

$$(p)^2 = 1 \qquad p_0 \geqq 0$$

On this hyperboloid we can define an invariant arc distance $\alpha$ by

$$\cosh \alpha = pp'$$

Since Schwartz's inequality implies

$$pp' \geqq p_0 p_0' - |p| |p'| \geqq 1$$

[we use the notation $|p| = (\sum_{k=1}^{3} p_k^2)^{1/2}$] this angle $\alpha$ is in fact real and can be chosen positive. The differential element of distance squared

$$ds^2 = -dp_0^2 + \sum_{t=1}^{3} dp_k^2 = d\alpha^2$$

is positive definite. This can be seen from

$$p_0 \, dp_0 - \sum_{k=1}^{3} p_k \, dp_k = 0$$

and

$$p_0^2 \, ds^2 = p_0^2 \sum_{k=1}^{3} dp_k^2 - \left( \sum_{k=1}^{3} p_k \, dp_k \right)^2$$

$$\geqq (p_0^2 - |p|^2) \sum_{k=1}^{3} dp_k^2$$

Spacelike orbits can be regarded as imaginary Lobachevskian spaces if we identify opposite points $p$ and $-p$ on them. We consider the copy $\mu^2 = 1$. Again we introduce an arc distance $\alpha$ by

$$\cosh \alpha = |pp'|$$

However, it is now easy to find points with an imaginary arc distance. We choose a reference point $p^R$

$$p^R = (0, 0, 0, 1)$$

and regard those points on the orbit with a fixed distance from $p^R$. The points
(Fig. 2–1)

$$p_3 \geqslant 1$$

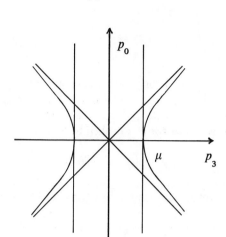

second real→ ←imaginary→ ← real

**Figure 2–1.**    Real and imaginary domains of a spacelike orbit

have a real distance from $p^R$. We call this part the real domain of the orbit
relative to $p^R$. The mirror image

$$p_3 \leqslant -1$$

which is identical to the real domain if we regard the orbit as a
Lobachevskian space, is denoted the second real domain of the orbit relative
to $p^R$. The points with

$$|p_3| < 1$$

form the imaginary domain. Their distance from $p^R$ is imaginary and lies
between 0 and $\tfrac{1}{2}i\pi$. Hence for spacelike orbits the element of distance squared
is indefinite. In fact, through each point $p$ of the orbit go isotropic lines

$$p + \lambda\pi \qquad (\pi)^2 = 0 \qquad -\infty < \lambda < \infty$$

which lie in the orbit and form a two-dimensional cone. Any pair of points on
one such isotropic line has a vanishing arc distance. The sign of the element of
distance squared changes from the interior to the exterior of the cone.

## 2–2    LITTLE GROUPS

We can ascribe a Hermitian $2 \times 2$ matrix $\mathbf{p}$ to any momentum vector $p$ by

$$\mathbf{p} = p_0 e + \sum_{k=1}^{3} p_k \sigma_k$$

Since we are always able to distinguish between momenta (characters) and coordinate vectors, we are in fact free to make the most convenient choice independent of $(1-5)$. The scalar product of $p_\mu$ with a contravariant coordinate or translation vector $y^\mu$ can then be obtained from

$$py = p_\mu y^\mu = \tfrac{1}{2} \text{Tr}(\mathbf{py})$$

if the matrix $\mathbf{y}$ is formed from the vector $y^\mu$ as in $(1-5)$. This matrix $\mathbf{y}$ transforms as

$$\mathbf{y}' = a\mathbf{y}a^\dagger$$

The matrix $\mathbf{p}$ transforms like a character $(1-37)$, namely, we require

$$\text{Tr}(\mathbf{py}') = \text{Tr}(\mathbf{p}'\mathbf{y})$$

which implies

$$\mathbf{p}' = \mathbf{p}_a = a^\dagger \mathbf{p} a \qquad\qquad (2\text{-}6)$$

The little groups consist of those elements $a \in SL(2, C)$ which obey

$$\mathbf{p} = a^\dagger \mathbf{p} a$$

For the reference momentum

$$p^R = (M, 0, 0, 0) \qquad M > 0$$

on positive timelike orbits the little group is SU(2). The same holds for negative timelike orbits with reference momentum

$$p^R = (-M, 0, 0, 0) \qquad M > 0$$

On spacelike orbits we choose the reference momentum

$$p^R = (0, 0, 0, \mu) \qquad \mu > 0$$

The little group consists of elements $a$ satisfying

$$\sigma_3 = a^\dagger \sigma_3 a$$

This condition coincides with the definition $(1-18)$ of the subgroup SU(1, 1) of SL(2, C). On the positive light cone we take as reference momentum

$$p^R = (1, 0, 0, 1)$$

The little group consists, consequently, of those elements which fulfill the condition

$$(\sigma_3 + e) = a^\dagger(\sigma_3 + e)a$$

From this constraint we find the little group $U(1)' \times T_2$ (Section 1–1d). For the negative light cone the reference momentum

$$p^R = (-1, 0, 0, -1)$$

yields the same little group. In the case of the orbit $p = 0$ (null orbit) there is no constraint on the little group at all; the little group is identical with the group SL(2, C).

All these little groups have the elements of the form

$$u(\psi) = \exp \frac{i}{2} \psi \sigma_3$$

in common. We shall see later in these lectures that the basis $\{\varphi\}$ in a Hilbert space carrying a unitary irreducible representation of any little group can always be chosen such that the operators representing $u(\psi)$ are diagonal. Except in the case of the little group SL(2, C), the basis elements are completely characterized by the eigenvalues $\exp iq\psi$, that is, we have

$$U_{u(\psi)} \varphi_q = \sum_{q'} D_{q'q}(u(\psi))\varphi_{q'}$$
$$= e^{iq\psi} \varphi_q$$

and each $\varphi_q$ spans a one-dimensional subspace; $2q$ is an integer. The discrete label $q$ describes the "spin degree of freedom." It bears different names for different choices of boosts (Section 2–3).

Taking the quotient of any little group of SL(2, C) with respect to $Z_2$ yields the corresponding little group of the orthochronous Lorentz group; the orbits do not change under this division. The representation theory for the little group of the light cones is so elementary (this holds also for harmonic analysis on this group dealing with Fourier-Bessel transformations on the real line) that we will leave it out.

## 2–3    BOOSTS FOR TIMELIKE ORBITS

If we would a priori know covariant functions $\varphi(a)$, $a \in SL(2, C)$, we could go ahead constructing unitary representations of the inhomogeneous $SL(2, C)$. In general, however, we have to use the auxiliary device of boosts to find such covariant functions. The manner in which the boosts are defined is to a high degree a matter of convenience. In general they have to solve the equation

$$\mathbf{p} = a(p)^{\dagger}\mathbf{p}^{R}a(p) \tag{2-7}$$

or

$$p_{\mu} = \Lambda_{\mu}{}^{\nu}(p)p_{\nu}{}^{R} \tag{2-8}$$

as follows from $(1-39)$ and $(2-6)$. Here we use the notation

$$\Lambda(p) = \Lambda(\pm a(p))$$

where $\Lambda(a)$ was defined by $(1-7)$ and $(1-8)$.

We consider first positive timelike orbits with the reference momentum

$$p^{R} = (M, 0, 0, 0)$$

If we make the additional requirements

$$a(p)^{\dagger} = a(p) \qquad \mathrm{Tr}\, a(p) > 0 \tag{2-9}$$

we obtain "Wigner's rotation-free boost." The condition $(2-7)$ can then be solved by the square root formula

$$a = \pm \frac{a^{2} + e}{[2 + \mathrm{Tr}(a^{2})]^{1/2}} \tag{2-10}$$

which follows from

$$a^{2} = a_{0}{}^{2}e + 2a_{0}\sum_{k=1}^{3} a_{k}\sigma_{k} + \left(\sum_{k=1}^{3} a_{k}{}^{2}\right)e$$

and

$$\mathrm{Tr}(a^{2}) = (\mathrm{Tr}\, a)^{2} - 2$$

If

$$\mathrm{Tr}(a^2) + 2 \neq 0.$$

(2–10) gives all square roots within SL(2, C), but there might be others outside SL(2, C). In this fashion we get the solution for (2–7), (2–9)

$$a(p) = [2M(M + p_0)]^{-1/2}\left\{(p_0 + M)e + \sum_{k=1}^{3} p_k \sigma_k\right\} \qquad (2\text{-}11)$$

Since the matrix (2–11) is Hermitian, it possesses real eigenvalues and is a pure Lorentz transformation. In addition it has the property to leave the plane spanned by $p$ and $p^R$ invariant as a whole (in general not pointwise). This statement suggests the problem of finding the matrix $o(p, p')$ which leaves the plane spanned by general vectors $p$ and $p'$ on the same orbit invariant and transforms $p'$ into $p$

$$\mathbf{p} = a(p, p')^\dagger \mathbf{p}' a(p, p')$$

Since invariance of a plane is itself an invariant property, the matrix $a(p, p')$ can be obtained from $\pm a(p)$ (2–11) by a change of basis in Minkowski space. Let $a$ be any element of SL(2, C) such that

$$\mathbf{p}' = a^\dagger \mathbf{p}^R a$$

We denote

$$\mathbf{p}'' = (a^{-1})^\dagger \mathbf{p} a^{-1}$$

and get

$$a(p, p') = a^{-1} a(p'') a$$

In explicit terms the solution is

$$a(p, p') = \pm [2M^2(M^2 + pp')]^{-1/2}$$
$$\times \left\{(M^2 + pp')e + \sum_{k=1}^{3}\left(p_k p_0' - p_k' p_0 + i \sum_{l,m=1}^{3} \varepsilon^{klm} p_l p_m'\right)\sigma_k\right\}$$
$$(2\text{-}12)$$

The matrix $\Lambda(p, p) = \Lambda(\pm a(p, p'))$ is correspondingly

$$\Lambda_\mu{}^\nu(p, p') = \delta_\mu{}^\nu - (M^2 + pp')^{-1}(p_\mu + p_\mu')(p^\nu + p'^\nu)$$
$$+ 2M^{-2} p_\mu p'^\nu \qquad (2\text{-}13)$$

It has the property

$$p_\mu = \Lambda_\mu{}^\nu(p, p')p'_\nu$$

If we put $p' = p^R$ we get back to

$$\Lambda_\mu{}^\nu(p, p^R) = \Lambda_\mu{}^\nu(\pm a(p, p^R)) = \Lambda_\mu{}^\nu(\pm a(p)) = \Lambda_\mu{}^\nu(p)$$

On negative timelike orbits we can define a rotation-free boost by $a(-p)$ if $a(p)$ is the matrix (2–11).

Another convenient boost has been defined by Jacob and Wick; it leads to the helicity formalism. They introduce a rotation $u(p)$ with the property

$$u(p)\mathbf{p}u(p)^{-1} = \mathbf{p}' \qquad p' = (p_0, 0, 0, |p|) \qquad |p| = \left(\sum_{k=1}^{3} p_k{}^2\right)^{1/2} \qquad (2\text{-}14)$$

The solution $u(p)$ can be made unique by imposing a condition on the *a priori* arbitrary phase $\psi$ in

$$u(p) = \exp\left\{\frac{i}{2}\psi\sigma_3\right\}\exp\left\{\frac{i}{2}\alpha(p)\sigma_2\right\}\exp\left\{\frac{i}{2}\beta(p)\sigma_3\right\}$$

This being done we define the Jacob-Wick boost by

$$a_{JW}(p) = a(p')u(p) = u(p)a(p) \qquad (2\text{-}15)$$

where $a(p)$ is a rotation-free boost (2–11). Wigner's rotation (1–43) for an element $u \in SU(2)$ and for Jacob-Wick boosts (2–15)

$$r(u, p) = u(p)a(p)ua(p_u)^{-1}u(p_u)^{-1}$$
$$= u(p)uu(p_u)^{-1}$$

is easily seen to be a diagonal element of $SU(2)$ of the form $u(\psi)$, for example, from

$$\mathbf{p}' = r(u, p)^{-1}\mathbf{p}'r(u, p)$$

where $p'$ is as in (2–14). This is a characteristic property of the Jacob-Wick boosts.

The unitary irreducible representations of the little group $SU(2)$ for timelike orbits are finite dimensional and in a basis $\{\phi_q\}$ (Section 2–2) can be described by Wigner's matrices $D^S_{q_1 q_2}(u)$, where $S$ denotes the "spin" and $q_{1,2}$ vary over

$$-S \leqq q_{1,2} \leqq S$$

These matrices are given explicitly in (2–30). If we use the rotation-free boosts (2–11), $q$ is called the "third component of spin"; in the case of the Jacob-Wick boost (2–15), $q$ is denoted "helicity." In both cases we have from (1–46)

$$U_{(a, y)} \varphi_q(p) = e^{ipy} \sum_{q'=-S}^{S} D^S_{qq'}(r(a, p)) \varphi_{q'}(p_a) \tag{2-16}$$

where $\varphi_q(p)$ is assumed to be measurable on a positive or negative timelike orbit for each $q$, and the norm

$$\|\varphi\| = \left\{ \int \sum_{q=-S}^{S} |\varphi_q(p)|^2 \, d\mu(p) \right\}^{1/2}$$

is finite. We used the notations (2–6) and Wigner's rotation (1–43)

$$r(a, p) = a(p)aa(p_a)^{-1}$$

We call $\varphi_q(p)$ a "wave function in momentum space," in brief a "wave function." Using (1–45) instead of (1–46) we have

$$U_{(a,y)} \varphi_q(a_1) = \exp i(p^R \Lambda(a_1)y) \varphi_q(a_1 a) \tag{2-17}$$

where the "covariant wave function" $\varphi_q(a)$ can be constructed from a wave function $\varphi_q(p)$ for any type of boost by

$$\varphi_q(ua(p)) = \sum_{q'} D^S_{qq'}(u) \varphi_{q'}(p) \tag{2-18}$$

The definition (2–18) implies the covariance constraint for the wave function $\varphi_q(a)$

$$\varphi_q(ua) = \sum_{q'} D^S_{qq'}(u) \varphi_{q'}(a) \tag{2-19}$$

## 2–4   BOOSTS FOR OTHER ORBITS

In the case of spacelike orbits it seems appealing to use also the matrix $a(p, p^R)$ (2–12) as a boost after a continuation in $M^2$ until $-\mu^2$. To gain an insight into what happens we study the eigenvalues of $\Lambda(p, p')$ and put $p' = p^R$ with the reference momentum fixed as in Section 2–2. Two eigenvalues must be equal to one, since the vectors orthogonal to the $pp^R$-plane are eigenvectors. If we operate with $\Lambda(p, p^R)$ on $p$ we get

$$\Lambda(p, p^R)p = -p^R - 2\frac{pp^R}{\mu^2}\, p$$

On the basis $p$ and $p^R$, $\Lambda(p, p^R)$ takes therefore the matrix form

$$\begin{pmatrix} -\dfrac{2pp^R}{\mu^2} & 1 \\[2mm] -1 & 0 \end{pmatrix}$$

from which we read off the remaining two eigenvalues

$$\lambda_{3,4} = \frac{p_3}{\mu} \pm \left[\left(\frac{p_3}{\mu}\right)^2 - 1\right]^{1/2}$$

In the real domain $p_3 \geq \mu$ (Section 2–1) the eigenvalues are real positive and $\Lambda(p, p^R)$ is a pure Lorentz transformation. In the imaginary domain $-\mu \leq p_3 \leq \mu$ the eigenvalues $\lambda_{3,4}$ are complex of modulus 1 and $\Lambda(p, p^R)$ is a pure rotation. In the second real domain both eigenvalues $\lambda_{3,4}$ are negative real, so that $\Lambda(p, p^R)$ is a pure Lorentz transformation times a total inversion $I_{st}$ times a rotation with angle $\pi$. The matrix $\Lambda(p, p^R)$ is no longer an element of the proper orthochronous Lorentz group. We note finally that the arc distance $\alpha$ between $p$ and $p^R$ and the eigenvalues $\lambda_{3,4}$ are related by

$$\lambda_{3,4} = e^{\pm\alpha}$$

for any type of orbit.

The behavior of $a(p, p^R)$ is similar. Continuing in $M^2$ we get

$$a(p, p^R) = a(p) = \pm[2\mu(\mu + p_3)]^{-1/2}\{\mu + p_3 + p_0\sigma_3 \tag{2-20}$$
$$- ip_2\sigma_1 + ip_1\sigma_2\}$$

which is a pseudo-Hermitian matrix in the sense

$$a(p)^\dagger \sigma_3 = \sigma_3\, a(p) \tag{2-21}$$

in the real and imaginary domain and solves there the correct equation (2–7)

$$a(p)^2 = \mu^{-1}\sigma_3\, \mathbf{p}$$

In the second real domain $\mu + p_3 \leqq 0$ the denominator of (2–20) becomes imaginary and the matrix $a(p)$ satisfies the wrong equation

$$\mathbf{p} = -a(p)^\dagger \mathbf{p}^R a(p)$$

The matrix $a(p)$ (2–20) is therefore no boost in this domain. It is, however, possible to define a boost in the second real domain by

$$a(p)' = i\sigma_2 a(-p)$$

where $a(p)$ is as in (2–20). With this definition the boost on the spacelike orbit consists of two pieces of analytic expressions. This is a general property of boosts on spacelike orbits and a consequence of the peculiar geometry on these orbits.

It is sometimes convenient to define a boost on spacelike orbits by a matrix of the triangular shape

$$a(p) = \begin{pmatrix} \lambda & 0 \\ \rho & \lambda^{-1} \end{pmatrix} \qquad \lambda \text{ real positive}$$

$$\lambda = \mu[\mu(p_3 - p_0)]^{-1/2} \qquad \rho = -\lambda\mu^{-1}(p_1 + ip_2) \qquad (2\text{-}22)$$

In this case there exists also a domain, namely, $p_3 - p_0 \leq 0$, in which the boost must be replaced by another expression, for example, by $i\sigma_2 a(-p)$, $a(p)$ as in (2–22).

On the positive light cone, boosts can be chosen in the form of triangular matrices (2–22). We have

$$a(p) = \begin{pmatrix} \lambda^{-1} & \rho \\ 0 & \lambda \end{pmatrix} \qquad \lambda \text{ real positive}$$

$$\lambda = 2[2(p_0 + p_3)]^{-1/2} \qquad \rho = \tfrac{1}{2}\lambda(p_1 - ip_2) \qquad (2\text{-}23)$$

This boost (2–23) is singular on the line $p_3 = -p_0, p_1 = p_2 = 0$. On the negative light cone we can again take $a(-p)$. In (2–22) and (2–23) the reference momenta are chosen as in Section 2–2.

## 2–5   SPINORS

The most important class of unitary representations of the Poincaré group consists of representations on timelike orbits. In applications these appear very often in the form of spinor wave functions or as solutions of classical free field equations. We shall sketch in brief how such quantities can be deduced from the wave functions $\varphi_q(p)$ which result from the theory of induced representations.

Doing this we first have to study finite-dimensional nonunitary representations of the group SL(2, C). We call these representations spinor representations; the vectors in the linear spaces carrying these representations are the spinors.

We consider complex two-vectors

$$\xi = (\xi^1, \xi^2)$$

which transform as

$$\xi' = \xi a \qquad \xi'^i = \sum_{k=1, 2} \xi^k a_{ki} \tag{2-24}$$

under application of a $2 \times 2$ matrix $a \, \epsilon \, SL(2, C)$. We construct a linear vector space of complex polynomials in $\xi^1, \xi^2$,

$$F(\xi) = F(\xi^1, \xi^2)$$

which are homogeneous in $\xi^1, \xi^2$ of degree $2S$. $2S$ is any nonnegative integer. In this space of dimension $2S + 1$ we define transformations $T_a$ for any $a \, \epsilon \, SL(2, C)$ by

$$T_a F(\xi) = F(\xi') = F(\xi a) \tag{2-25}$$

These transformations establish a $2S + 1$-dimensional representation of SL(2, C). The functions $F(\xi)$ are denoted covariant spinors to the spin $S$.

In order to obtain a more conventional notation we expand the polynomial $F$ into powers of $\xi^1$ and $\xi^2$

$$F(\xi) = \sum_{A=-S}^{S} \chi_A^{(S)} N_A^{(S)} (\xi^1)^{S+A} (\xi^2)^{S-A} \tag{2-26}$$

The $N_A^{(S)}$ are normalization constants

$$N_A^{(S)} = \left[ \frac{(2S)!}{(S+A)!(S-A)!} \right]^{1/2} \tag{2-27}$$

In terms of the expansion coefficients $\chi_A^{(S)}$ the transformation $T_a$ reads

$$(T_a \chi^{(S)})_A = \sum_{A'=-S}^{S} D_{AA'}^S(a) \chi_{A'}^{(S)} \tag{2-28}$$

Instead of using the functions $F$ we can also identify a covariant spinor with the

$2S + 1$-tupel of coefficients $\chi\binom{S}{A}$. From (2–25),(2–26), and (2–28) we read off easily

$$D_{AA'}^{S}(a) = \left[\frac{(S + A)!(S - A)!}{(S + A')!(S - A')!}\right]^{1/2} \sum_{n} \binom{S + A'}{n}\binom{S - A'}{S + A - n}$$
$$\times a_{11}^{n} a_{12}^{S+A-n} a_{21}^{S+A'-n} a_{22}^{n-A-A'} \qquad (2\text{-}29)$$

where the sum over $n$ extends over all integers for which neither of the binomial coefficients vanishes.

If we restrict the group SL(2, C) to the subgroup SU(2), we obtain the matrix $D_{AA'}^{S}(u)$ which defines a unitary irreducible representation of the rotation group SU(2) to the spin $S$. We may write for arbitrary $u \, \epsilon \,$ SU(2)

$$u = \exp\left\{\frac{i}{2}\psi_1\sigma_3\right\}\exp\left\{\frac{i}{2}\vartheta\sigma_2\right\}\exp\left\{\frac{i}{2}\psi_2\sigma_3\right\}$$

and get

$$D_{AA'}^{S}(u) = \exp i(A\psi_1 + A'\psi_2)\, d_{AA'}^{S}(\cos\vartheta)$$

with the function $d_{AA'}^{S}$ defined by

$$d_{AA'}^{S}(\cos\vartheta) = \left[\frac{(S + A)!(S - A)!}{(S + A')!(S - A')!}\right]^{1/2} \sum_{n}(-1)^{S+A'-n}$$
$$\times \binom{S + A'}{n}\binom{S - A'}{S + A - n}(\cos\tfrac{1}{2}\vartheta)^{2n-A-A'}(\sin\tfrac{1}{2}\vartheta)^{2S-2n+A+A'}$$

$$(2\text{-}30)$$

From (2–30) we derive the identities

$$d_{AA'}^{S}(\cos\vartheta) = (-1)^{A-A'}\, d_{A'A}^{S}(\cos\vartheta)$$
$$= (-1)^{A-A'}\, d_{-A,-A'}^{S}(\cos\vartheta)$$
$$= (-1)^{S+A'}\, d_{-A,A'}^{S}(\cos(\pi - \vartheta)) \qquad (2\text{-}31)$$

if we replace $n$ by $n$ itself, by $n + A + A'$, and by $S + A' - n$, respectively.

Each element $D_{AA'}^{S}(u)$ is a homogeneous polynomial of degree $2S$ in the matrix elements of $u$, and the coefficients of this polynomial are real. A substitution

$$u_{ij} \rightarrow a_{ij} \qquad i, j = 1, 2$$

leads us back to the matrix elements $D_{AA'}^S(a)$. If we write

$$u_{11} = \alpha + i\beta \qquad u_{12} = \gamma + i\delta$$
$$u_{21} = -\gamma + i\delta \qquad u_{22} = \alpha - i\beta$$

with real $\alpha, \beta, \gamma, \delta$ such that the constraint

$$\alpha^2 + \beta^2 + \gamma^2 + \delta^2 = 1$$

holds, we can interpret this substitution $u_{ij} \to a_{ij}$ as an extension of the real parameters $\alpha, \beta, \gamma, \delta$ into the complex domain. The constraint maintains its form under this extension. This is a complexification of SU(2) similar to that one studied in Section 1–1c, but it is free of singularities.

Because the coefficients of the polynomial $D_{AA'}^S(a)$ in $a_{ij}$ are real, we have

$$D_{AA'}^S(\bar{a}) = \overline{D_{AA'}^S(a)} \tag{2-32}$$

From the construction of these functions follows in addition

$$D_{AA'}^S(a^T) = D_{A'A}^S(a) \tag{2-33}$$

We define an element $\varepsilon$ of SU(2) by

$$\varepsilon = i\sigma_2 = \begin{pmatrix} 0 & 1 \\ -1 & 0 \end{pmatrix} \tag{2-34}$$

and have

$$(a^{-1})^T = \varepsilon a \varepsilon^{-1} \tag{2-35}$$

We introduce the matrix $C$ by

$$C_{AA'}^S = D_{AA'}^S(\varepsilon) \tag{2-36}$$

and have from (2–35)

$$D_{AA'}^S((a^{-1})^T) = (CD(a)C^{-1})_{AA'}^S$$

The definition (2–36) implies

$$C_{AA'}^S = (-1)^{S+A'} \delta_{A,-A'}$$

and

$$D_{AA'}^S((a^{-1})^T) = (-1)^{A-A'} D_{-A,-A'}^S(a) \tag{2-37}$$

Let us now consider a complex two-vector

$$\eta = (\eta^1, \eta^2)$$

which transforms as

$$\eta' = \eta\bar{a}$$

We can analogously construct a linear space of homogeneous polynomials in $\eta^1, \eta^2$ which are of degree $2S$

$$F(\eta) = F(\eta^1, \eta^2) \qquad T_a F(\eta) = F(\eta\bar{a})$$

By the expansion

$$F(\eta) = \sum_{A=-S}^{S} \chi_{\dot{A}}^{(S)} N_A^{(S)}(\eta^1)^{S+A}(\eta^2)^{S-A}$$

we obtain coefficients $\chi_{\dot{A}}^{(S)}$ which transform as

$$(T_a\chi^{(S)})_{\dot{A}} = \sum_{A'=-S}^{S} D_{AA'}^S(\bar{a})\chi_{\dot{A}'}^{(S)} \tag{2-38}$$

Because of $(2-32)$ these representations transform as complex conjugates of the covariant spinors; we call them therefore complex conjugate covariant spinors.

One might think that starting from spaces of polynomials in the variables $\zeta$, respectively $\vartheta$, which transform as

$$\zeta' = \zeta(a^{-1})^T$$

respectively,

$$\vartheta' = \vartheta(a^{-1})^\dagger$$

one could find new representations of SL(2, C). These would act on the $2S + 1$-dimensional spinors $\chi^A, \chi^{\dot{A}}$ as

$$(T_a\chi)^A = \sum_{A'} D_{AA'}^S((a^{-1})^T)\chi^{A'} \tag{2-39}$$

$$(T_a\chi)^{\dot{A}} = \sum_{A'} D_{AA'}^S((a^{-1})^\dagger)\chi^{\dot{A}'} \tag{2-40}$$

Relations (2–28), (2–33), and (2–39) imply

$$\sum_A (T_a\chi)^A(T_a\chi')_A = \sum_{ABC} \chi^B D^S_{BA}(a^{-1})D^S_{AC}(a)\chi'_C$$
$$= \sum_A \chi^A\chi'_A$$

namely, $\chi^A$ transforms contravariantly compared with the covariant spinor $\chi^A$. We call $\chi^A$, therefore, a contravariant spinor. A similar relation holds between the contravariant conjugate spinor $\chi^{\dot{A}}$ and the conjugate covariant spinor $\chi_{\dot{A}}$. Due to (2–35) the contravariant spinors can, however, be identified with the covariant spinors by

$$\chi_A = \sum_B C^{-1}_{AB}\chi^B \tag{2-41}$$

$$\chi_A = \sum_B C^{-1}_{AB}\chi^{\dot{B}} \tag{2-42}$$

The corresponding representations are therefore equivalent.

The contravariant spinors $\chi^A$ can also be considered from another point of view. The chain of relations

$$\overline{D(u)} = D(\bar{u}) = D((u^{-1})^T) \to D((a^{-1})^T)$$

and (2–39) tell us that these spinors result from continuing the complex conjugate representation of SU(2) to spin $S$ more complex, similar to the case considered earlier in this section. In this complexification the real parameters of $u_{ij}$ are extended into the complex domain so as to give the matrix elements $a_{ij}$. The corresponding analytic continuation of $(u^{-1})^T$ leads to $(a^{-1})^T$ as can be seen from

$$(u^{-1})^T = \varepsilon u \varepsilon^{-1} \qquad (a^{-1})^T = \varepsilon a \varepsilon^{-1}$$

The spinors $\chi_{\dot{A}}$ and $\chi^{\dot{A}}$ can be regarded analogously as obtained by an antianalytic continuation from representations of SU(2).

The spinor representations of SL(2, C) are irreducible in the elementary sense of invariant subspaces. Each tensor product of a covariant spinor with a conjugate covariant spinor is also irreducible. If we take tensor products $(S,S')$ of all pairs of a covariant spinor of spin $S$ and a complex conjugate spinor of spin $S'$, we exhaust all finite-dimensional irreducible representations of SL(2, C). Two spinor representations, $(S_1, S'_1)$ and $(S_2, S'_2)$, are equivalent if and only if $S_1 = S_2$ and $S'_1 = S'_2$.

## 2–6  CLASSICAL SPINOR FIELDS

We now make use of the fact that the representations of the little group SU(2) can be extended to spinor representations of SL(2, C) by analytic or antianalytic continuation as explained in Section 2–5. We define the two quantities

$$\psi_A(p) = \sum_{q=-S}^{S} D^S_{Aq}(a(p)^{-1})\varphi_q(p) \tag{2-43}$$

$$\psi^{\dot{A}}(p) = \sum_{q=-S}^{S} D^S_{Aq}(a(p)^\dagger)\varphi_q(p) \tag{2-44}$$

with $\varphi_q(p)$ as in (2–16). Since we have from (2–16)

$$r(a, p) = [r(a, p)^{-1}]^\dagger = [a(p)^{-1}]^\dagger (a^{-1})^\dagger a(p_a)^\dagger$$

these objects (2–43), (2–44) transform as

$$U_{(a, y)} \psi_A(p) = e^{ipy} \sum_{A'=-S}^{S} D^S_{AA'}(a)\psi_{A'}(p_a) \tag{2-45}$$

$$U_{(a, y)} \psi^{\dot{A}}(p) = e^{ipy} \sum_{A'=-S}^{S} D^S_{AA'}((a^{-1})^\dagger)\psi^{\dot{A}'}(p_a) \tag{2-46}$$

With these quantities the norm in Hilbert space can be expressed in several equivalent fashions

$$\|\psi\|^2 = \int d\mu(p) \sum_A \overline{\psi^A(p)}\psi_A(p)$$

$$= \int d\mu(p) \sum_{A, B} \overline{\psi_{\dot{A}}(p)} P^{\dot{A}B}\psi_B(p) \tag{2-47}$$

where we make explicit that the complex conjugate of $\psi_A$ transforms as $\psi_{\dot{A}}$, and so on. The matrices $P$ and $P'$ are defined by

$$P^{\dot{A}B} = \sum_{q=-S}^{S} D^S_{Aq}(a(p)^\dagger)D^S_{qB}(a(p)) = D^S_{AB}(M^{-1}\mathbf{p}) \tag{2-48}$$

$$\sum_B P^{\dot{A}B}P'_{B\dot{C}} = \delta_{\dot{C}}^{\dot{A}} \tag{2-49}$$

Due to the fact that both functions $\psi_A$ and $\psi^{\dot{A}}$ are formed of the same wave function $\varphi_q(p)$ they are linearly related by

$$\psi_A(p) = \sum_B P'_{A\dot{B}}\psi^{\dot{B}}(p) \tag{2-50}$$

$$\psi^{\dot{A}}(p) = \sum_B P^{\dot{A}B}\psi_B(p) \tag{2-51}$$

A similar construction can be based on the covariant wave function $\varphi_q(a)$ (2–18). In agreement with (2–18), (2–43), and (2–44) we define the functions

$$\psi_A(a) = \sum_q D^S_{Aq}(a^{-1})\varphi_q(a) \tag{2-52}$$

$$\psi^{\dot{A}}(a) = \sum_q D^S_{Aq}(a^\dagger)\varphi_q(a) \tag{2-53}$$

such that

$$\psi_A(p) \equiv \psi_A(a(p))$$

and so on. Instead of obeying the covariance relation (2–19) the functions (2–52) and (2–53) are constant on right cosets of SU(2)

$$\psi_A(ua) = \psi_A(a) \tag{2-54}$$

$$\psi^{\dot{A}}(ua) = \psi^{\dot{A}}(a) \tag{2-55}$$

We denote the functions (2–43), (2–44), (2–52), and (2–53) spinor wave functions.

We consider the Fourier transforms of $\psi_A(p)$ and $\psi^{\dot{A}}(p)$

$$\psi_A(x) = (2\pi)^{-3/2} \int d\mu(p) e^{-ipx} \psi_A(p) \tag{2-56}$$

$$\psi^{\dot{A}}(x) = (2\pi)^{-3/2} \int d\mu(p) e^{-ipx} \psi^{\dot{A}}(p) \tag{2-57}$$

The fact that the index transformations of $\psi_A(p)$ and $\psi^{\dot{A}}(p)$ are independent of $p$ is now crucial. We get simply

$$U_{(a,\,y)}^{-1} \psi_A(x) = \sum_{A'} D^S_{AA'}(a^{-1})\psi_{A'}(x') \tag{2-58}$$

$$U_{(a,\,y)}^{-1} \psi^{\dot{A}}(x) = \sum_{A'} D^S_{AA'}(a^\dagger)\psi^{\dot{A}'}(x') \tag{2-59}$$

where

$$\mathbf{x}' = a\mathbf{x}a^\dagger + \mathbf{y} \qquad x'^\mu = \Lambda_\nu{}^\mu(a)x^\nu + y^\mu \tag{2-60}$$

This transformation behavior (2–58) – (2–60) is typical for "classical free fields."

As can be inspected from the expression (2–29) the matrices $P$ and $P'$ can be regarded as homogeneous functions of degree $2S$ in the components of $p$ and of degree $-2S$ in $M$. The dependence on $M$ factored out, we can Fourier transform these matrices

$$(2\pi)^4 \mathscr{F}(P)^{\dot{A}B} \delta^{(4)}(x) = M^{2S} \int d^4 p e^{-ipx} p^{\dot{A}B} \tag{2-61}$$

$$(2\pi)^4 \mathscr{F}(P')_{A\dot{B}} \delta^{(4)}(x) = M^{2S} \int d^4 p e^{-ipx} P'_{A\dot{B}} \tag{2-62}$$

The differential operators $\mathscr{F}(P)$ and $\mathscr{F}(P')$ of order $2S$ obey

$$\sum_B \mathscr{F}(P)^{\dot{A}B} \mathscr{F}(P')_{B\dot{C}} = \left[ -\frac{\partial^2}{\partial x_0{}^2} + \sum_{k=1}^{3} \frac{\partial^2}{\partial x_k{}^2} \right]^{2S} \delta_{\dot{C}}{}^{\dot{A}} \tag{2-63}$$

The quantities $\psi_A(x)$ and $\psi^{\dot{A}}(x)$ themselves solve the "Bargmann-Wigner differential equations" which follow from $(2-50)$ and $(2-51)$

$$\sum_B \mathscr{F}(P)^{\dot{A}B} \psi_B(x) = M^{2S} \psi^{\dot{A}}(x) \tag{2-64}$$

and vice versa, and the "Klein-Gordon equation"

$$\left[ -\frac{\partial^2}{\partial x_0{}^2} + \sum_{k=1}^{3} \frac{\partial^2}{\partial x_k{}^2} \right] \psi(x) = M^2 \psi(x) \tag{2-65}$$

The Klein-Gordon equation follows from the Bargmann-Wigner equations only for $S = \frac{1}{2}$.

　　This derivation of the Bargmann-Wigner equations from the representation theory of the inhomogeneous SL(2, C) group is now to be forgotten. Instead we want to regard the quantities $\psi_A(x)$ and $\psi^A(x)$ as defined by their transformation properties $(2-58) - (2-60)$ and by the differential equations $(2-64)$ and $(2-65)$. We call such objects classical free spinor fields. They correspond to a reducible unitary representation of the inhomogeneous SL(2, C) with two irreducible components. These are characterized by $M^2$, $S$, and both signs sign $(p_0) = \pm 1$.

## 2–7　REMARKS

　　The subject of the first three sections of Chapter 2 has been dealt with in Wigner's famous article [37]. A discussion of the representations of the Poincaré group including the discrete transformations $I_s$ and $I_t$ has also been given by Wigner [39]. For a shorter account of the discrete transformations see Wightman [36]. The helicity formalism has been developed by Jacob and Wick [17]. Spinorial representations of SL(2, C) and the spinor calculus are treated in Bade and Jehle [4]. The connection of group theory and wave equations was

first emphasized by Bargmann and Wigner [7]. Weinberg [35] discusses
local quantized fields using the formalism of classical free spinor fields.

   Other treatments of the theory of unitary representations of the
inhomogeneous SL(2, C) are often based on the use of "plane wave states."
These describe states of an elementary particle with mass $M$ and spin $S$ for a
fixed four-momentum. This four-momentum is a contravariant vector in
Minkowski space like the position vector. If the spin is zero, we denote by
$|p^R\rangle$ the state of the particle at rest, and by $|p\rangle$ an arbitrary plane
wave state with momentum $p = \{p^\mu\}$. We require

$$U_a \, |p\rangle = |\Lambda(a)p\rangle \tag{2-66}$$

It is easy to see that such plane wave states can be identified with "improper"
wave functions

$$|p'\rangle \leftrightarrow \varphi(p) = 2p_0 \, \delta^{(3)}(p - p') \tag{2-67}$$

From (2–67), (2–16), and

$$2(p_a)_0 \, \delta^{(3)}(p_a - p') = 2p_0 \, \delta^{(3)}(p - p'_{a^{-1}})$$

$$(p'_{a^{-1}})^\mu = \Lambda_\nu{}^\mu(a)p'^\nu$$

follows (2–66) and

$$U^{-1}_{a(p)} |p^R\rangle = |p\rangle \tag{2-68}$$

In this fashion we can make our formulas agree completely with other
approaches based on plane wave states. We shall, however, never need such plane
wave states in these lectures.

# Chapter 3

# Representations of the Group SL(2, C)

Though it seems that the group SU(1, 1) plays a more important role in the theory of representations of the group SL(2, C) $\times$ T$_4$, it is for mathematical reasons more convenient to first study the representations of the group SL(2, C). The group SL(2, C) is in fact the most elementary example of a simple non-compact Lie group. The methods which we are going to present in dealing with the group SL(2, C) are later, and after some modifications, applied to the group SU(1, 1). In general we will, however, have physical applications in mind rather than the possibility of generalizing the results to other noncompact groups.

Unitary representations are in general realized in Hilbert spaces. Such Hilbert spaces are familiar to most theoretical physicists because of the use made of them in quantum mechanics. Although in standard textbooks of quantum mechanics Hilbert spaces are the only spaces called by a name, Dirac's formalism with bras and kets represents a generalization of the concept of Hilbert spaces insofar as it deals with nonnormalizable states. These can be considered as linear functionals over a dense subspace of a Hilbert space. Similar dense subspaces of a Hilbert space are of great importance in the theory of not necessarily unitary representations of noncompact groups like SL(2, C) and SU(1, 1).

These dense subspaces are themselves closed topological vector spaces (strictly speaking they are countably normed spaces) in a topology which is necessarily stronger than that of the Hilbert spaces in which they are embedded. They are introduced in Section 3–1 as spaces of functions which are homogeneous in certain variables. The definition of group operations in these spaces (Section 3–2) shows us how to relate these variables with cosets of a subgroup K in SL(2, C). This permits us to draw the connection to the theory of induced representations. Since there is a one-to-one correspondence between the homogeneous spaces SL(2, C)/K and SU(2)/U(1), a unique Banach space norm (which in fact turns out to be a Hilbert space norm) can be selected among an infinite number of different choices that allow us to complete the topological spaces. We need only embed our topological spaces in the Hilbert space of square integrable functions on SU(2)/U(1), which in turn can be regarded as a subspace of the Hilbert space $\mathscr{L}^2(U)$ of square integrable functions on SU(2). This

51

embedding procedure is denoted "canonical." Sections 3–3 to 3–5 are devoted to these problems.

In order to find out which of the representations constructed in this way are unitary and irreducible, we introduce the concepts of bilinear invariant functionals and intertwining operators in Sections 3–6 to 3–8. The important question of whether the unitary irreducible representations obtained are, up to equivalence, all that exist, is, however, left open. The matrix form of the intertwining operators is deduced in Section 3–9 by very simple arguments only from the existence of these operators. We need not use the explicit integral (or differential) operator form found for them in Section 3–8. In Section 3–10 we discuss a generalization of the concept of bilinear invariant functionals.

## 3–1　LINEAR SPACES OF HOMOGENEOUS FUNCTIONS

Infinite-dimensional representations of SL(2, C) can be constructed on spaces of homogeneous functions. These can be regarded as generalizations of the spaces of polynomials which are homogeneous with integral degrees $2S$, $2S'$ in the pairs of variables $(\xi^1, \xi^2)$ and $(\eta^1, \eta^2)$ (Section 2–4). We used such polynomial spaces to build up spinor representations of SL(2, C). Since it is possible to regard the second pair as the complex conjugate of the first, we denote these variables $(z_1, z_2)$, $(\bar{z}_1, \bar{z}_2)$ in the sequel. Functions of a complex argument, be they analytic or not, are written $f(z)$ or in the case of a pair of complex arguments, $F(z_1, z_2)$ or the like. We turn now to the basic definitions.

We call a function $F(z_1, z_2)$ homogeneous of degrees $\lambda$ and $\mu$ in the variables $z_1$ and $z_2$ if for any complex number $\alpha \neq 0$

$$F(\alpha z_1, \alpha z_2) = \alpha^\lambda \, \bar{\alpha}^\mu F(z_1, z_2) \tag{3-1}$$

with complex exponents $\lambda$ and $\mu$. In order to make the function $F$ single valued, the homogeneity condition

$$F(e^{i\omega}z_1, e^{i\omega}z_2) = e^{i(\lambda-\mu)\omega}F(z_1, z_2)$$

must reduce to an identity for $\omega = 2n\pi$ and any integer $n$. This forces us to require in addition that the difference

$$\mu - \lambda = m$$

be an integer. We shall make this assumption throughout. Instead of the degrees $\lambda$ and $\mu$ we shall characterize homogeneous functions by the labels

$$\chi = \{n_1, n_2\} = (m, \rho)$$

where

$$n_1 = \lambda + 1 = -\frac{1}{2}m + \frac{i}{2}\rho$$

$$n_2 = \mu + 1 = +\frac{1}{2}m + \frac{i}{2}\rho$$

(3-2)

In addition we use the notation

$$-\chi = \{-n_1, -n_2\} \quad \text{if} \quad \chi = \{n_1, n_2\}$$

It will soon become clear why these notations are convenient. We assume of course that $n_2 - n_1 = m$ also be an integer.

The linear spaces $\mathscr{D}_\chi$ of homogeneous functions are defined by the following postulates.

(1) $\mathscr{D}_\chi$ is a linear vector space with homogeneous functions of type $\chi$ as elements.

(2) Any element is infinitely differentiable in $z_1, z_2, \bar{z}_1, \bar{z}_2$ with the possible exception of the point $z_1 = z_2 = 0$.

(3) It is a topological vector space of the following kind. A sequence $\{F_q(z_1, z_2)\}$ is a null sequence if and only if it converges to zero uniformly together with all its derivatives on any compact set in the $z_1 z_2$-plane, which does not contain the point $z_1 = z_2 = 0$.

We shall not show that this indeed defines a topology.

It is easy to see that the spaces $\mathscr{D}_\chi$ are closed with this topology. The importance of the spaces $\mathscr{D}_\chi$ is due to the fact that we can define linear continuous operators $T_a{}^\chi$ in them for all elements $a \in SL(2, C)$ such that the group law

$$T^\chi_{a_1} T^\chi_{a_2} = T^\chi_{a_1 a_2}$$

is fulfilled. This is done explicitly in the subsequent section. These spaces are also in the domain of many unbounded operators of particular importance such as the group generators, the Casimir operators of the group and its subgroups, and the covariant operators.

## 3–2 THE GROUP OPERATIONS

We define an operator $T^\chi_a$ for any element of SL(2, C) in the space $\mathscr{D}_\chi$ exactly as in the case of spinor representations by

$$T_a{}^\chi F(z_1, z_2) = F(z'_1, z'_2) = F(z_1 a_{11} + z_2 a_{21}, z_1 a_{12} + z_2 a_{22}) \quad (3-3)$$

Since the origin $z_1 = z_2 = 0$ goes into itself, namely, $z'_1 = z'_2 = 0$, it is clear that

$T_a{}^\chi F$ is in $\mathscr{D}_\chi$ if $F$ itself is in $\mathscr{D}_\chi$. It is also a simple matter to show that $T_a{}^\chi$ is continuous in the group element $a$ and in the vector $F$ of the space $\mathscr{D}_\chi$. For the latter assertion we need only prove that any compact set in $z_1, z_2$ not containing the origin goes into a compact set which again does not contain the origin. This follows, however, from the continuity of the mapping $(z_1, z_2) \rightarrow (z_1', z_2')$ and the earlier mentioned trivial fact that the origin goes into itself under this mapping. Since we did not yet introduce a norm into the space $\mathscr{D}_\chi$ we cannot speak of bounded operators. Because boundedness was one of the postulates made in the definition of a representation, the operators do not establish a representation yet, though the group law

$$T_{a_1}^\chi T_{a_2}^\chi = T_{a_1 a_2}^\chi$$

is obviously fulfilled.

Instead of using the spaces of homogeneous functions it is sometimes very useful to consider other equivalent realizations of the spaces $\mathscr{D}_\chi$. We define a function

$$f(z) = F(z, 1)$$

Due to the homogeneity of $F(z_1, z_2)$ knowledge of the function $f(z)$ suffices to reconstruct $F(z_1, z_2)$,

$$F(z_1, z_2) = z_2^{n_1 - 1} \bar{z}_2^{n_2 - 1} f\left(\frac{z_1}{z_2}\right) \tag{3-4}$$

The function $f(z)$ depends only on one complex variable $z$. It is infinitely differentiable at all points $z$, the infinite far point included. At this infinite far point $f(z)$ possesses therefore an asymptotic series expansion

$$f(z) \cong z^{n_1 - 1} \bar{z}^{n_2 - 1} \sum_{j, k = 0}^{\infty} d_{jk} z^{-j} \bar{z}^{-k} \tag{3-5}$$

The topology of the spaces $\mathscr{D}_\chi$ of functions $F(z_1, z_2)$ can be carried over to the new space of functions $f(z)$. The space thus obtained will also be denoted $\mathscr{D}_\chi$. In fact, we do not make a difference between these realizations.

In the space of functions $f(z)$ the operator $T_a{}^\chi$ reads

$$T_a^\chi f(z) = \alpha(z, a) f(z_a)$$

$$z_a = \frac{a_{11} z + a_{21}}{a_{12} z + a_{22}} \tag{3-6}$$

$$\alpha(z, a) = (a_{12} z + a_{22})^{n_1 - 1} \overline{(a_{12} z + a_{22})}^{n_2 - 1}$$

Here we adopted the notations used for multiplier representations in Section 1–4a. In the following fashion we can show indeed that the complex number $z$ appearing as the argument of $f(z)$ represents a certain coset of a subgroup in SL(2, C).

We put the pair of variables $(z_1, z_2)$ of the function $F(z_1, z_2)$ into the bottom row of a matrix of SL(2, C). The transformation (3–3) of these variables can be formulated as a matrix multiplication

$$\begin{pmatrix} \cdots & \cdots \\ z_1 & z_2 \end{pmatrix} \begin{pmatrix} a_{11} & a_{12} \\ a_{21} & a_{22} \end{pmatrix} = \begin{pmatrix} \cdots & \cdots \\ z'_1 & z'_2 \end{pmatrix}$$

The matrix containing $z_1$ and $z_2$ can be decomposed into a product

$$\begin{pmatrix} \cdots & \cdots \\ z_1 & z_2 \end{pmatrix} = \begin{pmatrix} z_2^{-1} & \cdots \\ 0 & z_2 \end{pmatrix} \begin{pmatrix} 1 & 0 \\ \dfrac{z_1}{z_2} & 1 \end{pmatrix} \tag{3-7}$$

such that the matrix to the right can be identified with the argument $z = z_1/z_2$ of the functions $f$. Since we want the matrix

$$\begin{pmatrix} 1 & 0 \\ \dfrac{z_1}{z_2} & 1 \end{pmatrix}$$

to describe a right coset, we consider the subgroup of SL(2, C) which contains all the left factors of the product decomposition (3–7), that is, the subgroup K of SL(2, C) consisting of triangular matrices $k$,

$$k = \begin{pmatrix} \lambda^{-1} & \mu \\ 0 & \lambda \end{pmatrix} \qquad \lambda, \mu \text{ complex}$$

Since each element of SL(2, C) can be decomposed into the product

$$a = k \begin{pmatrix} 1 & 0 \\ z & 1 \end{pmatrix} \tag{3-8}$$

provided only $a_{22} \neq 0$, the right cosets SL(2, C)/K can be mapped one-to-one on complex numbers $z$ if $a_{22} \neq 0$, and the single coset corresponding to $a_{22} = 0$ is mapped on the infinite far point $z = \infty$. In particular the matrix

$$\begin{pmatrix} \cdots & \cdots \\ z_1 & z_2 \end{pmatrix}$$

belongs to the coset described by

$$\begin{pmatrix} 1 & 0 \\ \dfrac{z_1}{z_2} & 1 \end{pmatrix}$$

as desired. The mapping $(z_1, z_2) \to (z'_1, z'_2)$ can therefore be looked upon as a transformation of the set of right cosets SL(2, C)/K and can be formulated as

$$\begin{pmatrix} 1 & 0 \\ z & 1 \end{pmatrix}\begin{pmatrix} a_{11} & a_{12} \\ a_{21} & a_{22} \end{pmatrix} = \begin{pmatrix} \lambda(z, a)^{-1} & \mu \\ 0 & \lambda(z, a) \end{pmatrix}\begin{pmatrix} 1 & 0 \\ z_a & 1 \end{pmatrix} \tag{3-9}$$

The function $\alpha(z, a)$ of (3–6) is a multiplier and can be expressed by $\lambda(z, a)$ as

$$\alpha(z, a) = \lambda(z, a)^{n_1 - 1}\overline{\lambda(z, a)}^{n_2 - 1} \tag{3-10}$$

We can use $Dz$ as a quasi-invariant measure on SL(2, C)/K and have

$$\left(\frac{Dz_a}{Dz}\right)^{1/2} = |\lambda(z, a)|^{-2}$$

which implies (Section 1–4a)

$$\beta(z, a) = \left(\frac{Dz_a}{Dz}\right)^{-1/2} \alpha(z, a) = \lambda(z, a)^{n_1}\overline{\lambda(z, a)}^{n_2}$$

and

$$\beta(a) = a_{22}{}^{n_1} \overline{a_{22}}{}^{n_2}$$

This function $\beta(a)$ reduces to a one-dimensional, though not necessarily unitary, representation $\beta(k)$ of K,

$$\beta(k) = \lambda^{n_1}\overline{\lambda}^{n_2} \qquad k = \begin{pmatrix} \lambda^{-1} & \mu \\ 0 & \lambda \end{pmatrix}$$

The representation $\beta(k)$ is unitary if the parameter $\rho$ defined in (3–2) is real. In this case the theory of induced representations implies that $\mathscr{D}_\chi$ can be embedded in a Hilbert space $\mathscr{L}^2(Z)$ of measurable functions $f(z)$ with the finite norm $\|f\| = (f, f)^{1/2}$, where $(f_1, f_2)$ denotes the scalar product

$$(f_1, f_2) = \int \overline{f_1(z)}f_2(z)Dz$$

such that (3–6) can be extended on $\mathscr{L}^2(Z)$ and defines a unitary representation. These unitary representations are denoted "representations of the principal series"; we shall prove their irreducibility in Section 3–8.

## 3–3   THE CANONICAL BASIS

There exists another realization of the spaces $\mathscr{D}_\chi$ which is of great practical and theoretical importance. We start from the result of the preceding section which told us that $\mathscr{D}_\chi$ can be regarded as a space of functions $f(z)$ which are infinitely differentiable on the whole compact complex plane, each point $z$ of which has to be understood as a coset of SL(2, C)/K. We need only search for other possible representatives of these cosets.

It is known that all elements of SL(2, C) can be decomposed into the product

$$a = ku \quad k \in K \quad u \in SU(2)$$

The unitary matrix $u$ appearing in this decomposition can only be determined up to a phase $\gamma$,

$$\gamma = \begin{pmatrix} e^{i\omega} & 0 \\ 0 & e^{-i\omega} \end{pmatrix} \tag{3-11}$$

Instead of functions depending on $z$ we may therefore introduce functions $\varphi(u)$; however, in order to eliminate the ambiguous phase $\gamma$, we postulate an appropriate covariance

$$\varphi(\gamma u) = e^{im\omega}\varphi(u) \tag{3-12}$$

on the right cosets of U(1). Here, $m$ should be an integer to guarantee single-valuedness of $\varphi(u)$. A convenient choice for $m$ is found below.

We could as well have obtained the functions $\varphi(u)$ from the homogeneous functions $F(z_1, z_2)$. Because of the homogeneity, any function $F(z_1, z_2)$ is completely determined if its values on the sphere

$$|z_1|^2 + |z_2|^2 = 1$$

are given. We denote a point on this sphere by

$$z_1 = u_{21} \quad z_2 = u_{22}$$

Then

$$\varphi(u) = F(u_{21}, u_{22}) \tag{3-13}$$

defines a function on SU(2) if we put the matrix $u$ equal to

$$u = \begin{pmatrix} u_{11} & u_{12} \\ u_{21} & u_{22} \end{pmatrix} \qquad u_{11} = \overline{u_{22}} \qquad u_{12} = -\overline{u_{21}}$$

Due to the homogeneity of $F$ we get the additional constraint

$$\varphi(\gamma u) = F(e^{-i\omega}u_{21}, e^{-i\omega}u_{22}) = \exp\{i\omega(n_2 - n_1)\}\varphi(u)$$

The integer $m$ which was arbitrary in (3–12) now takes the definite value $n_2 - n_1$ in accord with (3–2). This is a very convenient definition of the covariance of $\varphi(u)$.

From the homogeneity we have

$$F(z_1, z_2) = (|z_1|^2 + |z_2|^2)^{(1/2)(n_1 + n_2) - 1}\varphi(u)$$

and from (3–4)

$$\varphi(u) = (|z_1|^2 + |z_2|^2)^{-(1/2)(n_1 + n_2) + 1} z_2^{n_1 - 1} \bar{z}_2^{n_2 - 1} f\left(\frac{z_1}{z_2}\right)$$

$$= (1 + |z|^2)^{-(1/2)(n_1 + n_2) + 1} e^{im\psi} f(z) \tag{3-14}$$

where we used the notations

$$z = \frac{z_1}{z_2} \qquad \psi = -\arg z_2 = -\arg u_{22}$$

If we parametrize $u$ by

$$u_{11} = t^{1/2} \exp\{i\theta_1\} \qquad u_{12} = (1 - t)^{1/2} \exp\{i\theta_2\} \tag{3-15}$$

$$0 \le t \le 1 \qquad 0 \le \theta_{1,2} \le 2\pi$$

we obtain the useful formulas

$$t = (1 + |z|^2)^{-1} \qquad \theta_1 = \psi \qquad \theta_2 = \pi - \arg z + \psi$$

$$d\mu(u) = (2\pi)^{-2} dt \, d\theta_1 \, d\theta_2$$

$$= 2(2\pi)^{-2}(1 + |z|^2)^{-2} Dz \, d\psi \tag{3-16}$$

which allow us to translate between $u$ and $z$.

The functions $\varphi(u)$ of $\mathscr{D}_\chi$ possess derivatives of all orders on SU(2). A coset $u$ of the homogeneous space SL(2, C)/K goes into $u_a$ via

$$ua = ku_a \qquad k = \begin{pmatrix} \lambda(u, a)^{-1} & \mu \\ 0 & \lambda(u, a) \end{pmatrix} \qquad (3\text{-}17)$$

The operator $T_a^\chi$ reads, correspondingly,

$$T_a^\chi \varphi(u) = \alpha(u, a)\varphi(u_a) \qquad (3\text{-}18)$$

with

$$\alpha(u, a) = \lambda(u, a)^{n_1 - 1}\overline{\lambda(u, a)}^{n_2 - 1} \qquad (3\text{-}19)$$

The Plancherel theorem for the group SU(2) can be applied to the functions $\varphi(u)$ of $\mathscr{D}_\chi$. According to this theorem the functions

$$(2j + 1)^{1/2}D^j_{q_1 q_2}(u) \qquad j = 0, \tfrac{1}{2}, 1, \tfrac{3}{2}, \ldots \qquad -j \leq q_{1, 2} \leq j$$

where $D^j_{q_1 q_2}(u)$ was defined in (2–29) as a matrix element of a representation of SU(2), are a complete and orthonormal system in the Hilbert space $\mathscr{L}^2(U)$ of measurable functions $\varphi(u)$ on SU(2) with finite norm $\|\varphi\| = (\varphi, \varphi)^{1/2}$ where $(\varphi_1, \varphi_2)$ denotes the scalar product

$$(\varphi_1, \varphi_2) = \int \overline{\varphi_1(u)}\varphi_2(u) \, d\mu(u)$$

Due to the covariance constraint on $\varphi(u)$, the subset of functions

$$\varphi_q^{\,j}(u) = (2j + 1)^{1/2}D^j_{(1/2)m, \, q}(u)$$
$$j = |\tfrac{1}{2}m| + n \qquad n = 0, 1, 2, \ldots \qquad -j \leq q \leq j \qquad (3\text{-}20)$$

lies in $\mathscr{D}_\chi$, and it is complete in $\mathscr{D}_\chi$. We call this set the canonical basis of $\mathscr{D}_\chi$. Any function $\varphi(u)$ of $\mathscr{D}_\chi$ can be expanded into a series

$$\varphi(u) = \sum_{j=|(1/2)m|}^{\infty} \sum_{q=-j}^{j} d_q^{\,j}\varphi_q^{\,j}(u)$$

with the Fourier coefficient $d_q^{\,j}$ defined by

$$d_q^{\,j} = \int \overline{\varphi_q^{\,j}(u)}\varphi(u) \, d\mu(u)$$

This coefficient has the asymptotic behavior

$$\lim_{j \to \infty} \sup_q (|d_q{}^j| j^n) = 0 \qquad \text{for all} \quad n$$

We call this assertion an analog of the Paley-Wiener theorem. The proof is easy, we postpone it to our remarks in Section 3–11.

By means of the formula (3–14) we can also give the canonical basis in terms of functions $f_q{}^j(z)$. These functions $f_q{}^j(z)$ are orthogonal in the sense

$$\frac{1}{\pi} \int \overline{f_{q_1}^{j_1}(z)} f_{q_2}^{j_2}(z)(1 + |z|^2)^{\operatorname{Im} \rho} Dz = \delta_{j_1 j_2} \delta_{q_1 q_2}$$

as can be seen from (3–14) and (3–16). We can correspondingly construct Hilbert spaces $\mathscr{L}_{\rho_1}^2(Z)$, $\rho_1 = \operatorname{Im} \rho$, consisting of measurable functions $f(z)$ with finite norm $\|f\| = (f, f)^{1/2}$, where $(f_1, f_2)$ is the scalar product

$$(f_1, f_2) = \frac{1}{\pi} \int \overline{f_1(z)} f_2(z)(1 + |z|^2)^{\rho_1} Dz$$

in which the functions $f_q{}^j(z)$ are orthonormal and complete. In the case of the principal series where $\operatorname{Im} \rho = 0$, we are led back this way to the Hilbert space $\mathscr{L}^2(Z)$ with invariant norm found by the method of induction in Section 3–2. The functions $f_q{}^j(z)$ are explicitly given in Section 3–9.

The subspace of $\mathscr{L}^2(U)$ which we obtain from a completion of $\mathscr{D}_\chi$ is called $\mathscr{L}_m{}^2(U)$. It consists of those elements $\varphi(u)$ of $\mathscr{L}^2(U)$ which are covariant in the sense (3–12) for almost all $u$. The fact that all spaces $\mathscr{D}_\chi$ allow us to introduce a norm in a unique way, which after a completion leads to a Hilbert space, is certainly rather important. If we would have admitted arbitrary norms, we would have obtained an infinite set of different Banach spaces with common dense subspace $\mathscr{D}_\chi$. The single "nucleus of a representation" on $\mathscr{D}_\chi$ would have implied an infinity of representations on the different Banach spaces. The canonical norm supplied by $\mathscr{L}_m{}^2(U)$ allows us to select a unique extension of this nucleus onto a normed space.

Moreover the Hilbert space $\mathscr{L}_m{}^2(U)$ depends only on the difference $m$ of the two numbers $n_1$ and $n_2$. The sum $n_1 + n_2$ is an arbitrary complex number. The set of representations with a fixed parameter $m$ possesses operators in the same Hilbert space $\mathscr{L}_m{}^2(U)$. These operators are therefore operator functions of $n_1 + n_2 = i\rho$. their matrix elements in the canonical basis are analytic in $\rho$. This enables us therefore to speak of analytic continuations of a representation.

In these statements we anticipated that $T_a{}^\chi$ could be extended from $\mathscr{D}_\chi$ onto $\mathscr{L}_m{}^2(U)$. For this to be possible it suffices to show that $T_a{}^\chi$ is bounded on $\mathscr{D}_\chi$ with respect to the canonical norm. We compute the operator norm of $T_a{}^\chi$ in the subsequent section.

## 3–4    THE OPERATOR NORM OF THE GROUP ELEMENTS

The operator $T_u{}^\chi$ can be extended from $\mathscr{D}_\chi$ onto $\mathscr{L}_m^2(U)$ and yields a unitary operator. Indeed, from the definition (3–17)

$$T_u{}^\chi \varphi(u_1) = \varphi(u_1 u)$$

we deduce the isometry of $T_u{}^\chi$ on $\mathscr{D}_\chi$

$$\|T_u{}^\chi \varphi\|^2 = \|\varphi\|^2 \qquad \varphi \in \mathscr{D}_\chi$$

$T_u{}^\chi$ can then be extended onto $\mathscr{L}_m^2(U)$ and gives an isometric operator. The same can be done with $T_{u^{-1}}^{\chi}$. Since under this extension the group law is conserved, $T_u{}^\chi$ possesses the inverse $T_{u^{-1}}^{\chi}$ and is unitary.

An arbitrary element $a \in SL(2, C)$ can be decomposed as

$$a = u_1 \, d u_2 \tag{3-21}$$

where $u_{1,2}$ are in SU(2) and $d$ is diagonal

$$d = \begin{pmatrix} e^{\eta/2} & 0 \\ 0 & e^{-\eta/2} \end{pmatrix} \qquad \eta \geqq 0 \tag{3-22}$$

When we have proved that the norm of $T_d{}^\chi$ is finite over $\mathscr{D}_\chi$, we can show easily that $T_a{}^\chi$ has also finite norm equal to that of $T_d{}^\chi$. In fact it is

$$\|T_a{}^\chi\|^2 = \|T_a{}^\chi T_a{}^{\chi\dagger}\| = \|T_{u_1}^{\chi} T_d{}^\chi T_{u_2}^{\chi} \, T_{u_2}^{\chi\dagger} T_d{}^{\chi\dagger} T_{u_1}^{\chi\dagger}\| = \|T_a{}^\chi\|^2$$

We compute therefore the norm of $T_d{}^\chi$. We use the definition

$$\|T_d{}^\chi\|^2 = \sup_{\varphi \in \mathscr{D}_\chi} \frac{\|T_d{}^\chi \varphi\|^2}{\|\varphi\|^2} \tag{3-23}$$

into which we insert

$$\|T_d{}^\chi \varphi\|^2 = \int |\lambda(u, d)|^{2\,\mathrm{Re}(n_1 + n_2) - 4} |\varphi(u_d)|^2 \, d\mu(u)$$

where we may replace

$$d\mu(u) = |\lambda(u, d)|^4 \, d\mu(u_d)$$

The parameters (3–15) give

$$|\lambda(u, d)|^2 = (1 - t)e^\eta + te^{-\eta}$$

We get finally

$$\| T_d{}^\chi \| = \max_{0 \le t \le 1} \left[ (1 - t)e^\eta + te^{-\eta} \right]^{(\mathrm{Re}(n_1 + n_2))/2} \tag{3-24}$$

$$= \exp\{\tfrac{1}{2}\eta \, |\mathrm{Re}(n_1 + n_2)|\}$$

With the notation (3–2) we can rewrite this result as

$$\| T_d{}^\chi \| = \exp\{\tfrac{1}{2}\eta |\mathrm{Im}\,\rho|\} \tag{3-25}$$

If $\rho$ is real, the norm of the operators $T_a{}^\chi$ is 1. This is the case of the principal series.

We introduce some further notations. For any element $a \in \mathrm{SL}(2, \mathrm{C})$ we define a matrix norm $|a|$ by

$$|a|^2 = \mathrm{Tr}\,(aa^\dagger) = \sum_{ij} |a_{ij}|^2 \tag{3-26}$$

This norm has the properties

$$|a| = |a^{-1}| = |a^\dagger| \qquad |\alpha a| = |\alpha||a|$$

$$|a_1 + a_2| \le |a_1| + |a_2| \qquad |a_1 a_2| \le |a_1||a_2|$$

where $\alpha$ is any complex number. Since any element of $\mathrm{SL}(2, \mathrm{C})$ admits of the decomposition (3–21), (3–22), we can express the norm $|a|$ by $\eta$ alone,

$$|a|^2 = 2 \cosh \eta \tag{3-27}$$

The norm $|a|$ has a positive minimum $2^{1/2}$ on $\mathrm{SL}(2, \mathrm{C})$.

We say that a function $f(a)$ on $\mathrm{SL}(2, \mathrm{C})$ is *polynomially bounded* if real numbers $s$ exist for which

$$|a|^{-s} f(a)$$

is bounded. We call $\inf(s)$ the polynomial order of the function $f(a)$. With these notations we can formulate our result on the operator norm of $T$ as:

$$\| T_a{}^\chi \| = \left[ \tfrac{1}{2}(|a|^2 + \sqrt{|a|^4 - 4}) \right]^{|\mathrm{Im}\,\rho|/2}$$

A weaker formulation of our result is: The polynomial order of $\|T\|$ is $|\mathrm{Im}\,\rho|$.

## 3–5   AN INTEGRAL REPRESENTATION FOR THE MATRIX ELEMENTS OF THE GROUP ELEMENTS IN THE CANONICAL BASIS

For the matrix elements in the canonical basis (3–20) we use the notations

$$D^\chi_{j_1 q_1 j_2 q_2}(a) = \langle j_1 q_1 | T_a^\chi | j_2 q_2 \rangle = (\varphi^{j_1}_{q_1}, T_a^\chi \varphi^{j_2}_{q_2})$$

Since

$$T_u^\chi \varphi(u_1) = \varphi(u_1 u)$$

the space spanned by $\varphi_q^{\ j}$ with fixed $j$ is invariant against transformations of the subgroup SU(2). The matrix elements are

$$D^\chi_{j_1 q_1 j_2 q_2}(u) = \delta_{j_1 j_2} D^{j_1}_{q_1 q_2}(u)$$

The space $\mathscr{L}_m^2(U)$ decomposes therefore into a direct orthogonal sum of $(2j + 1)$-dimensional Hilbert spaces $\mathscr{H}_j$, each of which carries an irreducible representation of SU(2) with spin $j$,

$$\mathscr{L}_m^2(U) = \sum_{j=|(1/2)m|}^{\infty} \oplus \quad \mathscr{H}_j \qquad \dim \mathscr{H}_j = 2j + 1 \qquad (3\text{-}28)$$

Once again we make use of the possibility of decomposing elements of SL(2, C) as in (3–21), (3–22). This decomposition implies

$$D^\chi_{j_1 q_1 j_2 q_2}(a) = \sum_q D^{j_1}_{q_1 q}(u_1) D^{j_2}_{q q_2}(u_2) D^\chi_{j_1 q j_2 q}(d) \qquad (3\text{-}29)$$

where we anticipated that the matrix of $T_d^\chi$ is diagonal as

$$D^\chi_{j_1 q_1 j_2 q_2}(d) = \delta_{q_1 q_2} D^\chi_{j_1 q_1 j_2 q_1}(d) \qquad (3\text{-}30)$$

We need therefore compute only the matrix elements of $T_d^\chi$ which from now on we denote by

$$d^\chi_{j_1 j_2 q}(\eta) = D^\chi_{j_1 q j_2 q}(d) \qquad (3\text{-}31)$$

By definition we have

$$d^\chi_{j_1 j_2 q}(\eta) = \int \overline{\varphi^{j_1}_q(u)} \lambda(u, d)^{n_1 - 1} \overline{\lambda(u, d)}^{n_2 - 1} \varphi^{j_2}_q(u_d) \, d\mu(u)$$

In the parameters (3–15) we obtain

$$\lambda(u, d) = [te^{-\eta} + (1 - t)e^\eta]^{1/2}$$

$$t_d = te^{-\eta}[te^{-\eta} + (1 - t)e^\eta]^{-1}$$

$$(\theta_1)_d = \theta_1 \qquad (\theta_2)_d = \theta_2$$

and

$$D^j_{(1/2)m,\,q}(u) = \exp\left\{\frac{i}{2}\,m(\theta_1 + \theta_2) + iq(\theta_1 - \theta_2)\right\}d^j_{(1/2)m,\,q}(2t - 1)$$

with $d^j_{q_1 q_2}$ as in (2–30). In this fashion we get the integral representation

$$d^\chi_{j_1 j_2 q}(\eta) = (2j_1 + 1)^{1/2}(2j_2 + 1)^{1/2}\int_0^1 dt d^{j_1}_{(1/2)m,\,q}(2t - 1)$$

$$\times\, d^{j_2}_{(1/2)m,\,q}(2t_d - 1)[te^{-\eta} + (1 - t)e^\eta]^{(i/2)\rho - 1} \qquad (3\text{-}32)$$

It is easy to see that this expression (3–32) yields an elementary function of $\eta$. We can generalize this result immediately and assert that all matrix elements of any representation in the canonical basis are elementary functions of the elements of the 2 × 2 matrix $a \in SL(2, C)$. However in general these functions are rather complicated aggregates of elementary functions. In addition we notice that $d^{(m,\rho)}_{j_1 j_2 q}(\eta)$ is entire in $\rho$. This is due to the fact that the square bracket under the integral sign in (3–32) has a positive minimum on the interval $0 \leqq t \leqq 1$.

The following symmetries can easily be read off the integral representation (3–32) for $d^{(m,\rho)}_{j_1 j_2 q}(\eta)$. If $\rho$ is real, the representation belongs to the principal series and the matrix elements constitute a unitary matrix. This implies

$$d^{(m,\,-\rho)}_{j_1 j_2 q}(\eta) = \overline{d^{(m,\,\rho)}_{j_1 j_2 q}(\eta)} = d^{(m,\,\rho)}_{j_2 j_1 q}(-\eta)$$

or

$$d^{(m,\,\rho)}_{j_1 j_2 q}(\eta) = d^{(m,\,-\rho)}_{j_2 j_1 q}(-\eta) \qquad (3\text{-}33)$$

In the last form the relation can be continued analytically in $\rho$ such that it holds for all complex $\rho$. The label $m$ goes into the integral representation (3–32) only through the rotation functions $d^j_{q_1 q_2}$. These functions satisfy the identities (2–31), in particular

$$d^j_{(1/2)m,\,q}(2t - 1) = (-1)^{j+q}d^j_{-(1/2)m,\,q}(-2t + 1)$$

where $t$ has been replaced by $1 - t$ in the argument. Under the same replacement and if we change the sign of $\eta$ simultaneously, $t_d$ goes into $1 - t_d$ and the square bracket

$$[te^{-\eta} + (1 - t)e^\eta]^{(i/2)\rho - 1}$$

is invariant. We obtain in this fashion

$$d^{(m,\,\rho)}_{j_1 j_2 q}(\eta) = (-1)^{j_1 - j_2}d^{(-m,\,\rho)}_{j_1 j_2 q}(-\eta) \qquad (3\text{-}34)$$

and if we use (3–33) to replace $-\eta$ by $\eta$

$$d_{j_1 j_2 q}^{(m; \rho)}(\eta) = (-1)^{j_1 - j_2} d_{j_2 j_1 q}^{(-m, -\rho)}(\eta) \tag{3-35}$$

Two further relations can be derived by trivial arguments from the other two identities (2–31)

$$d_{j_1 j_2 q}^{(m; \rho)}(\eta) = d_{j_1 j_2, -q}^{(-m, \rho)}(\eta) = d_{j_1 j_2, (1/2)m}^{(2q, \rho)}(\eta)$$

A less trivial relation connecting $d_{j_1 j_2 q}^{(m,\rho)}(\eta)$ with $d_{j_1 j_2 q}^{(m,\rho)}(\eta)$ is derived in Section 3–9. A series expansion of $d_{j_1 j_2 q}^{\chi}(\eta)$ is given in Section 4–5.

Finally we give the simplest example of a function $d_{j_1 j_2 q}^{\chi}(\eta)$ explicitly, namely,

$$d_{000}^{(0, \rho)}(\eta) = \frac{2 \sin (1/2) \eta \rho}{\rho \sinh \eta} \tag{3-36}$$

which can be computed easily from the integral (3–32).

## 3–6  INVARIANT BILINEAR FUNCTIONALS

In the preceding sections the discussion of representations was based on an elementary use of the spaces of homogeneous functions. Though we have accomplished already the explicit matrix form of the representation operators, some more fundamental problems remain unsolved, for example, proving the irreducibility, the equivalence, and the unitarity of such representations. It turns out that one further concept, namely, that of bilinear invariant functionals, enables us to settle the issue of establishing equivalence and unitarity completely, whereas irreducibility can be proved only for unitary representations. As mentioned in Section 1–3a, we will be content with such a limited solution. The method of constructing bilinear invariant functionals is of great practical interest itself because it can be generalized immediately to derive multilinear invariant functionals, in particular covariant operators, in a very elegant fashion. We can, however, touch these questions of generalizations only briefly in Section 3–10.

The bilinear functional $B(h, f)$ is defined for any pair $f \epsilon \mathcal{D}_\chi$ and $h \epsilon \mathcal{D}_{\chi'}$ and is separately continuous on both $\mathcal{D}_\chi$ and $\mathcal{D}_{\chi'}$. Bilinearity means

$$B(\alpha_1 h_1 + \alpha_2 h_2, \beta_1 f_1 + \beta_2 f_2) = \sum_{ij} \alpha_i \beta_j B(h_i, f_j) \tag{3-37}$$

It is moreover called invariant with respect to the representations $\chi$ and $\chi'$ if

$$B(T_a^{\chi'} h, T_a^{\chi} f) = B(h, f) \tag{3-38}$$

for all $a$. It is easy to see that any element of SL(2, C) can be decomposed into a product of matrices of the following types:

(i) Translations $\zeta$,

$$\zeta = \begin{pmatrix} 1 & 0 \\ z & 1 \end{pmatrix} \qquad T_\zeta^\chi f(z_1) = f(z_1 + z) \tag{3-39}$$

(ii) Dilatations $\delta$,

$$\delta = \begin{pmatrix} \lambda^{-1} & 0 \\ 0 & \lambda \end{pmatrix} \tag{3-40}$$

$$T_\delta^\chi f(z) = \lambda^{n_1 - 1} \bar{\lambda}^{n_2 - 1} f(\lambda^{-2} z)$$

(iii) The inversion $\varepsilon$ (2–34)

$$\varepsilon = i\sigma_2 = \begin{pmatrix} 0 & 1 \\ -1 & 0 \end{pmatrix}$$

$$T_\varepsilon^\chi f(z) = z^{n_1 - 1} \bar{z}^{n_2 - 1} f\left(-\frac{1}{z}\right) \tag{3-41}$$

If we succeed in constructing a bilinear functional that is invariant with respect to transformations of these three types, we can be sure of full invariance.

The very powerful method which we are going to use is based on distributions. We consider realizations of the spaces $\mathscr{D}_\chi$ as spaces of functions $f(z)$ where the complex number $z$ is a representative of a coset of SL(2, C)/K. As usual we introduce a space of test functions. We define the space $\mathscr{C}^\infty$ to consist of all functions $f(z)$ of $\mathscr{D}_\chi$ which vanish outside a compact set in the finite $z$-plane. (In this course we shall always restrict ourselves to spaces of test functions which are infinitely differentiable and vanish outside a compact set. We denote such spaces $\mathscr{C}^\infty$. The linear continuous functionals on $\mathscr{C}^\infty$ are called distributions.) The space $\mathscr{C}^\infty$ is a common subspace (not dense) of all $\mathscr{D}_\chi$. The convergence in $\mathscr{C}^\infty$ is as usual: A sequence $f_n(z)$ is a null sequence if and only if a compact set D in the $z$-plane exists such that $f_n(z) = 0$ outside D for all $n$, and if $f_n(z)$ is a null sequence in the topology of $\mathscr{D}_\chi$.

We consider two elements $f \in \mathscr{D}_\chi$ and $h \in \mathscr{D}_{\chi'}$ which are in $\mathscr{C}^\infty$. Let

$$\chi = \{n_1, n_2\} \qquad \chi' = \{N_1, N_2\}$$

Taking into account invariance against translations we make the following ansatz for the bilinear functional

$$B(h, f) = \int Dz_2 \, M(z_2) \int Dz_1 h(z_1 + z_2) f(z_1) \tag{3-42}$$

$M(z)$ is assumed to be a distribution. In fact, the inner integral of (3–42) defines an element of $\mathscr{C}^\infty$. One can prove that this ansatz (3–42) is the most general one that is in accord with translational invariance and the premises of linearity and continuity. The integral (3–42) may also formally be written

$$\int M(z_2 - z_1)h(z_2)f(z_1)Dz_1 Dz_2$$

We impose dilatational invariance on this integral (3–42). Inserting the particular dilatation $\delta = e_-, \lambda = -1$, yields the constraint

$$n_1 + N_1 \cong n_2 + N_2 \bmod 2 \tag{3-43}$$

For general matrices $\delta$ we get

$$\int Dz_2\, M(z_2) \int Dz_1 h(z_1 + z_2)f(z_1)$$

$$= \int Dz_2\, M(z_2) \int Dz_1 \lambda^{n_1 + N_1 - 2} \bar{\lambda}^{n_2 + N_2 - 2} h(\lambda^{-2}(z_1 + z_2))f(\lambda^{-2}z_1)$$

$$= \int Dz_2\, M(z_2)\lambda^{n_1 + N_1}\bar{\lambda}^{n_2 + N_2} \int Dz_1 h(z_1 + \lambda^{-2}z_2)f(z_1)$$

Invariance requires therefore that $M(z)$ be a homogeneous distribution of degree

$$-\tfrac{1}{2}(n_1 + N_1) - 1 = s_1 - 1 \quad \text{in} \quad z$$

and

$$-\tfrac{1}{2}(n_2 + N_2) - 1 = s_2 - 1 \quad \text{in} \quad \bar{z}$$

All homogeneous distributions are known. They have the form

$$M(z) = cz^{s_1 - 1}\bar{z}^{s_2 - 1} \tag{3-44}$$

($c$ is a constant) except the singular case in which $s_1$ and $s_2$ are both negative integers or zero. In this case the homogeneous distribution is

$$M(z) = c\left(\frac{\partial}{\partial z}\right)^{-s_1}\left(\frac{\partial}{\partial \bar{z}}\right)^{-s_2}\delta(z) \tag{3-45}$$

The nonsingular alternative (3–44) leads to the functional

$$B(h, f) = \int (z_2 - z_1)^{s_1 - 1}(\bar{z}_2 - \bar{z}_1)^{s_2 - 1}h(z_2)f(z_1)Dz_2 Dz_1 \tag{3-46}$$

and the singular alternative (3—45) to

$$B(h, f) = \int h(z)\left(\frac{\partial}{\partial z}\right)^{-s_1}\left(\frac{\partial}{\partial \bar{z}}\right)^{-s_2} f(z)Dz \qquad (3\text{-}47)$$

We emphasize that $M(z)$ has to be understood as a distribution. This involves the fact that integrals have to be regularized tacitly if necessary. For instance, if

$$-v < \text{Re}(s_1 + s_2) < -v + 1 \qquad v > 0 \qquad \text{an integer}$$

the integral (3—46) is meant in the sense

$$B(h, f) = \int Dz_2\, z_2^{s_1-1}\bar{z}_2^{s_2-1} \int D(z_1)\left[h(z_1 + z_2)\right.$$

$$\left. - \sum_{i,\,j=0}^{i+j=v-1} \frac{z_2^i \bar{z}_2^j}{i!\,j!}\left(\frac{\partial}{\partial z_1}\right)^i\left(\frac{\partial}{\partial \bar{z}_1}\right)^j h(z_1)\right] f(z_1) \qquad (3\text{-}48)$$

which converges at $z_2 = 0$ and at $z_2 = \infty$.

We assume now in addition that $f$ and $h$ vanish in a neighborhood of $z = 0$. Then we impose on (3—46) and (3—47) invariance under inversion. In the nonsingular alternative (3—46) we have the condition

$$B(h, f) = \int (z_2 - z_1)^{s_1-1}(\bar{z}_2 - \bar{z}_1)^{s_2-1} z_1^{n_1-1} z_2^{N_1-1}$$

$$\times\; \bar{z}_1^{n_2-1}\bar{z}_2^{N_2-1} h(-z_2^{-1}) f(-z_1^{-1})\, Dz_1\, Dz_2$$

$$= \int (z_2 - z_1)^{s_1-1}(\bar{z}_2 - \bar{z}_1)^{s_2-1} z_1^{-s_1-n_1} z_2^{-s_1-N_1}$$

$$\times\; \bar{z}_1^{-s_2-n_2}\bar{z}_2^{-s_2-N_2} h(z_2) f(z_1)\, Dz_1\, Dz_2$$

Invariance is possible only if

$$s_1 + n_1 = s_1 + N_1 = 0$$

$$s_2 + n_2 = s_2 + N_2 = 0$$

which means

$$n_1 = N_1 = -s_1$$

$$n_2 = N_2 = -s_2$$

We formulate this result for the nonsingular alternative explicitly: For two representations $\chi$ and $\chi'$

$$\chi = \{n_1, n_2\} \qquad \chi' = \{N_1, N_2\}$$

where

$$-s_1 = \tfrac{1}{2}(n_1 + N_1) \qquad -s_2 = \tfrac{1}{2}(n_2 + N_2)$$

are not simultaneously nonnegative integers, a bilinear invariant functional can exist only if

$$\chi = \chi' \qquad n_1 = N_1 \qquad n_2 = N_2$$

If $f$ and $h$ are in $\mathscr{C}^\infty$ and if both vanish in a neighborhood of $z = 0$, the functional exists and has the form

$$B(h, f) = \int (z_2 - z_1)^{-n_1 - 1}(\bar{z}_2 - \bar{z}_1)^{-n_2 - 1} h(z_2) f(z_1)\, Dz_1\, Dz_2 \quad (3\text{-}49)$$

which eventually has to be understood in the sense of a regularized integral [see (3–48)].

It is possible to show that this expression appropriately regularized can be extended onto the whole spaces $\mathscr{D}_\chi$ and $\mathscr{D}_{\chi'}$, and thereby remains a bilinear invariant functional. In fact, the functions in these spaces are regular enough to allow for eventually necessary subtractions, and the topology of $\mathscr{D}_\chi$ is strong enough to make the integral (3–49) a continuous functional. Though the proof of these assertions is easy, we do not want to go through the lengthy arguments here. Instead we consider the singular alternative (3–47).

If $f$ and $h$ vanish in a neighborhood of $z = 0$, the integral (3–47) yields under inversion

$$B(h, f) = \int h(-z^{-1}) z^{N_1 - 1} \bar{z}^{N_2 - 1} \left(\frac{\partial}{\partial z}\right)^{-s_1} \left(\frac{\partial}{\partial \bar{z}}\right)^{-s_2}$$

$$\times \{z^{n_1 - 1}\bar{z}^{n_2 - 1} f(-z^{-1})\}\, Dz$$

$$= \int h(z) z^{-N_1 - 1} \bar{z}^{-N_2 - 1} \left(z^2 \frac{\partial}{\partial z}\right)^{-s_1}$$

$$\times \left(\bar{z}^2 \frac{\partial}{\partial \bar{z}}\right)^{-s_2} \{z^{-n_1 + 1}\bar{z}^{-n_2 + 1} f(z)\}\, Dz$$

To perform the differentiations it is most convenient to use a combinatorial approach. We make an ansatz

$$\left(x^2 \frac{d}{dx}\right)^p \{x^{-q+1} f(x)\} = \sum_{n=0}^{p} a_p^{\,n}(q) x^{p-q+n+1} f^{(n)}(x) \qquad (3\text{-}50)$$

where $p$ is a nonnegative integer and $q$ is arbitrary. This ansatz can be justified by showing that it implies the consistent recursion relation for $a_p{}^n(q)$

$$a_{p+1}^n(q) = (p - q + n + 1)a_p{}^n(q) + a_p^{n-1}(q) \tag{3-51}$$

with the initial condition $a_p{}^p(q) = 1$. The recursion relation (3–51) possesses the unique solution

$$a_p{}^n(q) = \binom{p}{n} \frac{\Gamma(p - q + 1)}{\Gamma(n - q + 1)} \tag{3-52}$$

In order that the right-hand side of (3–50) reduce to

$$x^{2p-q+1}f^{(p)}(x)$$

we must require that

$$a_p{}^n(q) = 0 \qquad \text{for} \quad 0 \leq n < p$$

This is possible only if $p = 0$ (in which case the requirement is void) or if $p = q$.

    Inserting this result into the integral under consideration, we see that only the following four possibilities are in accord with invariance against inversion:

(1)    $-s_1 = n_1$,    $-s_2 = n_2$,    i.e.,   $n_1 = N_1$,   $n_2 = N_2$;

(2)    $-s_1 = n_1$,    $s_2 = 0$,    i.e.,   $n_1 = N_1$,   $n_2 = -N_2$;

(3)    $s_1 = 0$,    $-s_2 = n_2$,    i.e.,   $n_1 = -N_1$,   $n_2 = N_2$;

(4)    $s_1 = s_2 = 0$,    i.e.,   $n_1 = -N_1$,   $n_2 = -N_2$.

We formulate this result for the singular alternative as follows:

    If two representations $\chi$ and $\chi'$ are such that

$$-s_1 = \tfrac{1}{2}(n_1 + N_1) \qquad -s_2 = \tfrac{1}{2}(n_2 + N_2)$$

are both nonnegative integers, a bilinear invariant functional can exist only in the following four cases (which differ a little bit from the preceding cases; we shall later refer to this and not to the preceding classification):

(1)   $n_1 = N_1$, $n_2 = N_2$ ($n_1$ and $n_2$ are both positive integers)

$$B(h, f) = \int h(z) \left(\frac{\partial}{\partial z}\right)^{n_1} \left(\frac{\partial}{\partial \bar{z}}\right)^{n_2} f(z) \, Dz \tag{3-53}$$

(2)  $n_1 = N_1$, $n_2 = -N_2$ ($n_1$ is a positive integer, $n_2$ an arbitrary integer)

$$B(h, f) = \int h(z) \left( \frac{\partial}{\partial z} \right)^{n_1} f(z) \, Dz \qquad (3\text{-}54)$$

(3)  $n_1 = -N_1$, $n_2 = N_2$ ($n_2$ is a positive integer, $n_1$ an arbitrary integer)

$$B(h, f) = \int h(z) \left( \frac{\partial}{\partial \bar{z}} \right)^{n_2} f(z) \, Dz \qquad (3\text{-}55)$$

(4)  $n_1 = -N_1$, $n_2 = -N_2$ ($n_1, n_2$ are arbitrary numbers)

$$B(h, f) = \int h(z) f(z) \, Dz \qquad (3\text{-}56)$$

The functionals given here explicitly exist if $f$ and $h$ are in $\mathscr{C}^{\infty}$ and if they both vanish in a neighborhood of $z = 0$. Recall that $n_2$ is an integer if $n_1$ is, and vice versa, because $n_2 - n_1 = m$ is an integer.

It is again possible to show that these integrals can be extended to the spaces $\mathscr{D}_\chi$ and $\mathscr{D}_{\chi'}$ and thereby remain invariant bilinear functionals. The integrals need not even be regularized on these spaces. In all cases the integrals can be seen to converge at $z = \infty$ due to the asymptotic property (3–5) of the elements of $\mathscr{D}_\chi$ and $\mathscr{D}_{\chi'}$.

Before we end this section, we want to illustrate the usefulness of these bilinear invariant functionals by one elementary application. Given a bilinear invariant functional $B(h, f)$, we may consider all those elements $f \in \mathscr{D}_\chi$ which make the functional $B(h, f)$ vanish on the whole space $\mathscr{D}_{\chi'}$ of elements $h$. These elements $f$ form a closed invariant subspace of $\mathscr{D}_\chi$. The closure of this space follows from the continuity of the functional, the invariance of the subspace from the invariance of the functional. First we study the singular alternative, that is, the case in which $\frac{1}{2}(n_2 + N_1)$ and $\frac{1}{2}(n_2 + N_2)$ are both nonnegative integers.

In case (1), $B(h, f)$ (3–53) is zero if $f$ has the form

$$f(z) = \sum_{j=0}^{n_1-1} z^j a_j(\bar{z}) + \sum_{k=0}^{n_2-1} \bar{z}^k a_k(z)$$

with

$$\frac{\partial}{\partial z} a_j(\bar{z}) = 0 \qquad \frac{\partial}{\partial \bar{z}} a_k'(z) = 0$$

These are the Cauchy-Riemann conditions for antianalytic functions $a_j(\bar{z})$ and analytic functions $a_k'(z)$. Since $f(z)$ is in $\mathscr{D}_\chi$, $a_j(\bar{z})$ is antiholomorphic for all $z$ with a pole at most of order $n_2 - 1$ at $z = \infty$. Analogously $a_k'(z)$ is holomorphic with a pole at most of order $n_1 - 1$ at $z = \infty$. In other words, $f(z)$ is itself a polynomial

$$f(z) = \sum_{j=0}^{n_1-1} \sum_{k=0}^{n_2-1} d_{jk} z^j \bar{z}^k \tag{3-57}$$

These polynomials span an invariant subspace of $\mathscr{D}_\chi$ which we denote $\mathscr{E}_\chi$, and whose dimension is $n_1 n_2$. The representation carried by $\mathscr{E}_\chi$ is a spinor representation with spins $(S, S') = (\frac{1}{2}(n_1 - 1), \frac{1}{2}(n_2 - 1))$ (compare the notation of Section 2–5).

If we assume that in the cases (2) (3–54) and (3) (3–55) both $n_1$ and $n_2$ are positive integers, the same arguments as before lead us to the conclusion that $\mathscr{D}_\chi$ possesses an invariant spinor space $\mathscr{E}_\chi$ of dimension $n_1 n_2$. This subspace consists of exactly the same polynomials (3–57) as in case (1).

Finally we consider the nonsingular alternative (3–49) and assume that $n_1$ and $n_2$ are simultaneously negative integral. We realize immediately that the functional vanishes if and only if $f(z)$ possesses vanishing moments up to order $-n_1 - 1, -n_2 - 1$, namely,

$$\int f(z) z^j \bar{z}^k \, Dz = 0 \tag{3-58}$$

for all $0 \leq j \leq -n_1 - 1, 0 \leq k \leq -n_2 - 1$. Such functions $f(z)$ form an infinite-dimensional invariant subspace $\mathscr{F}_\chi$. The quotient space $\mathscr{D}_\chi/\mathscr{F}_\chi$ can be seen to have dimension $n_1 n_2$.

## 3–7   SCALAR PRODUCTS

Using the results of Section 3–6 we now want to investigate the conditions under which an invariant scalar product $(h, f)$ can be introduced in $\mathscr{D}_\chi$. This problem can partly be reduced to the issue studied in Section 3–6 because the scalar product as a Hermitian invariant form can be identified with a bilinear invariant functional

$$(h, f) = B(\bar{h}, f) \tag{3-59}$$

Since both $f$ and $h$ are in $\mathscr{D}_\chi$, $\bar{h}$ lies in $\mathscr{D}_{\chi'}$, if $\chi' = \{\bar{n}_2, \bar{n}_1\}$, whereas $\chi = \{n_1, n_2\}$. We have therefore to insert $N_1 = \bar{n}_2$ and $N_2 = \bar{n}_1$ in the formulas of Section 3–6. In addition the scalar product has to be positive definite; this requires further investigations.

The nonsingular alternative is characterized by the fact that the number

$$-s_1 = -\bar{s}_2 = \tfrac{1}{2}(n_1 + \bar{n}_2)$$

is neither a positive integer nor zero. A Hermitian functional exists if and only if

$$n_1 = \bar{n}_2$$

Since $n_1 = n_2 - m$, this condition implies $n_1 = \bar{n}_1 + m$, which shows that $m = 0$ and $n_1 = n_2$ is real. The functional $B(\bar{h}, f)$ (3–49) takes on the form

$$(h, f) = \int |z_2 - z_1|^{-2n_1 - 2} \overline{h(z_2)} f(z_1) \, Dz_1 \, Dz_2 \tag{3-60}$$

The singular alternative leads to two possibilities

$$n_1 = \bar{n}_2 = n_2 \quad n_1 \text{ a positive integer}$$

corresponding to case (1), and

$$n_1 = -\bar{n}_2 \quad n_1 \text{ arbitrary}$$

corresponding to case (4). Cases (2) and (3) do not occur. If $n_1 = n_2$ is a positive integer, the Hermitian form is (3–53)

$$(h, f) = \int \overline{h(z)} \left(\frac{\partial}{\partial z}\right)^{n_1} \left(\frac{\partial}{\partial \bar{z}}\right)^{n_1} f(z) \, Dz \tag{3-61}$$

In the case $n_1 = -\bar{n}_2$ we have

$$n_1 = -\bar{n}_1 - m \quad \text{Re } n_1 = -\tfrac{1}{2}m$$

and therefore

$$n_1 = -\tfrac{1}{2}m + \tfrac{i}{2}\rho \quad n_2 = \tfrac{1}{2}m + \tfrac{i}{2}\rho \quad \rho \text{ real}$$

The Hermitian form is obtained from (3–56)

$$(h, f) = \int \overline{h(z)} f(z) \, Dz \tag{3-62}$$

This form (3–62) is obviously positive definite. The representation on $\mathscr{D}_\chi$ can be extended onto the Hilbert space $\mathscr{L}^2(Z)$ of square integrable functions $f(z)$. In this fashion we get once more the unitary representations of the principal series, which we obtained from unitary one-dimensional representations of the subgroup K by induction in Section 3–2.

It remains to investigate the case that

$$n_1 = n_2 \quad n_1 \text{ real}$$

where $n_1$ is either a positive integer (this case belongs to the singular alternative) or neither a positive integer nor zero (this is the nonsingular alternative, the case $n_1 = n_2 = 0$ is already included in the principal series). The respective expressions for the Hermitian forms, (3–61) and (3–60), can be united into the single term

$$(h, f) = \frac{1}{\Gamma(-n_1)} \int |z_2 - z_1|^{-2n_1-2} \overline{h(z_2)} f(z_1) \, Dz_1 \, Dz_2 \qquad (3\text{-}63)$$

if we use the relation

$$\left. \frac{|z|^{-2n-2}}{\Gamma(-n)} \right|_{n \geq 0, \text{ integer}} = (-1)^n \frac{\pi}{n!} \left(\frac{\partial}{\partial z}\right)^n \left(\frac{\partial}{\partial \bar{z}}\right)^n \delta(z) \qquad (3\text{-}64)$$

known from the theory of distributions.

We consider the case $n_1 < 0$. We show in Section 3–8 that the case $n_1 > 0$ can be reduced to $n_1 < 0$, and that we therefore need not study $n_1 > 0$ separately. The Fourier transforms $\mathscr{F}_1(w)$, $\mathscr{F}_2(w)$ of two functions $f_1(z)$, $f_2(z)$ of $\mathscr{D}_\chi$ exist as can be seen from the asymptotic behavior (3–5) of these functions. The Fourier transform of the distribution

$$\frac{|z|^{-2n-2}}{\Gamma(-n)}$$

is the distribution

$$\frac{\pi |w|^{2n}}{2^{2n}\Gamma(n+1)}$$

Consequently the Hermitian form (3–63) is

$$(f_1, f_2) = \frac{1}{2^{2n_1+2}\pi\Gamma(n_1+1)} \int \overline{\mathscr{F}_1(w)} \mathscr{F}_2(w) |w|^{2n_1} \, Dw \qquad (3\text{-}65)$$

This integral has to be understood as a regularized integral for $n_1 \leq -1$. The form (3–65) is obviously positive definite if $-1 < n_1 < 0$. In this case the integral converges absolutely. This positive definite Hermitian form gives rise to the unitary representations of the supplementary series.

If $n_1$ lies in the open interval between $-v-1$ and $-v$ with $v$ a nonnegative integer, the regularization of the integral (3–65) can be performed as

$$(f, f) = \frac{1}{2^{2n_1+2}\pi\Gamma(n_1+1)} \int \left[ |\mathscr{F}(w)|^2 \right.$$

$$\left. - \sum_{j+k=0}^{2v-2} \frac{w^j \bar{w}^k}{j! \, k!} \left(\frac{\partial}{\partial w}\right)^j \left(\frac{\partial}{\partial \bar{w}}\right)^k |\mathscr{F}(w)|^2 \Big|_{w=0} \right]$$

$$\times |w|^{2n_1} \, Dw \qquad (3\text{-}66)$$

such that the integral converges at $w = 0$ and at $w = \infty$. If $n_1 = -\nu - 1$, $\nu$ a nonnegative integer, we have instead

$$(f, f) = \frac{(-1)^\nu}{2^{-2\nu}\nu!} \left(\frac{\partial}{\partial w}\right)^\nu \left(\frac{\partial}{\partial \overline{w}}\right)^\nu |\mathcal{F}(w)|^2 \Bigg|_{w=0} \tag{3-67}$$

It is easy to see that for nonintegral $n_1 < -1$ the Hermitian form (3–66) is indefinite. Negative integral $n_1$ yield Hermitian forms (3–67) that vanish on the invariant subspaces $\mathcal{F}_\chi$ and are indefinite on the quotient spaces (spinor spaces) $\mathscr{D}_\chi / \mathcal{F}_\chi$ except in the trivial case $n_1 = -1$ where $\mathscr{D}_\chi / \mathcal{F}_\chi$ is one dimensional.

## 3–8    EQUIVALENCE AND IRREDUCIBILITY

Now we pose the problem of finding all operators from $\mathscr{D}_\chi$ into $\mathscr{D}_{\chi'}$ which are continuous and intertwine the two representations in the sense

$$T_a^{\chi'} A = A T_a^\chi \qquad \text{for all } a \in \text{SL}(2, \text{C})$$

We call $A$ an intertwining operator. In particular we are interested in that operator $A$ which maps $\mathscr{D}_\chi$ one-to-one on $\mathscr{D}_{\chi'}$ and is bicontinuous (that means it has also a continuous inverse). If such an operator $A$ exists we call the representations $\chi$ and $\chi'$ equivalent. If $\chi$ is identical with $\chi'$, we say that $A$ commutes with the representation $\chi$ or that $A$ is a commuting operator on $\mathscr{D}_\chi$. Analogous definitions are used for representations in Hilbert spaces, and $A$ is then assumed to be a bounded operator.

Such intertwining operators can serve as very powerful tools. Let us consider in particular the case in which the representations are unitary. If $A$ intertwines the unitary representations $\chi$ and $\chi'$, and is one-to-one and bi-continuous, a scalar product $(h, f)_{\chi'}$ (3–59) in $\mathscr{D}_{\chi'}$ induces a Hermitian invariant form in $\mathscr{D}_\chi$

$$(Ah, Af)_{\chi'} \qquad h, f \in \mathscr{D}_\chi$$

In fact, it is the intertwining property of $A$ which guarantees the invariance of this form. Because of the uniqueness of the Hermitian invariant form, this form must be proportional to the scalar product in $\mathscr{D}_{\chi'}$,

$$(Ah, Af)_{\chi'} = c(h, f)_\chi \qquad c' = \text{const} > 0$$

This proves that $A$ is proportional to an isometric operator that can be extended to map the two Hilbert spaces of the unitary representations $\chi$ and $\chi'$ onto each other. These representations are then equivalent in the standard sense (Section 1–3a).

If for a unitary representation $\chi$ in a Hilbert space $\mathscr{H}$ any commuting operator $A$ in $\mathscr{H}$ is necessarily proportional to the unit operator, then the representation is irreducible. In fact, if the Hilbert space $\mathscr{H}$ of the representation contains an invariant subspace $\mathscr{H}' = P\mathscr{H}$, the orthogonal projection operator $P$ has the property

$$PT_a^{\chi}P = T_a^{\chi}P \qquad (3\text{-}68)$$

for all $a \,\epsilon\, SL(2, C)$. Because the representation is unitary we have

$$(T_a^{\chi}P)^{\dagger} = PT_{a^{-1}}^{\chi} \qquad (PT_a^{\chi}P)^{\dagger} = PT_{a^{-1}}^{\chi}P$$

Replacing $a$ by $\bar{a}^{-1}$ in the last relation and comparing with (3–68) yields

$$T_a^{\chi}P = PT_a^{\chi}P = PT_a^{\chi} \qquad (3\text{-}69)$$

The orthogonal projection operator $P$ is therefore a commuting operator which is submitted to our premise, that is, $P$ is proportional to the unit operator. Consequently $P$ is the unit operator itself or the null operator, and the representation is irreducible.

We turn now to the construction of the intertwining operators $A$. Our discussion in Section 3–6 showed that for any pair of representations $\chi = \{n_1, n_2\}$ and $-\chi = \{-n_1, -n_2\}$ there exists a bilinear functional (3–56)

$$[h, f] = \int h(z)f(z)\, Dz \qquad (3\text{-}70)$$

where $f \,\epsilon\, \mathscr{D}_{\chi}$ and $h \,\epsilon\, \mathscr{D}_{-\chi}$. An intertwining operator $A$ which maps $\mathscr{D}_{\chi}$ into $\mathscr{D}_{\chi'}$ permits us to construct the bilinear functional

$$B(h, f) = [\,h, Af\,] \qquad f \in \mathscr{D}_{\chi} \qquad h \in \mathscr{D}_{-\chi'} \qquad (3\text{-}71)$$

where the square bracket form is as in (3–70). Invariance of the functional (3–71) follows from

$$[T_a^{-\chi'}h,\, T_a^{\chi}f] = [h, f]$$

and

$$\begin{aligned}
B(T_a^{-\chi'}h,\, T_a^{\chi}f) &= [T_a^{-\chi'}h,\, AT_a^{\chi}f]\\
&= [T_a^{-\chi'}h,\, T_a^{\chi'}Af]\\
&= [h, Af]
\end{aligned}$$

We can turn this argument around. If $A$ is continuous from $\mathscr{D}_{\chi}$ into $\mathscr{D}_{\chi'}$ such that the functional $B(h, f)$ (3–71) is invariant, then $A$ intertwines the two representations. In the proof we need only make use of the fact that $f \in \mathscr{D}_\chi$ is uniquely determined if $[h, f]$ is known for all elements $h \in \mathscr{D}_{-\chi}$, in other words that

$$[h, f] = 0 \qquad \text{for all} \quad h \in \mathscr{D}_{-\chi}$$

implies $f = 0$, which is obvious.

　　　With the help of this construction we have reduced the issue of finding an intertwining operator $A$ from $\mathscr{D}_\chi$ into $\mathscr{D}_{\chi'}$ to the problem of constructing a bilinear invariant functional on the spaces $\mathscr{D}_\chi$ and $\mathscr{D}_{-\chi'}$. We can then apply our results on the bilinear invariant functionals (Section 3–6) and get (the operators $A$ are determined only up to a constant factor, of course):

　　(i)　$\chi = \{n_1, n_2\}$, $\chi' = \{-n_1, -n_2\}$. This is the nonsingular alternative (3–49) where $n_1$ and $n_2$ are not simultaneously positive integers or zero.

$$Af(z) = \int (z - z_1)^{-n_1-1}(\bar{z} - \bar{z}_1)^{-n_2-1} f(z_1)\, Dz_1 \qquad (3\text{-}72)$$

　　(ii)　$\chi = \{n_1, n_2\}$, $\chi' = \{-n_1, -n_2\}$. This is case (1) of the singular alternative (3–53) where both $n_1$ and $n_2$ are positive integers.

$$Af(z) = \left(\frac{\partial}{\partial z}\right)^{\tilde{n}_1}\left(\frac{\partial}{\partial \bar{z}}\right)^{n_2} f(z) \qquad (3\text{-}73)$$

　　(iii)　$\chi = \{n_1, n_2\}$, $\chi' = \{-n_1, n_2\}$. This is case (2) of the singular alternative (3–54) where $n_1$ is a positive integer and $n_2$ is an arbitrary integer.

$$Af(z) = \left(\frac{\partial}{\partial z}\right)^{n_1} f(z) \qquad (3\text{-}74)$$

　　(iv)　$\chi = \{n_1, n_2\}$, $\chi' = \{n_1, -n_2\}$. This is case (3) of the singular alternative (3–55) where $n_1$ is an arbitrary integer and $n_2$ is a positive integer.

$$Af(z) = \left(\frac{\partial}{\partial \bar{z}}\right)^{n_2} f(z) \qquad (3\text{-}75)$$

　　(v)　$\chi = \{n_1, n_2\}$, $\chi' = \{n_1, n_2\}$, This is case (4) of the singular alternative (3–56) where $n_1$ and $n_2$ are arbitrary numbers.

$$Af(z) = f(z) \qquad (3\text{-}76)$$

　　Only in case (v) is $\chi = \chi'$. A continuous operator commuting with a representation on $\mathscr{D}_\chi$ is therefore proportional to the unit operator. In the case of unitary representations one can show that the domain of the intertwining operators $A$ found above can be extended onto the Hilbert spaces and that no additional intertwining operators exist which map $\mathscr{D}_\chi$ into the Hilbert space completion of $\mathscr{D}_\chi$. This and (3–76) complete the proof of the irreducibility of the principal and supplementary series of representations.

Provided $n_1$ and $n_2$ are, in addition, not simultaneously negative integers or zero, case (i) shows that there exists an operator $A'$ which in turn maps $\mathscr{D}_{\chi'}$ into $\mathscr{D}_\chi$. It has the form

$$A'f(z) = \int (z - z_1)^{n_1 - 1}(\bar{z} - \bar{z}_1)^{n_2 - 1}f(z_1)\, Dz_1 \tag{3-77}$$

The product $A'A$ maps $\mathscr{D}_\chi$ into $\mathscr{D}_\chi$. Evaluating the convolution integrals (e.g., by means of Fourier transformations) we get

$$A'Af(z) = \frac{4\pi^2(-1)^m}{m^2 + \rho^2}\, f(z) \tag{3-78}$$

We see that $A$ and $A'$ are one-to-one and bicontinuous mappings. We have thus obtained the result that two representations $\chi$ and $-\chi$ are equivalent provided $n_1$ and $n_2$ are neither integers of the same sign nor is one integral and the other zero.

Next we consider the interesting cases (ii), (iii), and (iv). In case (ii) both $n_1$ and $n_2$ are positive integers. We know that $\mathscr{D}_\chi$ possesses an invariant spinor space $\mathscr{E}_\chi$. This space consists of the polynomials

$$\sum_{j=0}^{n_1 - 1} \sum_{k=0}^{n_2 - 1} d_{jk}\, z^j \bar{z}^k$$

and is annihilated by $A$. We know also that $\mathscr{D}_{\chi'}$ possesses an invariant subspace $\mathscr{F}_{\chi'}$ consisting of the functions whose moments

$$\int g(z)z^j\bar{z}^k\, Dz \qquad 0 \leq j \leq n_1 - 1 \qquad 0 \leq k \leq n_2 - 1$$

vanish. If we insert $g(z) = Af(z)$, $A$ as in (3–73), into these moment integrals and recall the combinatorial formula

$$z^j\bar{z}^k\left(\frac{\partial}{\partial z}\right)^{n_1}\left(\frac{\partial}{\partial \bar{z}}\right)^{n_2} f(z)$$

$$= \sum_{\substack{0 \leq r \leq j \\ 0 \leq s \leq k}} (-1)^{r+s} r!\, s!\, \binom{j}{r}\binom{n_1}{r}\binom{k}{s}\binom{n_2}{s}$$

$$\times \left(\frac{\partial}{\partial z}\right)^{n_1 - r}\left(\frac{\partial}{\partial \bar{z}}\right)^{n_2 - s} z^{j-r}\bar{z}^{k-s} f(z)$$

where $f(z)$ can be replaced on either side by the function $f_1(z) \in \mathscr{D}_\chi / \mathscr{E}_\chi$ defined by

$$f_1(z) = f(z) - \sum_{p=0}^{n_1-1} \sum_{q=0}^{n_2-1} d_{pq} z^p \bar{z}^q$$

$$\lim_{|z| \to \infty} f_1(z) = 0$$

we notice that the moments of $Af(z)$ vanish and that $A$ maps $\mathscr{D}_\chi$ into $\mathscr{F}_{\chi'}$. On the other hand, we can find a primitive function $f$ for any element $g \in \mathscr{F}_{\chi'}$ such that $A$ applied to $f$ yields $g$. This primitive function can be chosen from $\mathscr{D}_\chi$ and is then unique up to a component in $\mathscr{E}_\chi$.

To prove the last assertion we define an element $f(z) \in \mathscr{D}_\chi$ for an arbitrary function $g(z)$ from $\mathscr{F}_{\chi'}$ by the (convergent) integral

$$f(z) = \frac{2}{\pi(n_1 - 1)!(n_2 - 1)!} \int (z - z_1)^{n_1 - 1} (\bar{z} - \bar{z}_1)^{n_2 - 1} \log |z - z_1|$$

$$\times g(z_1) Dz_1 \tag{3-79}$$

If we differentiate $f(z)$ with respect to $z$, the logarithmic factor is not affected as long as there is at least one power of $z - z_1$, since the moments of $g(z)$ vanish. The same statement holds for differentiation with respect to $\bar{z}$. Thus we get

$$Af(z) = \left(\frac{\partial}{\partial z}\right)^{n_1} \left(\frac{\partial}{\partial \bar{z}}\right)^{n_2} f(z)$$

$$= \frac{2}{\pi} \int \frac{\partial}{\partial z} \frac{\partial}{\partial \bar{z}} \log |z - z_1| \, g(z_1) \, Dz_1$$

$$= \frac{1}{2\pi} \int \Delta \log |z - z_1| \, g(z_1) \, Dz_1$$

$$= g(z)$$

using a well-known property of the Laplacian $\Delta$.

In this fashion we have proved that $A$ maps the quotient space $\mathscr{D}_\chi / \mathscr{E}_\chi$ one-to-one and bicontinuous on $\mathscr{F}_{\chi'}$. The representation induced by $\chi$ on the quotient space $\mathscr{D}_\chi / \mathscr{E}_\chi$ is therefore equivalent to the representation on the invariant subspace $\mathscr{F}_{\chi'}$.

So far we have studied the continuous mapping of $\mathscr{D}_\chi$ into $\mathscr{D}_{\chi'}$, $\chi$ and $\chi'$ as in case (ii), by means of the intertwining operator $A$ given by (3-73). In turn it is also possible to map $\mathscr{D}_{\chi'}$ into $\mathscr{D}_\chi$ using an intertwining operator $A'$ as defined in (3–72)

$$A'f(z) = \int (z - z_1)^{n_1-1}(\bar{z} - \bar{z}_1)^{n_2-1} f(z_1)\, Dz_1$$

It has the kernel $\mathscr{F}_{\chi'}$ and an image in $\mathscr{E}_\chi$. It is not difficult to verify that the image is the whole of $\mathscr{E}_\chi$. The representation induced by $\chi'$ in the quotient space $\mathscr{D}_{\chi'}/\mathscr{F}_{\chi'}$ is finite dimensional and therefore equivalent to the spinor representation in $\mathscr{E}_\chi$.

We get similar results in the remaining cases (iii) and (iv). If in case (iii) both $n_1$ and $n_2$ are positive integers, the operator $A$ of (3–74) annihilates the invariant subspace $\mathscr{E}_\chi$. The representation on the quotient space $\mathscr{D}_\chi / \mathscr{E}_\chi$ is equivalent to the representation on $\mathscr{D}_{\chi'}$. If $n_2$ is negative integral, $\mathscr{D}_{\chi'}$ has an invariant subspace $\mathscr{F}_{\chi'}$, and the representation on $\mathscr{F}_{\chi'}$ is equivalent to the representation on $\mathscr{D}_\chi$. If $n_1$ is positive integral and $n_2$ is zero, neither $\mathscr{D}_\chi$ nor $\mathscr{D}_{\chi'}$ possesses invariant subspaces of the type $\mathscr{E}$ or $\mathscr{F}$. Both representations $\chi$ and $\chi'$ are in fact equivalent since (3–72) supplies us with the continuous inverse operator $A'$ of the operator $A$ (3–74). The situation is analogous in case (iv).

## 3–9　GENERATORS AND THE MATRIX FORM OF THE INTERTWINING OPERATORS IN THE CANONICAL BASIS

The operator $A$ which intertwines the representations $\chi$ and $-\chi$ will later be needed in an explicit matrix form. To derive its matrix elements it is most convenient to make use of the group generators. An infinitesimal group element of SL(2, C) can be parametrized by six real numbers $\varepsilon_k$, $\eta_k$:

$$a = e + i \sum_{k=1}^{3} \left( \varepsilon_k \frac{1}{2}\sigma_k + \eta_k \frac{i}{2}\sigma_k \right) + O(\varepsilon_k^2, \eta_k^2) \tag{3-80}$$

Correspondingly we expand an operator $T_a^\chi$ on $\mathscr{D}_\chi$ as

$$T_a^\chi = E + i \sum_{k=1}^{3} (\varepsilon_k H_k + \eta_k F_k) + O(\varepsilon_k^2, \eta_k^2) \tag{3-81}$$

If we realize $\mathscr{D}_\chi$ as a space of functions $f(z)$, the generators $H_k$ and $F_k$ become differential operators with range in $\mathscr{D}_\chi$. With the notations

$$H_\pm = H_1 \pm iH_2 \qquad\qquad F_\pm = F_1 \pm iF_2$$

we have

$$H_+ = -z^2 \frac{\partial}{\partial z} - \frac{\partial}{\partial \bar{z}} + (n_1 - 1)z$$

$$H_- = + \frac{\partial}{\partial z} + \bar{z}^2 \frac{\partial}{\partial \bar{z}} - (n_2 - 1)\bar{z}$$

$$H_3 = +z\frac{\partial}{\partial z} - \bar{z}\frac{\partial}{\partial \bar{z}} + \tfrac{1}{2}(n_2 - n_1)$$

$$F_+ = -iz^2\frac{\partial}{\partial z} + i\frac{\partial}{\partial \bar{z}} + i(n_1 - 1)z$$

$$F_- = +i\frac{\partial}{\partial z} - i\bar{z}^2\frac{\partial}{\partial \bar{z}} + i(n_2 - 1)\bar{z}$$

$$F_3 = +iz\frac{\partial}{\partial z} + i\bar{z}\frac{\partial}{\partial \bar{z}} - \frac{i}{2}(n_1 + n_2 - 2) \tag{3-82}$$

The matrix form of the generators in the canonical basis $\{f_q{}^j(z)\}$ can be obtained using the explicit form of these basis functions. If we normalize the basis with respect to the measure

$$(1 + |z|^2)^{\rho_1}\, Dz \qquad \rho_1 = \operatorname{Im}\rho$$

such that it is orthonormal in a space $\mathcal{L}^2_{q_1}(Z)$ (Section 3–3) we have from (3–20), (3–14), and (2–30)

$$f_q{}^j(z) = \left[\frac{2j + 1}{\pi}\frac{(j + (1/2)m)!\,(j - (1/2)m)!}{(j+q)!\,(j-q)!}\right]^{1/2}(1 + z\bar{z})^{(i/2)\rho - j - 1}G_q{}^j(z) \tag{3-83}$$

where $G_q{}^j(z)$ are certain hypergeometric polynomials

$$G_q{}^j(z) = \begin{cases} \binom{j+q}{q-\frac{1}{2}m}z^{q-m/2}{}_2F_1(-j+q, -j-\tfrac{1}{2}m, q-\tfrac{1}{2}m+1; -z\bar{z}) \\ \qquad\qquad\qquad\qquad\qquad\qquad \text{for}\quad q \geq \tfrac{1}{2}m \\[2mm] \binom{j-q}{\frac{1}{2}m-q}(-\bar{z})^{m/2-q}{}_2F_1(-j-q, -j+\tfrac{1}{2}m, \tfrac{1}{2}m-q+1; -z\bar{z}) \\ \qquad\qquad\qquad\qquad\qquad\qquad \text{for}\quad \tfrac{1}{2}m \geq q \end{cases} \tag{3-84}$$

The matrix elements of $H_\pm$ and $H_3$ are those known from the rotation group SU(2)

$$H_\pm f_q{}^j = [(j \mp q)(j + 1 \pm q)]^{1/2}f_{q\pm1}^j$$
$$H_3 f_q{}^j = q f_q{}^j \tag{3-85}$$

From (3–82),(3–83), (3–84) we derive the matrix element of the generator $F_+$

$$F_+ f_j^j = 2i\left(\frac{i}{2}\rho - j - 1\right)\left[\frac{(j + 1)^2 - m^2/4}{(2j + 2)(2j + 3)}\right]^{1/2}f_{j+1}^{j+1} \tag{3-86}$$

As is well known, the group SL(2, C) admits of two Casimir invariants. One of these is a scalar, namely, it does not change sign under the "parity" substitution $H_k \to H_k$, $F_k \to -F_k$. We normalize it as

$$I_1 = F_+ F_- + F_- F_+ + 2F_3{}^2 - H_+ H_- - H_- H_+ - 2H_3{}^2 \qquad (3\text{-}87)$$

The other invariant is pseudoscalar, it changes sign under the "parity" substitution mentioned,

$$I_2 = H_+ F_- + H_- F_+ + F_+ H_- + F_- H_+ + 4 H_3 F_3 \qquad (3\text{-}88)$$

These operators $I_1$ and $I_2$ are proportional to the unit operator with eigenvalues

$$i_1 = -\tfrac{1}{2}(m^2 - \rho^2 - 4) \qquad (3\text{-}89)$$

respectively,

$$i_2 = m\rho \qquad (3\text{-}90)$$

if they act on a function space $\mathscr{D}_\chi$. It is easy to check this result with the differential operator forms (3–82) of the generators.

From the definition of the intertwining operator

$$T_a^{-\chi} A = A T_a^\chi \qquad \text{for all} \quad a \in \text{SL}(2, \text{C})$$

follows by differentiation with respect to $a$

$$H_k^{-\chi} A = A H_k^\chi \qquad F_k^{-\chi} A = A F_k^\chi$$

for all generators $H_k$ and $F_k$. A vector $f_q^{\,j}(z)$ of the canonical basis can be characterized in the Hilbert space $\mathscr{L}^2_{\rho_1}(Z)$ as an eigenvector of the operators $H_3$ and $H^2$

$$H_3 f_q^{\,j} = q f_q^{\,j} \qquad H^2 f_q^{\,j} = j(j+1) f_q^{\,j} \qquad (3\text{-}91)$$

where $H^2$ is the Casimir operator of SU(2)

$$H^2 = \tfrac{1}{2}[H_+ H_- + H_- H_+] + H_3{}^2 \qquad (3\text{-}92)$$

Under application of the operator $A$, the vector $f_q^{\,j(+)}$ of the basis in $\mathscr{D}_\chi$ goes into the vector $f_q^{\,j(-)}$ of the basis in $\mathscr{D}_{-\chi}$ since we have

$$H_3^{-\chi}(A f_q^{\,j(+)}) = A H_3^\chi f_q^{\,j(+)} = q(A f_q^{\,j(+)})$$

and a similar relation for $H^2$. We may therefore write

$$A f_q^{\,j(+)} = \beta_q^{\,j} f_q^{\,j(-)} \qquad (3\text{-}93)$$

The operator $A$ intertwines the generator $H_+$ whose matrix elements (3–85) do not depend on $\chi$. The coefficients $\beta_q{}^j$ can therefore not depend on $q$ either,

$$\beta_q{}^j = \beta^j$$

The matrix elements of $F_+$, however, depend on $\chi$. This allows us to compute the numbers $\beta^j$ explicitly. If we intertwine $A$ with $F_+$ we get

$$AF_+{}^\chi f_j^{j(+)} = \beta^{j+1}\tau_{j+1}(\chi)f_{j+1}^{j+1(-)}$$

and simultaneously

$$AF_+{}^\chi f_j^{j(+)} = F_+^{-\chi}Af_j^{j(+)}$$
$$= \beta^j F_+^{-\chi}f_j^{j(-)}$$
$$= \beta^j \tau_{j+1}(-\chi)f_{j+1}^{j+1(-)}$$

where $\tau_j(\chi)$ is the matrix element (3–86) of $F_+$

$$\tau_j(\chi) = 2i\left(\frac{i}{2}\rho - j\right)\left[\frac{j^2 - m^2/4}{2j(2j+1)}\right]^{1/2}$$

Consequently we have

$$\frac{\beta^j}{\beta^{j-1}} = \frac{\tau_j(-\chi)}{\tau_j(\chi)} = \frac{j + (i/2)\rho}{j - (i/2)\rho}$$

This relation can be multiplied up from $j_0 = \tfrac{1}{2}|m|$ till $j$,

$$\frac{\beta^j}{\beta^{j_0}} = \frac{\Gamma(j + 1 + (i/2)\rho)\Gamma(j_0 + 1 - (i/2)\rho)}{\Gamma(j + 1 - (i/2)\rho)\Gamma(j_0 + 1 + (i/2)\rho)}$$

Since the operator $A$ may be normalized arbitrarily, we can define

$$\beta^j = \frac{\Gamma(j + 1 + (i/2)\rho)}{\Gamma(j + 1 - (i/2)\rho)} \tag{3-94}$$

This is the result desired. Finally we note that the matrix elements of $T_a^\chi$ and $T_a^{-\chi}$ are related by

$$\beta^{j_1} D_{j_1 q_1 j_2 q_2}^{\chi}(a) = \beta^{j_2} D_{j_1 q_1 j_2 q_2}^{-\chi}(a) \tag{3-95}$$

This relation implies in particular

$$\beta^{j_1} d^{(m;\,\rho)}_{j_1 j_2 q}(\eta) = \beta^{j_2} d^{(-m;\,-\rho)}_{j_1 j_2 q}(\eta) \tag{3-96}$$

Relation (3–96) completes the set of symmetry relations for the function $d^{(m,\rho)}_{j_1 j_2 q}(\eta)$ derived in Section 3–5.

A similar consideration yields the matrix elements of the operator $T_a{}^\chi$ on the quotient space $\mathscr{D}_\chi / \mathscr{E}_\chi$ when $\chi = \{n_1, n_2\}$ and $n_1$ and $n_2$ are both positive integers. Case (iii) of Section 3–8 tells us that an intertwining operator $A$ exists which relates these matrix elements with those of the operator $T_a{}^{\chi*}$ on the space $\mathscr{D}_{\chi*}$ where $\chi^* = \{-n_1, n_2\}$. In this case

$$n_1 + n_2 = i\rho$$

is a positive integer. Let us denote

$$j_0 = \frac{1}{2}|n_1 - n_2|, \qquad j'_0 = \frac{i}{2}\rho$$

The canonical bases in $\mathscr{D}_\chi$ and $\mathscr{D}_{\chi*}$ consist of the vectors

$$\mathscr{D}_\chi : f_q{}^j, \quad j = j_0 + \text{nonnegative integer}$$

$$\mathscr{D}_\chi{}^*: f_q{}^{j*}, \quad j = j'_0 + \text{nonnegative integer}$$

where $j_0 < j_0'$. The intertwining operator maps these bases onto each other such that

$$A f_q{}^j = \begin{cases} \delta^j f_q{}^{j*} & \text{for } j \geq j'_0 \\ 0 & \text{for } j < j'_0 \end{cases}$$

To compute the coefficient $\delta^j$ we proceed as before and get

$$\frac{\delta^j}{\delta^{j-1}} = \frac{\tau_j(\chi^*)}{\tau_j(\chi)} = \left[\frac{(j - m/2)(j + j'_0)}{(j + m/2)(j - j'_0)}\right]^{1/2}$$

Multiplying these relations we have

$$\frac{\delta^j}{\delta^{j'_0}} = \left[\frac{(j - m/2)!(j'_0 + m/2)!(j + j'_0)!}{(j + m/2)!(j'_0 - m/2)!(2j'_0)!(j - j'_0)!}\right]^{1/2} \tag{3-97}$$

for $j \geq j_0'$. Here $\delta^{j_0}$ can be fixed arbitrarily. With the help of these coefficients (3–97) we may write

$$\delta^{j_1} D^\chi_{j_1 q_1 j_2 q_2}(a) = \delta^{j_2} D^{\chi*}_{j_1 q_1 j_2 q_2}(a) \tag{3-98}$$

All relations existing between the matrix elements in the four representations $\chi = \{\pm n_1, \pm n_2\}$, $n_{1,2}$ integers, can be derived by combining the results (3–95) and (3–98).

Formula (3–95) can be viewed also in another respect. We remember that the matrix elements of the operator $T_a{}^\chi$ in the canonical basis are entire functions of $\rho$. If the $\Gamma$-functions contained in the coefficients $\beta^j$ in (3–95) have poles or zeros these must be cancelled by zeros of the matrix elements. We consider the example where $i\rho$ is a positive integer

$$j_0' = \frac{i}{2}\rho \qquad j_0 = \frac{1}{2}|m| \qquad j_0' > j_0$$

such that the two numbers

$$n_{1,2} = \mp \frac{1}{2}m + \frac{i}{2}\rho$$

are both positive integers. In this case relation (3–95) has the following form:

$$\frac{\Gamma(j_2 + 1 - j_0')}{\Gamma(j_2 + 1 + j_0')} D^\chi_{j_1 q_1 j_2 q_2}(a) = \frac{\Gamma(j_1 + 1 - j_0')}{\Gamma(j_1 + 1 + j_0')} D^{-\chi}_{j_1 q_1 j_2 q_2}(a)$$

It implies

$$D^\chi_{j_1 q_1 j_2 q_2}(a) = 0 \qquad \text{for} \quad j_0 \leq j_2 < j_0' \qquad j_1 \geq j_0'$$

$$D^{-\chi}_{j_1 q_1 j_2 q_2}(a) = 0 \qquad \text{for} \quad j_0 \leq j_1 < j_0' \qquad j_2 \geq j_0'$$

This means that the basis elements $f_q^{j(+)}$ in $\mathscr{D}_\chi$ with $j_0 \leq j < j_0'$ span an invariant subspace $\mathscr{E}_\chi$, whereas the basis elements $f_q^{j(-)}$ in $\mathscr{D}_{-\chi}$ with $j \geq j_0'$ span an invariant subspace $\mathscr{F}_{-\chi}$. The dimension of $\mathscr{E}_\chi$ is

$$\dim \mathscr{E}_\chi = \sum_{j=j_0}^{j_0'-1} (2j+1) = j_0'^2 - j_0^2 = n_1 n_2$$

as expected. One checks easily that the subspaces $\mathscr{E}_\chi$ and $\mathscr{F}_{-\chi}$ found this way from the matrix elements of the intertwining operators coincide with the spaces which we know to exist from the study of the abstract form of the intertwining operators.

## 3–10   COVARIANT OPERATORS

We consider continuous linear operators $T$ from the space $\mathscr{D}_\chi$ into the space $\mathscr{D}_{\chi'}$. We call such operators $T$ covariant operators if they are operator functions of a complex variable $x$, if for any pair $f \in \mathscr{D}_\chi$ and $g \in \mathscr{D}_{-\chi'}$

$$h(x) = \int g(z)T(x)f(z)\, Dz$$

is an element of a space $\mathscr{D}_{\chi^*}$, $\chi^* = \{\lambda, \mu\}$, and if the operator equation

$$T^{\chi'}_{a-1} T(x) T^{\chi}_a = (a_{12} x + a_{22})^{\lambda - 1} \overline{(a_{12} x + a_{22})}^{\mu - 1} T(x_a) \qquad (3\text{-}99)$$

is fulfilled. The constraint (3–99) guarantees the right-transformation property of $h \,\epsilon\, \mathscr{D}_{\chi^*}$. We call $\chi^*$ the covariance of $T(x)$.

These covariant operators are a more general notion than the intertwining operators to which they reduce if $\lambda = \mu = 1$ and if $h$ is restricted to $\mathscr{E}_{\chi^*}$. The general theory of covariant operators can be built up the same way as the theory of intertwining operators: it can be based on the theory of trilinear invariant functionals. We do not want to go into the details of this formalism, but give only some results for covariant operators which transform as Lorentz tensors. In certain applications, for example, the theory of Lorentz invariant equations due to Gelfand and Yaglom, these tensor operators are of considerable importance. For Lorentz tensors the numbers $\lambda$ and $\mu$ are positive integers such that their difference $\lambda - \mu$ is even. We note finally that the problem of decomposing tensor products of representations into irreducible components can also be traced back to trilinear invariant functionals.

Covariant operators have a property which enables us to build higher tensors out of lower rank tensors. If $T^{(1)}(x)$ is continuous from $\mathscr{D}_\chi$ into $\mathscr{D}_{\chi'}$ with covariance $\chi^{(1)} = \{\lambda_1, \mu_1\}$, and if $T^{(2)}(x)$ is continuous from $\mathscr{D}_{\chi'}$ into $\mathscr{D}_{\chi''}$ with covariance $\chi^{(2)} = \{\lambda_2, \mu_2\}$ then the product $T^{(2)}(x)T^{(1)}(x)$ is continuous from $\mathscr{D}_\chi$ into $\mathscr{D}_{\chi''}$ with a covariance described by $\{\lambda_1 + \lambda_2 - 1, \mu_1 + \mu_2 - 1\}$. This assertion follows immediately from the definition (3–99).

We are therefore mainly interested in the elementary cases

$\lambda = \mu = 2$, where $T$ is a four vector;

$\lambda = 3, \mu = 1$ and $\lambda = 1, \mu = 3$, where $T$ are antisymmetric tensor operators.

Let us denote the representations on $\mathscr{D}_\chi$ and $\mathscr{D}_{\chi'}$ by

$$\chi = \{n_1, n_2\} \quad \text{respectively} \quad \chi' = \{N_1, N_2\}$$

$$n_2 - n_1 = m \qquad N_2 - N_1 = m'$$

$$n_1 + n_2 = i\rho \qquad N_1 + N_2 = i\rho'$$

We use realizations of $\mathscr{D}_\chi$ and $\mathscr{D}_{\chi'}$ in terms of spaces of functions $f(z)$ and denote the operators $T(x)$ correspondingly by $T(x|z)$. The following operators T satisfy all requirements:

$\lambda = \mu = 2$, first possibility: $N_1 = n_1 - 1, N_2 = n_2 + 1, m' = m + 2, \rho' = \rho$,

$$T(x \,|\, z) = (z - x)^{n_1} (\bar{z} - \bar{x}) \frac{\partial}{\partial z} (z - x)^{-n_1 + 1} \qquad (3\text{-}100)$$

$\lambda = \mu = 2$, second possibility: $N_1 = n_1 + 1, N_2 = n_2 - 1, m' = m - 2, \rho' = \rho$,

$$T(x \mid z) = (\bar{z} - \bar{x})^{n_2}(z - x) \frac{\partial}{\partial \bar{z}} (\bar{z} - \bar{x})^{-n_2 + 1} \qquad (3\text{-}101)$$

$\lambda = \mu = 2$, third possibility: $N_1 = n_1 + 1, N_2 = n_2 + 1, m' = m, \rho' = \rho - 2i$,

$$T(x \mid z) = (\bar{z} - \bar{x})(z - x) \qquad (3\text{-}102)$$

$\lambda = \mu = 2$, fourth possibility: $N_1 = n_1 - 1, N_2 = n_2 - 1, m' = m, \rho' = \rho + 2i$,

$$T(x \mid z) = (\bar{z} - \bar{x})^{n_2}(z - x)^{n_1} \frac{\partial}{\partial z} \frac{\partial}{\partial \bar{z}} (\bar{z} - \bar{x})^{-n_2 + 1}(z - x)^{-n_1 + 1} \quad (3\text{-}103)$$

$\lambda = 3, \mu = 1: N_1 = n_1, N_2 = n_2$,

$$T(x \mid z) = (z - x)^{n_1 + 1} \frac{\partial}{\partial z} (z - x)^{-n_1 + 1} \qquad (3\text{-}104)$$

$\lambda = 1, \mu = 3: N_1 = n_1, N_2 = n_2$,

$$T(x \mid z) = (\bar{z} - \bar{x})^{n_2 + 1} \frac{\partial}{\partial \bar{z}} (\bar{z} - \bar{x})^{-n_2 + 1} \qquad (3\text{-}105)$$

For the proof of the covariance we need only verify the operator identity

$$T(x \mid z)(a_{12} z + a_{22})^{n_1 - 1}\overline{(a_{12} z + a_{22})}^{n_2 - 1}$$
$$= (a_{12} x + a_{22})^{\lambda - 1}\overline{(a_{12} x + a_{22})}^{\mu - 1}(a_{12} z + a_{22})^{N_1 - 1}$$
$$\times \overline{(a_{12} z + a_{22})}^{N_2 - 1} T(x_a \mid z_a) \qquad (3\text{-}106)$$

which is easily done if we make use of the relations

$$z_a - x_a = (z - x)(a_{12} z + a_{22})^{-1}(a_{12} x + a_{22})^{-1}$$

$$\frac{\partial}{\partial z_a} = (a_{12} z + a_{22})^2 \frac{\partial}{\partial z}$$

We suspect that the antisymmetric tensor operators (3–104), (3–105) are equivalent to the group generators. This can indeed be verified as follows. Expanding $T(x)$ (3–104) in powers of $x$ and $\bar{x}$

$$T(x) = \frac{1}{\sqrt{2}} [x^2 T_{+1} + \sqrt{2} x T_0 + T_{-1}] \qquad (3\text{-}107)$$

yields a covariant spinor operator $T_A$, $A = \pm 1, 0$. This expansion corresponds to the expansion (2–26) of a homogeneous polynomial $F(\xi)$ in powers of $\xi^1$ and $\xi^2$ [recall that $x = \xi^1/\xi^2$, $N_0^{(1)} = 2^{1/2}$, $N_{\pm 1}^{(1)} = 1$]. We introduce the contravariant spinor $T^A$ by (2–41)

$$T^A = (-1)^{1-A} T_{-A}$$

This contravariant spinor operator $T^A$ can be expressed by the generators as

$$T^{\pm 1} = \mp \frac{1}{\sqrt{2}} (H_\pm - iF_\pm) \qquad T^0 = H_3 - iF_3 \qquad (3\text{-}108)$$

where the differential operator forms (3–82) of the generators have to be inserted.

Similarly we expand the operator $T(x)$ (3–105) and get

$$T(x) = \frac{1}{\sqrt{2}} [\bar{x}^2 T_{+i} + \sqrt{2} \bar{x} T_{\dot{0}} + T_{-i}] \qquad (3\text{-}109)$$

where the components of the conjugate spinor $T^{\dot{A}}$ are connected with the generators by

$$T_{\pm i} = \mp \frac{1}{\sqrt{2}} (H_\pm + iF_\pm) \qquad T_{\dot{0}} = H_3 + iF_3 \qquad (3\text{-}110)$$

In the cases (3–100) to (3–103) the four-vector nature of the operators $T(x)$ can also be made explicit. We expand $T(x)$ in powers of $x$ and $\bar{x}$

$$T(x) = x\bar{x}\Gamma_{1/2,\,i/2} + x\Gamma_{1/2,\,-i/2} + \bar{x}\Gamma_{-1/2,\,-i/2} + \Gamma_{-1/2,\,i/2} \qquad (3\text{-}111)$$

This yields a $2 \times 2$ matrix $\Gamma_{A\dot{B}}$ that transforms as

$$T^{\chi'}_{a-1} \Gamma T^{\chi}_a = a\Gamma a^\dagger$$

If we set

$$\Gamma = \Gamma^0 e + \sum_{k=1}^{3} \Gamma^k \sigma_k$$

we obtain

$$T^{\chi'}_{a-1} \Gamma^\mu T^{\chi}_a = \Lambda_\nu{}^\mu(a) \Gamma^\nu \qquad (3\text{-}112)$$

namely, a contravariant vector $\Gamma^\mu$. The components of this vector are for (3–100)

$$\Gamma^0 = \frac{1}{2}\left[(1 + z\bar{z})\frac{\partial}{\partial z} - (n_1 - 1)\bar{z}\right]$$

$$\Gamma^1 = \frac{1}{2}\left[-(z + \bar{z})\frac{\partial}{\partial z} + (n_1 - 1)\right]$$

$$\Gamma^2 = \frac{1}{2}\left[i(z - \bar{z})\frac{\partial}{\partial z} - i(n_1 - 1)\right] \tag{3-113}$$

$$\Gamma^3 = \frac{1}{2}\left[(1 - z\bar{z})\frac{\partial}{\partial z} + (n_1 - 1)\bar{z}\right]$$

The components $\Gamma^\mu$ for (3–101) can be obtained from (3–113) by the following replacements: $z$ by $\bar{z}$ and vice versa, $n_1$ by $n_2$, and a sign change of $\Gamma^2$. It is straightforward to give the vector components also for (3–102) and (3–103).

Let us denote the two vector operators (3–100) and (3–101) $T$ and $T'$, respectively. Symmetric tensor operators of rank $n$ can be obtained as a product of totally $n$ factors $T$ and $T'$. This yields differential operators of $n$th order. For all these differential operators the invariant $\rho$ is the same for $\mathscr{D}_\chi$ and $\mathscr{D}_{\chi'}$ whereas $m$ may change by an even amount. This is illustrated in Fig. 3–1.

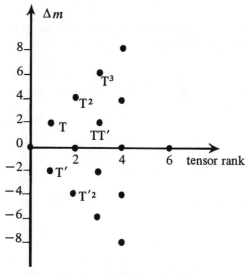

**Figure 3–1.** Differential operators transforming like symmetric tensors

We denote $\Delta m = m'-m$. Each point of the figure defines a single operator since the order of the factors $T$ and $T'$ in the product does not matter. Among symmetric tensor operators of even rank there is always exactly one with $\Delta m = 0$, and it maps $\mathscr{D}_\chi$ into itself. The space $\mathscr{D}_\chi$ can be mapped into itself also if the covariant operator is followed by an intertwining operator $A$. For the differential operators (3–100) and (3–101) this is possible if $\rho = 0$ and $m' = -m$, that is, if $\chi = \pm\{\frac{1}{2}, -\frac{1}{2}\}$. This is the first of two cases found by Majorana, where an irreducible representation of SL(2, C) is mapped into itself by a vector operator. This representation $\chi = \{\frac{1}{2}, -\frac{1}{2}\}$ belongs to the principal series. The second case of Majorana's is the representation $\chi = \{-\frac{1}{2}, -\frac{1}{2}\}$ which goes under $T(x)$ (3–102) into $\chi' = -\chi$. This representation belongs to the supplementary series.

In order to find all symmetric tensor operators we have also to admit powers of the operators (3–102) and (3–103). This forces us to enlarge the scheme presented in Fig. 3–1 by adding a third dimension which takes account of the changes in the invariant $\rho$. In general there are $(n + 1)^2$ different symmetric tensor operators of rank $n$. One can show that there are no further symmetric tensor operators, in particular there are no further vector operators beyond those given in (3–100) to (3–103). Not necessarily symmetric tensors are obtained by multiplying the symmetric tensors with the generators. One can again show that this procedure exhausts all possible tensor operators.

## 3–11   REMARKS

Our presentation of spaces of homogeneous functions and the construction of bilinear invariant functionals goes back to the work of Gelfand *et al.* [12 (Volume 5, Chapter III)] .

In Section 3–3 we mentioned two theorems, the Plancherel and the Paley-Wiener theorems. The Plancherel theorem asserts that for any function $\psi(x)$ which is square integrable on the real line $-\infty < x < +\infty$, the Fourier transform

$$\varphi(y) = \int_{-\infty}^{+\infty} \psi(x)e^{ixy}\, dx$$

exists as an $l^2$-limit is itself square integrable, and can be inverted to give

$$\psi(x) = \frac{1}{2\pi} \int_{-\infty}^{+\infty} \varphi(y)e^{-ixy}\, dy$$

almost everywhere. Moreover we have

$$\int_{-\infty}^{+\infty} |\varphi(y)|^2 \, dy = 2\pi \int_{-\infty}^{+\infty} |\psi(x)|^2 \, dx$$

which is known as Parseval's formula. In this course we shall use the notations Plancherel theorem and Parseval's formula for analogous generalizations in the theory of Fourier transforms on Lie groups. In the case of compact Lie groups like SU(2), the Plancherel theorem is contained in the theorem of Peter and Weyl; see Pontrjagin [27 (§ 29)]. The analog of the Plancherel theorem for the groups SL(2, C) and SU(1, 1) is one of the topics of our Chapters 4 and 6.

We call Paley-Wiener theorems certain assertions about the asymptotic behavior of the Fourier transform $\varphi(y)$ at $y = \infty$ which characterize classes of functions $\psi(x)$. For example, $\varphi(y)$ is the Fourier transform of an infinitely differentiable function with compact support,

$$\psi(x) = 0 \qquad \text{for} \quad |x| > a > 0$$

if and only if it exhibits the following properties:

(1)  it is entire analytic in $y$;
(2)  the inequality

$$|y^q \varphi(y)| \leqq c_q \, e^{a |\text{Im } y|}$$

holds for all $q = 0, 1, 2, \ldots$.

The proof is elementary, it is based on partial integrations; see, for example, Gelfand and Shilov [12 (Volume 1, Chapter II, Section 1)]. The constant $c_q$ can be estimated by

$$c_q \leqq 2a \, \max \left| \left(\frac{d}{dx}\right)^q \psi(x) \right|$$

If instead we assume that $\psi(x)$ decreases only rapidly, that is to say, faster than any power of $x$ if $x$ tends to infinity, the Fourier transform $\varphi(y)$ can in general only be defined on the real axis and has the same asymptotic behavior as $\psi(x)$ itself.

Either of these assertions can be generalized to many Lie groups. What we call Paley-Wiener theorems in this course is sometimes only the statement that a certain asymptotic property of $\varphi(y)$ is necessary. The proof for the assertion made in Section 3–3 makes use of the self-adjointness of the Casimir operator $H^2$ (3–92) of SU(2), which replaces the partial integrations in the proof of the classical theorem.

The ansatz (3–42) for the translational invariant functional is proved to be the most general one possible in Gelfand and Vilenkin [12 (Volume 4, Chapter II, Section 3.5)]. Distributions of a complex variable, in particular homogeneous distributions and the Fourier transformations of such distributions are treated in Gelfand and Shilov [12 (Volume 1, Appendix B)]. The proof that the necessary conditions on the bilinear invariant functionals found by us in Section 3–6 are also sufficient can be looked for in Gelfand *et al.* [ 12 (Volume 5, Chapter III)]. Covariant operators of the vector type as infinite-dimensional matrices appear first in Majorana [23]; this article contains also the two unitary irreducible representations which admit four-vector operators. The general theory of vector operators is dealt with in Naimark [24]. The theory of Lorentz invariant equations has been developed by Gelfand and Yaglom [13, 14] and is also presented in Naimark [24]. The problem of decomposing tensor products of unitary representations of SL(2, C) has been treated by Naimark [26]. His article contains the trilinear invariant functionals for three unitary irreducible representations.

# Chapter 4

# Harmonic Analysis of Polynomially Bounded Functions on the Group SL(2, C)

    The main result in the theory of Fourier transformations on a group is the Plancherel theorem. In Section 4–1 we introduce the notion of Fourier transformations on SL(2, C) and discuss their connection with representations of the group algebra. To prepare the proof of the Plancherel theorem we define characters of the group representations. The Plancherel theorem is derived in Section 4–2. In general the Plancherel theorem and its corollary, the Parseval formula, assert that the Fourier transformation is an isometric mapping of two $\mathscr{L}^2$-spaces onto each other. One of these spaces is the Hilbert space of square integrable functions on the group; it carries the right- or left-regular representation. The other space can be presented as a direct integral over Hilbert spaces each of which carries a factorial representation of the group. In this sense the Plancherel theorem solves the problem of decomposing the right-regular representation into factorial representations. This aspect is discussed in detail in Section 4–3. In Section 4–4 we define distributions (or generalized functions) and their Fourier transforms on SL(2, C). We introduce the concept of analytic functionals, which we regard as Fourier transforms of a certain class of distributions on SL(2, C). The most important problem in this context is to find out which classes of generalized functions possess which kinds of analytic functionals as Fourier transforms. To the knowledge of the author this problem has not yet been solved in general, and therefore we refrain from attacking it here also. It is, however, of great interest even for applications in physics.

    In the remaining sections we discuss more elementary topics. There are the representation functions of the second kind in Section 4–5. The method by which we introduce them does not make use of the theory of differential equations, but of the known explicit form of the functions of the first kind as given, for example, by the integral representation (3–32). For the reader to become acquainted with the concept of analytic functionals and the technique of continuing a distribution analytically in a parameter, we discuss two examples explicitly in Section 4–6. Formal aspects of an important class of distributions,

93

SU(2) bicovariant distributions, are investigated in Section 4–7. The scattering amplitude for elastic scattering of elementary particles in forward direction as studied in elementary particle physics belongs to this class of distributions (Chapter 8).

## 4–1    FOURIER TRANSFORMS AND REPRESENTATIONS OF THE GROUP ALGEBRA

The group algebra $\mathscr{R}(A)$ of SL(2, C) [A is a shorthand for SL(2, C)] consists of integrable functions $x(a)$ on SL(2, C), which themselves form an algebra $\mathscr{L}^1(A)$, and of a formal unit element. The multiplication operation in these algebras is defined by the convolution

$$x_1 \cdot x_2(a) = \int x_1(a_1) x_2(a_1^{-1} a) \, d\mu(a_1) \tag{4-1}$$

$\mathscr{L}^1(A)$ contains a linear subspace $\mathscr{C}^\infty$ which we define by the requirements:

($\alpha$)   The elements $x(a)$ of $\mathscr{C}^\infty$ possess derivatives of all orders.

($\beta$)   They have compact support, namely, there is a number $N$ depending on $x$ such that $x(a) = 0$ for $|a| > N$.

($\gamma$)   A sequence $x_n(a)$ is a null sequence if there is a common number $N$ such that all $x_n(a)$ vanish for $|a| > N$ and if $x_n(a)$ converges to zero uniformly together with all its derivatives.

For an element $x(a) \, \epsilon \, \mathscr{C}^\infty$ and a representation $\chi$ we define an operator $T_a^\chi$ in the Hilbert space $\mathscr{L}_m^2(U)$ or the equivalent space $\mathscr{L}_{\rho_1}^2(Z)$ (Section 3–3) by

$$T_x^\chi = \int x(a) T_a^\chi \, d\mu(a) \tag{4-2}$$

This operator is bounded because [see (3–25)]

$$\|T_x^\chi\| \leq \int |x(a)| \, \|T_a^\chi\| \, d\mu(a)$$

$$\leq N^{|\operatorname{Im} \rho|} \int |x(a)| \, d\mu(a)$$

where $N$ is such that $x(a) = 0$ for $|a| > N$. It is easy to see that $\mathscr{C}^\infty$ is not only a subspace of $\mathscr{L}^1(A)$ but moreover an algebra. In fact, the convolution product of two elements of $\mathscr{C}^\infty$ gives again an element of $\mathscr{C}^\infty$, and this convolution product is continuous in each factor with respect to the convergence defined in $\mathscr{C}^\infty$. We have

$$T_{x_1 \cdot x_2}^\chi = \int \left[ \int x_1(a_1) x_2(a_1^{-1} a) \, d\mu(a_1) \right] T_a^\chi \, d\mu(a)$$

$$= \int x_1(a_1) T_{a_1}^\chi \, d\mu(a_1) \int x_2(a) T_a^\chi \, d\mu(a)$$

$$= T_{x_1}^\chi \, T_{x_2}^\chi \tag{4-3}$$

The operators $T_x{}^\chi$ establish a representation of the algebra $\mathscr{C}^\infty$. Because of the inequality

$$\| T_x^\chi \| \le N^{|\mathrm{Im}\,\rho|} \int |x(a)| \, d\mu(a) \tag{4-4}$$

this representation is continuous with respect to the convergence defined in $\mathscr{C}^\infty$.

We consider now a representation $\chi$ of the principal series. In this case we have $\rho_1 = \mathrm{Im}\,\rho = 0$, and the representation in $\mathscr{L}_m^2(U)$ can be extended to a representation of the full Banach algebra $\mathscr{L}^1(A)$. Because of the inequality

$$\| T_x^\chi \| \le \int |x(a)| \, d\mu(a) \tag{4-5}$$

this representation of $\mathscr{L}^1(A)$ is continuous in the topology of $\mathscr{L}^1(A)$. In this case it makes sense to define an adjoint element $x(a)^\dagger$ by

$$x(a)^\dagger = \overline{x(a)^{-1}} \tag{4-6}$$

Definition (4–6) implies

$$(T_x^\chi)^\dagger = \int \overline{x(a)} (T_a^\chi)^\dagger \, d\mu(a)$$

$$= \int \overline{x(a^{-1})} T_a^\chi \, d\mu(a) = T_{x^\dagger}^\chi \tag{4-7}$$

since the measure $d\mu(a)$ is also invariant against inversion. Two elements $x(a)$ and $x(a)^\dagger$ of $\mathscr{L}^1(A)$, which are adjoints of each other, are therefore represented by a pair of adjoint operators in the Hilbert space $\mathscr{L}_m^2(U)$. Such representations of the algebra $\mathscr{R}(A)$ are called symmetric. We do not consider the supplementary series in this context, since we do not want to abandon the Hilbert spaces $\mathscr{L}_m^2(U)$ or $\mathscr{L}^2(Z)$.

We show next that $T_x{}^\chi$ is an integral operator in the space $\mathscr{L}_m^2(U)$ such that

$$T_x^\chi \varphi(u_1) = \int K_x(u_1, u_2 \,|\, \chi) \varphi(u_2) \, d\mu(u_2) \tag{4-8}$$

for arbitrary $\chi$. We call $T_x{}^\chi$ or the corresponding kernel $K_x(u_1, u_2 \,|\, \chi)$ the Fourier transform of $x$. If $x(a)$ is in $\mathscr{C}^\infty$, the operator function $T_x{}^\chi$ can be defined on the whole complex $\rho$-plane and for each $m$. We shall find later that the kernel $K_x(u_1, u_2 \,|\, \chi)$ can be chosen as an entire function of $\rho$ for any fixed

$u_1$, $u_2$, and $m$. The matrix elements of the operator $T_x{}^\chi$ in the canonical basis are

$$\langle j_1 q_1 | T_x{}^\chi | j_2 q_2 \rangle = \int \overline{\varphi_{q_1}^{j_1}(u_1)} K_x(u_1, u_2 | \chi) \varphi_{q_2}^{j_2}(u_2) \, d\mu(u_1) \, d\mu(u_2)$$

$$= \int x(a) D_{j_1 q_1 j_2 q_2}^\chi(a) \, d\mu(a)$$

We see that the matrix elements of the Fourier transform of $x(a)$ are the Fourier coefficients of $x(a)$ with respect to the matrix elements $D_{j_1 q_1 j_2 q_2}^\chi(a)$ of $T_a{}^\chi$.

In order to compute the kernel $K_x(\chi)$, we recall how $T_a{}^\chi$ acts in the space of functions $\varphi(u)$ (3–18)

$$T_x{}^\chi \varphi(u) = \int x(a) T_a{}^\chi \varphi(u) \, d\mu(a)$$

$$= \int x(a) \alpha(u, a) \varphi(u_a) \, d\mu(a)$$

We use (3–17) and replace $a$ by $u^{-1} a$

$$T_x{}^\chi \varphi(u) = \int x(u^{-1} a) \alpha(e, a) \varphi(k^{-1} a) \, d\mu(a)$$

In this expression $k$ depends on $a$ only and is defined such that $k^{-1} a = u_1$ is in SU(2). We set $a = k u_1$ and get, with (3–19),

$$T_x{}^\chi \varphi(u) = \int x(u^{-1} k u_1) \lambda^{n_1 - 1} \bar{\lambda}^{n_2 - 1} \varphi(u_1) \, d\mu(a)$$

Finally we make use of the decomposition

$$d\mu(a) = \pi \, d\mu(u_1) \, d\mu_l(k) \tag{4-9}$$

for $a = k u_1$, where $d\mu_l(k)$ is the left-invariant measure on the group K,

$$d\mu_l(k) = (2\pi)^{-4} D\lambda D\mu, \qquad k = \begin{pmatrix} \lambda^{-1} & \mu \\ 0 & \lambda \end{pmatrix} \tag{4-10}$$

The Jacobian (4–9) is computed in Appendix A–1a. The measure (4–10) is not right invariant, since

$$d\mu_l(k k_1) = |\lambda_1|^4 \, d\mu_l(k) \tag{4-11}$$

Inserting the expression (4–9) for the measure $d\mu(a)$ yields the result (4–8) desired, where

$$K_x(u_1, u_2 \mid \chi) = \pi \int x(u_1^{-1} k u_2) \lambda^{n_1 - 1} \bar{\lambda}^{n_2 - 1} \, d\mu_l(k)$$

$$\chi = \{n_1, n_2\} = (m, \rho) \tag{4-12}$$

For $x(a) \in \mathscr{C}^\infty$ the integral kernel $K_x(u_1, u_2 \mid \chi)$ defined by (4–12) has the following properties:

(1)  It is entire analytic in $\rho$ for fixed $u_1, u_2, m$.
(2)  It possesses derivatives of all orders with respect to $u_1$ and $u_2$.
(3)  It is covariant in the sense [see (3–11), (3–12)]

$$\begin{aligned} K(\gamma u_1, u_2 \mid \chi) &= K(u_1, \gamma^{-1} u_2 \mid \chi) \\ &= e^{im\omega} K(u_1, u_2 \mid \chi) \end{aligned} \tag{4-13}$$

(4)  It represents the algebra $\mathscr{C}^\infty$

$$K_{x_1 \cdot x_2}(u_1, u_3 \mid \chi)$$

$$= \int K_{x_1}(u_1, u_2 \mid \chi) K_{x_2}(u_2, u_3 \mid \chi) \, d\mu(u_2) \tag{4-14}$$

and if $\chi$ is in the principal series, we have in addition

$$K_x \dagger (u_1, u_2 \mid \chi) = \overline{K_x(u_2, u_1 \mid \chi)} \tag{4-15}$$

(5)  Considered as a function of the two variables $u_1$ and $u_2$, the kernel $K_x(u_1, u_2 \mid \chi)$ has a finite function norm

$$\|\!\| K_x(\chi) \|\!\|^2 = \int |K_x(u_1, u_2 \mid \chi)|^2 \, d\mu(u_1) \, d\mu(u_2) \tag{4-16}$$

Assertion (5) follows from (2) and the fact that the set of variables $u_1, u_2$ is compact. The function norm of the kernel is not smaller than the operator norm

$$\| T_x^\chi \| \leq \|\!\| K_x(\chi) \|\!\|$$

This holds in general. Integral operators whose kernels have a finite function norm are called operators of Hilbert-Schmidt type. Properties (3) and (4) remain valid if the representation of $\mathscr{C}^\infty$ is extended to a representation of $\mathscr{L}^1(A)$, if $\chi$ belongs to the principal series.

So far we have stressed only one aspect of the Fourier transforms of a function $x(a)$, namely, their relation with the algebra $\mathscr{L}^1(A)$ as expressed by the equations (4–14), (4–15). However, it is also possible to establish a connection with the Hilbert space $\mathscr{L}^2(A)$, which, as we remember, carries the right- and left-regular representations. If we exert a right translation $T_a{}^r$ on $x$, $x' = T_a{}^r x$, we get [see (4–12)]

$$K_{x'}(u_1, u_2 \mid \chi) = \pi \int x(u_1^{-1} k u_2 a) \lambda^{n_1-1} \bar{\lambda}^{n_2-1} \, d\mu_l(k)$$

$$= \pi \int x(u_1^{-1} k k'(u_2)_a) \lambda^{n_1-1} \bar{\lambda}^{n_2-1} \, d\mu_l(k)$$

$$= \alpha^{-\chi}(u_2, a) K_x(u_1, (u_2)_a \mid \chi) \tag{4-17}$$

where $\alpha^{-\chi}(u, a)$ denotes the multiplier (3–17), (3–19) for the representation $-\chi$. A left translation yields similarly

$$K_{x'}(u_1, u_2 \mid \chi) = \alpha^{\chi}(u_1, a) K_x((u_1)_a, u_2 \mid \chi)$$

$$x' = T_a{}^l x \tag{4-18}$$

We postpone further discussion of (4–17), (4–18) until we turn to the problem of decomposing the regular representations (Section 4–3).

Instead of representing the operator $T_x^\chi$ by an integral operator in $\mathscr{L}_m^2(U)$, we can also consider an integral operator in $\mathscr{L}_{\rho_1}^2(Z)$ with the kernel $K_x(z_1, z_2 \mid \chi)$. For the derivation of this kernel we may proceed exactly as for the kernel $K_x(u_1, u_2 \mid \chi)$. For the derivation of this kernel we may proceed exactly as for the kernel $K_x(u_1, u_2 \mid \chi)$. When

$$a = k\zeta \qquad \zeta = \begin{pmatrix} 1 & 0 \\ z & 1 \end{pmatrix}$$

we can decompose the measure $d\mu(a)$ as

$$d\mu(a) = d\mu_l(k) \, Dz$$

as is shown in Appendix A–1b. In this fashion we obtain

$$T_x^\chi f(z_1) = \int K_x(z_1, z_2 \mid \chi) f(z_2) Dz_2 \tag{4-19}$$

with

$$K_x(z_1, z_2 \mid \chi) = \int x(\zeta_1^{-1} k \zeta_2) \lambda^{n_1-1} \bar{\lambda}^{n_2-1} \, d\mu_l(k)$$

$$\zeta_{1,2} = \begin{pmatrix} 1 & 0 \\ z_{1,2} & 1 \end{pmatrix} \tag{4-20}$$

From (4–17), (4–18), and analogous formulas for the kernel $K_x(z_1, z_2|\chi)$, we see that the integral kernels transform as an element of $\mathcal{D}_\chi$ in their first variables $u_1$ or $z_1$, and as an element of $\mathcal{D}_{-\chi}$ in their second variables $u_2$ or $z_2$. In the same sense we may compare the integral (4–19) with the bilinear functional (3–56). We can use relation (3–14) between $\varphi(u)$ and $f(z)$ to switch between the two integral kernels. Replacing $u_1$ by $z_1$ as for a representation $\chi$ and $u_2$ by $z_2$ as for a representation $-\chi$, we get with [see (3–15), (3–16)]

$$u_1 = k_1\,\zeta_1 \qquad (u_1)_{21} = \lambda_1\,z_1 \qquad (u_1)_{22} = \lambda_1$$

$$u_2 = k_2\,\zeta_2 \qquad (u_2)_{21} = \lambda_2\,z_2 \qquad (u_2)_{22} = \lambda_2$$

$$\psi_{1,2} = -\arg\,(u_{1,2})_{22} = -\arg\lambda_{1,2}$$

$$|\lambda_{1,2}|^2 = (1 + |z_{1,2}|^2)^{-1}$$

the relation [see (4–11)]

$$K_x(u_1, u_2\,|\,\chi) = \pi \int x(\zeta_1^{-1}k_1^{-1}kk_2\,\zeta_2)\lambda^{n_1-1}\bar{\lambda}^{n_2-1}\,d\mu_l(k)$$

$$= \pi(\lambda_1\lambda_2^{-1})^{n_1-1}(\bar{\lambda}_1\bar{\lambda}_2^{-1})^{n_2-1}|\lambda_2|^{-4}K_x(z_1, z_2\,|\,\chi)$$

$$= \pi(1 + |z_1|^2)^{-(1/2)(n_1+n_2)+1}(1 + |z_2|^2)^{(1/2)(n_1+n_2)+1}$$

$$\times \exp\{im(\psi_1 - \psi_2)\}K_x(z_1, z_2\,|\,\chi) \qquad (4\text{-}21)$$

as expected.

For any $x \in \mathscr{C}^\infty$ the quantity

$$\mathrm{Tr}(T_x^x) = \int K_x(u, u\,|\,\chi)\,d\mu(u) \qquad (4\text{-}22)$$

can be seen to exist because of the property (2) of our list given earlier and the compactness of SU(2). We denote it the trace of the operator $T_x^{\,x}$. With the kernels in the space $\mathscr{L}^2_{\rho_1}(Z)$ we have from (4–21), (3–16)

$$\mathrm{Tr}(T_x^x) = \int K_x(z, z\,|\,\chi)Dz \qquad (4\text{-}23)$$

With the notation (4–22) and

$$x_1(a) = x \cdot x^\dagger(a)$$

we can write the Hilbert-Schmidt norm (4–16) in the simple form

$$\|K_x(\chi)\|^2 = \mathrm{Tr}(T_{x_1}^x) \qquad (4\text{-}24)$$

To justify the notation "trace" we show that in $\mathscr{L}^2_{\rho_1}(Z)$ and with respect to the canonical basis $\{f_q{}^j(z)\}$ the trace of $T_x{}^\chi$ in the standard sense is identical with (4–23). We have indeed

$$\mathrm{Tr}(T_x^\chi) = \sum_{j,q} \langle jq| \, T_x^\chi \, |jq\rangle$$

$$= \sum_{j,q} \int \overline{f_q{}^j(z)} (T_x^\chi f_q{}^j(z))(1 + |z|^2)^{\rho_1} \, Dz$$

$$= \sum_{j,q} \int \overline{f_q{}^j(z_1)} K_x(z_1, z_2 \,|\, \chi) f_q{}^j(z_2)$$

$$\times (1 + |z_1|^2)^{\rho_1} \, Dz_1 \, Dz_2$$

and the completeness relation in $\mathscr{L}^2_{\rho_1}(Z)$

$$\sum_{j,q} \overline{f_q{}^j(z_1)} f_q{}^j(z_2)(1 + |z_1|^2)^{\rho_1} = \delta(z_1 - z_2) \tag{4-25}$$

which together imply (4–23).

The trace of $T_x{}^\chi$, $x \, \epsilon \, \mathscr{C}^\infty$, has the remarkable property of being linear and continuous in $x$ with respect to the convergence in $\mathscr{C}^\infty$, as follows from (4–22) and (4–12). The trace can therefore be regarded as a linear continuous functional on $\mathscr{C}^\infty$, and defines a distribution $\xi^\chi(a)$ such that

$$\mathrm{Tr}(T_x^\chi) = \int x(a)\xi^\chi(a) \, d\mu(a) \tag{4-26}$$

The distribution $\xi^\chi(a)$ is customarily called the character of the representation $\chi$. To compute this character we recall (4–23) and (4–20),

$$\mathrm{Tr}(T_x^\chi) = \int x(\zeta^{-1}k\zeta)\lambda^{n_1-1}\bar{\lambda}^{n_2-1} \, d\mu_t(k) \, Dz \tag{4-27}$$

Denoting

$$a = \zeta^{-1}k\zeta$$

and taking into account the result of Appendix A–1c, we get

$$\mathrm{Tr}(T_x^\chi) = \int x(a) \frac{\lambda^{n_1}\bar{\lambda}^{n_2} + \lambda^{-n_1}\bar{\lambda}^{-n_2}}{|\lambda - \lambda^{-1}|^2} \, d\mu(a) \tag{4-28}$$

where $\lambda$ is any solution of

$$\lambda + \lambda^{-1} = a_{11} + a_{22}$$

Equation (4–28) determines the character $\xi^x(a)$ uniquely.

## 4–2   THE INVERSE FOURIER TRANSFORMATION

Assuming that the Fourier transform of a function $x \in \mathscr{C}^{\infty}$ is given in the form of a kernel $K_x(\chi)$, we wish now to express $x$ by $K_x(\chi)$, that is, we want to invert the Fourier transformation. The solution of this problem can be formulated in several equivalent fashions, for example, by the formulas

$$x(a) = \tfrac{1}{2} \int_{-\infty}^{+\infty} d\rho \sum_{m=-\infty}^{+\infty} (m^2 + \rho^2) \int \alpha^{-x}(z, a)$$

$$\times K_x(z, z_a \,|\, \chi) \, Dz \tag{4-29}$$

or

$$x(a) = \tfrac{1}{2} \int_{-\infty}^{+\infty} d\rho \sum_{m=-\infty}^{+\infty} (m^2 + \rho^2) \int \alpha^{-x}(u, a)$$

$$\times K_x(u, u_a \,|\, \chi) \, d\mu(u) \tag{4-30}$$

where $\chi = (m, \rho)$. The function $\alpha^{-x}(z, a)$ $[\alpha^{-x}(u, a)]$ is the multiplier defined by (3–9), (3–10) [(3–17), (3–19)] for a representation $-\chi$. Setting $a = e$ in (4–29) or (4–30), we obtain from (4–22), (4–23)

$$x(e) = \tfrac{1}{2} \int_{-\infty}^{+\infty} d\rho \sum_{m=-\infty}^{+\infty} (m^2 + \rho^2) \, \mathrm{Tr}(T_x^x) \tag{4-31}$$

We may regain (4–30) from (4–31) by exerting a right translation $T_a{}^r$ on $x$ and recalling (4–17). Formula (4–31) is therefore another equivalent inversion formula, and it is this form which we shall prove in this section. A third form of the inversion formula can be obtained from (4–31) if we write the trace of $T_x{}^x$ as a sum of matrix elements in the canonical basis

$$\mathrm{Tr}(T_x^x) = \sum_{j=|(1/2)m|}^{\infty} \sum_{q=-j}^{j} \int x(a_1) D_{jqjq}^x(a_1) \, d\mu(a_1)$$

and make use of the fact that the Fourier coefficients for a principal series representation $\chi$ behave under right translations as

$$\int x(a_1 a) D_{jqjq}^x(a_1) \, d\mu(a_1)$$

$$= \sum_{j_1 q_1} \overline{D_{jqj_1 q_1}^x(a)} \int x(a_1) D_{jqj_1 q_1}^x(a_1) \, d\mu(a_1)$$

This gives

$$x(a) = \tfrac{1}{2} \int\limits_{-\infty}^{+\infty} d\rho \sum_{m=-\infty}^{+\infty} (m^2 + \rho^2) \sum_{j_1, j_2 = |(1/2)m|}^{\infty} \sum_{q_1 = -j_1}^{j_1} \sum_{q_2 = -j_2}^{j_2}$$

$$\times \overline{D^x_{j_1 q_1 j_2 q_2}(a)} \int x(a_1) D^x_{j_1 q_1 j_2 q_2}(a_1) \, d\mu(a_1) \qquad (4\text{-}32)$$

In order to prove (4–31) we recall (4–27) from which we obtain by partial integration

$$-\tfrac{1}{4}(m^2 + \rho^2) \, \mathrm{Tr}(T_x^x)$$

$$= \int \lambda^{-(1/2)m + (i/2)\rho} \bar{\lambda}^{(1/2)m + (i/2)\rho} \left\{ \frac{\partial}{\partial \lambda} \frac{\partial}{\partial \bar{\lambda}} \, x(\zeta^{-1} k \zeta) \right\} d\mu_l(k) \, Dz$$

Inserting this into (4–31) and introducing the shorthand

$$G(\lambda) = |\lambda|^2 \int \frac{\partial}{\partial \lambda} \frac{\partial}{\partial \bar{\lambda}} \, x(\zeta^{-1} k \zeta) \, D\mu \, Dz \qquad (4\text{-}33)$$

yields

$$x(e) = -2(2\pi)^{-4} \int\limits_{-\infty}^{+\infty} d\rho \sum_{m=-\infty}^{+\infty}$$

$$\times \int \lambda^{-(1/2)m + (i/2)\rho - 1} \bar{\lambda}^{(1/2)m + (i/2)\rho - 1} G(\lambda) \, D\lambda$$

The sum and the integrations can be performed and we get

$$x(e) = -2(2\pi)^{-2} G(1) \qquad (4\text{-}34)$$

Relations (4–31) and (4–34) are obviously equivalent, we need therefore only prove the simpler relation (4–34).

We consider a function $x^*(a)$ of four complex variables $a_{ij}$, $i, j = 1, 2$, which has compact support and possesses derivatives of all orders in the eight real variables $\mathrm{Re}(a_{ij})$ and $\mathrm{Im}(a_{ij})$. Moreover we require

$$x^*(a) = x(a) \qquad \text{if} \quad a_{11} a_{22} - a_{12} a_{21} = 1$$

The function $x^*(a)$ possesses the classical Fourier transform $f(b)$ defined such that

$$x^*(a) = \int \prod_{i, j = 1, 2} Db_{ij} \, f(b) \exp\left\{ i \, \mathrm{Re} \sum_{i, j} a_{ij} b_{ji} \right\} \qquad (4\text{-}35)$$

Inserting (4–35) into (4–33) gives

$$G(\lambda) = (2\pi)^2 \, |\lambda|^2 \int \prod_{i,\,j} Db_{ij} \, f(b)$$

$$\times \int Dz \, \delta(b_{21} + (b_{11} - b_{22})z - b_{12} z^2)$$

$$\times \frac{\partial}{\partial \lambda} \frac{\partial}{\partial \bar{\lambda}} \exp\{i \, \mathrm{Re}(b_{11}\lambda^{-1} + b_{22}\lambda + b_{12} z(\lambda - \lambda^{-1}))\}$$

and further

$$G(1) = -\tfrac{1}{4}(2\pi)^2 \int \prod_{i,\,j} Db_{ij} \, f(b) \exp\{i \, \mathrm{Re}(b_{11} + b_{22})\}$$

$$\times \int Dz \, |b_{22} - b_{11} + 2b_{12} z|^2$$

$$\times \, \delta(b_{21} + (b_{11} - b_{22})z - b_{12} z^2)$$

The argument of the delta-function has two zeros for almost all $b$. The integration over $z$ yields therefore a function of $b$ which is equal to 2 almost everywhere. Taking into account (4–35) and $x^*(e) = x(e)$ we get

$$G(1) = -\tfrac{1}{2}(2\pi)^2 x(e)$$

as desired. This completes the proof of the inversion formulas.

In the inverse Fourier transformation the summation and integration extend only over the principal series of representations. The measure

$$d\chi = (m^2 + \rho^2) \, d\rho$$

which is restricted to the principal series, is called the Plancherel measure. Instead of (4–31) we may, for example, write

$$x(e) = \tfrac{1}{2} \int \mathrm{Tr}(T_x^\chi) \, d\chi \qquad (4\text{-}36)$$

For a function

$$x_1(a) = x \cdot x^\dagger(a) \qquad x \in \mathscr{C}^\infty$$

which implies $x_1(a) \in \mathscr{C}^\infty$, we get from (4–1), (4–6), and (4–36)

$$x_1(e) = \int |x(a)|^2 \, d\mu(a)$$

$$= \tfrac{1}{2} \int \mathrm{Tr}(T_{x_1}^\chi) \, d\chi$$

and with (4—24)

$$\int |x(a)|^2 \, d\mu(a) = \int \||K_x(\chi)\||^2 \, d\chi \qquad (4\text{-}37)$$

This is Parseval's formula for SL(2, C) for a function $x \in \mathscr{C}^\infty$.

The linear space $\mathscr{C}^\infty$ is dense in the Hilbert space $\mathscr{L}^2(A)$. With the help of Parseval's formula (4—37) we may therefore extend the Fourier transformation from $\mathscr{C}^\infty$ onto $\mathscr{L}^2(A)$ with the following result. For any function $x(a) \in \mathscr{L}^2(A)$ the integral (4—20) converges in the mean and is a measurable function in the product space of the three variables $z_1', z_2, \chi$ with the measure $Dz_1 | Dz_2 \, d\chi$. The integral

$$\int |K_x(z_1, z_2 | \chi)|^2 \, Dz_1 \, Dz_2 \, d\chi$$

is finite. The Fourier transformation can be inverted by formula (4—29) such that the integral (4—29) converges in the mean and the equality holds for almost all $a \in$ SL(2, C). These assertions are the content of Plancherel's theorem for the group SL(2, C).

We close this section with a heuristic remark. In the inversion formula for Fourier transformations on compact Lie groups, the Plancherel measure is a measure on the discrete set of equivalence classes of representations $\chi$. Each class appears with the weight dim $\chi$; remember the group SU(2) where dim $\chi(j) = 2j + 1$ [see e.g., (3—20)]. An irreducible spinor representation of SL(2, C) in a space of homogeneous functions of degrees $n_1$ and $n_2$ has the dimension $n_1 n_2$,

$$n_1 n_2 = -\tfrac{1}{4}(m^2 + \rho^2)$$

It is this dimension factor which (with opposite sign) is contained in the Plancherel measure of SL(2, C).

## 4—3    THE DECOMPOSITION OF THE REGULAR REPRESENTATION

Next we attack the problem of decomposing the regular representation in the space $\mathscr{L}^2(A)$ into irreducible components. We consider the Hilbert space $\mathscr{L}^2(K^*)$ of functions $K(z_1, z_2 | \chi)$ which are measurable on the topological product of the three manifolds of variables $z_1, z_2, \chi$ ($\chi$ is restricted to the principal series) with respect to the measure $Dz_1 \, Dz_2 \, d\chi$, and whose norm squared

$$\||K\||^2 = \tfrac{1}{2} \int d\chi \||K(\chi)\||^2 = \tfrac{1}{2} \int d\chi \, Dz_1 \, Dz_2 \, |K_x(z_1, z_2 | \chi)|^2 \qquad (4\text{-}38)$$

is finite. It is obvious as to how one must define the scalar product in $\mathscr{L}^2(K^*)$. Parseval's formula (4—37) proves that the Fourier transformation is an isometric mapping of $\mathscr{L}^2(A)$ into $\mathscr{L}^2(K^*)$. The image of $\mathscr{L}^2(A)$ is certainly a Hilbert subspace of $\mathscr{L}^2(K^*)$; we denote it $\mathscr{L}^2(K)$. How can $\mathscr{L}^2(K)$ be characterized?

Let $\chi$ and $-\chi$ be two different representations of the principal series realized in Hilbert spaces $\mathscr{L}^2(Z_1)$ and $\mathscr{L}^2(Z_2)$, respectively. There exists an intertwining operator $A$ (3–72) with the property

$$T_a^{-\chi}A = AT_a^{\chi} \qquad \text{for all } a \in \text{SL}(2, \text{C})\qquad (4\text{-}39)$$

which after a proper normalization maps $\mathscr{L}^2(Z_1)$ isometrically on $\mathscr{L}^2(Z_2)$. We can write

$$g(z_2) = \int A(z_2, z_1 \,|\, \chi) f(z_1)\, Dz_1 \qquad (4\text{-}40)$$

$$f(z_1) \in \mathscr{L}^2(Z_1) \qquad g(z_2) \in \mathscr{L}^2(Z_2)$$

where [see (3–72) and (3–78)]

$$A(z_2, z_1 \,|\, \chi) = (2\pi)^{-1}(m^2 + \rho^2)^{1/2}(z_2 - z_1)^{(1/2)m - (i/2)\rho - 1}$$
$$\times (\bar{z}_2 - \bar{z}_1)^{-(1/2)m - (i/2)\rho - 1} \qquad (4\text{-}41)$$

and the integral converges in the mean. From (4–2), (4–39) we have

$$T_x^{-\chi}A = AT_x^{\chi}$$

such that (4–19) and (4–40) imply

$$\int K_x(z_1, z_2 \,|\, -\chi)A(z_2, z_3 \,|\, \chi)\, Dz_2 = \int A(z_1, z_2 \,|\, \chi)K_x(z_2, z_3 \,|\, \chi)\, Dz_2 \quad (4\text{-}42)$$

for any function $x(a) \in \mathscr{C}^{\infty}$. We denote the image of $\mathscr{C}^{\infty}$ in $\mathscr{L}^2(K)$ by $\mathscr{X}$. $\mathscr{X}$ is dense in $\mathscr{L}^2(K)$. The constraint (4–42) can therefore be extended from $\mathscr{X}$ onto $\mathscr{L}^2(K)$ in the familiar sense: For any element of $\mathscr{L}^2(K)$ both integrals (4–42) exist and the equality (4–42) holds for almost all $z_1$, $z_3$, $\chi$. On the other hand, one can show that the constraint (4–42) characterizes the subspace $\mathscr{L}^2(K)$ in $\mathscr{L}^2(K^*)$ completely. The proof of this assertion is elementary but lengthy, and we skip over it.

Now we construct Hilbert spaces $\mathscr{L}^2(Z_2 \,|\, \chi)_q{}^j$ for any $\chi = (m, \rho)$ of the principal series and

$$j = |\tfrac{1}{2}m| + n \qquad n = 0, 1, 2, \dots \qquad -j \leqq q \leqq j$$

They consist of measurable functions $f(z_2) \equiv f(z_2|\chi)_q^j$ with finite norm $\|f\| = (f, f)^{1/2}$, where $(f_1, f_2)$ denotes the scalar product

$$(f_1, f_2) = \int \overline{f_1(z_2)} f_2(z_2) \, Dz_2$$

We construct the direct orthogonal sum

$$\mathscr{L}^2(Z_2|\chi) = \sum_{j=|(1/2)m|}^{\infty \oplus} \sum_{q=-j}^{j \oplus} \mathscr{L}^2(Z_2|\chi)_q^j \tag{4-43}$$

and the direct integral

$$\mathscr{H} = \int_{\rho > 0}^{\oplus} d\chi \, \mathscr{L}^2(Z_2|\chi) \tag{4-44}$$

out of such spaces (Section 1–3). If we define

$$f_x(z_2|\chi)_q^j = \int \overline{f_q^{\,j}(z_1)} K(z_1, z_2| -\chi) \, Dz_1 \tag{4-45}$$

where $f_q^{\,j}(z)$ is an element of the canonical basis in $\mathscr{L}^2(Z)$, we have achieved an isometric mapping of $\mathscr{L}^2(K)$ on $\mathscr{H}$. In fact, for any $K_x(z_1, z_2|-\chi) \in \mathscr{L}^2(K)$, (4–45) is a measurable function in $z_2$ and $\chi$ for all $j$ and $q$, and

$$\int_{\rho > 0} d\chi \sum_{j=|(1/2)m|}^{\infty} \sum_{q=-j}^{j} \int Dz_2 \, |f_x(z_2|\chi)_q^j|^2 = \int_{\rho > 0} d\chi \, \|\!| K(-\chi)\|\!|^2 = \|\!|\!| K \|\!|\!|^2 \tag{4-46}$$

where we used (4–45), the completeness of the canonical basis, the isometry of the intertwining operator $A$ in (4–42), and (4–38).

A right translation $T_a^r$ in $\mathscr{L}^2(A)$,

$$x(a_1) \xrightarrow[\cdot \ T_a^r]{} x'(a_1) = x(a_1 a)$$

induces a transformation (4–17) in $\mathscr{L}^2(K)$

$$K_x(z_1, z_2| -\chi) \xrightarrow[T_a^r]{} K_{x'}(z_1, z_2| -\chi)$$

$$= \alpha^x(z_2, a) K_x(z_1, (z_2)_a| -\chi)$$

which in turn implies

$$f_x(z_2|\chi)_q^j \xrightarrow[T_a^r]{} f_{x'}(z_2|\chi)_q^j \tag{4-47}$$

$$= \alpha^x(z_2, a) f_x((z_2)_a|\chi)_q^j$$

because of (4–45). By means of the Fourier transformation and the map (4–45) we have therefore accomplished a direct integral and direct orthogonal sum decomposition of $\mathscr{L}^2(A)$ into Hilbert spaces $\mathscr{L}^2(Z_2|\chi)_q^j$ such that the right translations in $\mathscr{L}^2(A)$ generate irreducible, unitary representations $\chi$ of the principal series in $\mathscr{L}^2(Z_2|\chi)_q^j$. This is the complete solution of the task of decomposing the right-regular representation into irreducible components.

Finally we consider the Hilbert space $\mathscr{L}^2(Z_2|\chi)$ defined by (4–43) and (4–45). It carries a factorial representation of type one, that is, an infinite multiple of the representation $\chi$ of the principal series. On the other hand, we see easily that its elements can be identified with the kernels $K_x(z_1, z_2|\chi)$. The norm in $\mathscr{L}^2(Z_2|\chi)$ can then be expressed by the Hilbert-Schmidt norm $|||K_x(\chi)|||$ of $K_x(z_1, z_2|\chi)$, using (4–16), (4–21), and (3–16). The degeneracy of the representation $\chi$ in the factorial representation is labeled by the continuous parameter $z_1$ instead of $j$ and $q$. The fact that the group SL(2, C) is of type one guarantees that the decomposition (4–44) of the right-regular representation into factorial representations is unique, as asserted by Theorem 3 in Section 1–3c, and it allows us to decompose each factorial representation into a direct orthogonal sum of irreducible components. In order to obtain an analogous result for the left-regular representation on $\mathscr{L}^2(A)$, we need only replace (4–45) by

$$f_x(z_1|\chi)_q^j = \int K_x(z_1, z_2|\chi)f_q^j(z_2)\, Dz_2 \qquad (4\text{-}48)$$

## 4–4    GENERALIZED FUNCTIONS ON THE GROUP SL(2, C)

We consider again the space $\mathscr{C}^\infty$ of functions with compact support on SL(2, C). Any linear continuous functional $p$ on $\mathscr{C}^\infty$ is called a distribution or a generalized function on SL(2, C). We write it for any $x \in \mathscr{C}^\infty$

$$(p, x) \quad \text{or formally} \quad \int p(a)x(a)\, d\mu(a)$$

Such distributions have no particular properties themselves if compared with distributions over other multidimensional spaces. They exhibit features of interest only in connection with Fourier transformations.

Distributions form a closed linear space $\mathscr{C}^{\infty\,\prime}$. The Fourier transformation maps $\mathscr{C}^\infty$ into the space $\mathscr{L}^2(K)$, the image being denoted $\mathscr{Z}$. $\mathscr{C}^\infty$ is dense in $\mathscr{L}^2(A)$, and $\mathscr{Z}$ is correspondingly dense in $\mathscr{L}^2(K)$. We can introduce a convergence in $\mathscr{Z}$ by simply carrying over the convergence in $\mathscr{C}^\infty$. Any distribution $p \in \mathscr{C}^{\infty\,\prime}$ induces a linear functional $p^*$ on the space $\mathscr{Z}$ if we define

$$(p, x) = (p^*, K_x) \qquad (4\text{-}49)$$

As we shall see later, (4–49) can be regarded as an extension of Parseval's formula (4–37). The linear continuous functionals $p^*$ on $\mathscr{L}$ form a closed linear space $\mathscr{L}'$. The space $\mathscr{L}'$ is in one-to-one correspondence with the space of distributions $\mathscr{C}^{\infty\prime}$ through the Fourier transformation and this "Parseval's formula" (4–49).

If the distribution $p$ is regular, that is to say, given by a square integrable function $p(a)$ such that the functional $(p, x)$ is a proper integral

$$(p, x) = \int p(a)x(a)\,d\mu(a)$$

then the functional $p^*$ is similarly defined by a function, namely, the Fourier transform $K_{\bar{p}}(u_1, u_2|\chi)$ of $\overline{p(a)}$ with $\chi$ in the principal series. In fact, by means of the original Parseval's formula (4–37) we obtain

$$(p, x) = \tfrac{1}{2}\int\limits_{-\infty}^{+\infty} d\rho \sum_{m=-\infty}^{+\infty} (m^2 + \rho^2)\int K_{\bar{p}}(u_1, u_2|\chi)$$

$$\times\, K_x(u_1, u_2|\chi)\,d\mu(u_1)\,d\mu(u_2)$$

$$= (p^*, K_x) \tag{4-50}$$

This relation (4–50) justifies the notation "Parseval's formula" for (4–49).

Regular distributions are of course not very exciting. Nevertheless they may be used as heuristic guides in finding a bigger and more interesting class of distributions. Let C be an integration contour in the complex $\rho$-plane. We call it symmetric if it does not change under the replacement $\rho \to -\rho$. On a symmetric contour C and for all $m$ we give a function $K(u_1, u_2|\chi)$. This function is continuous in the variables $u_1$ and $u_2$, covariant as specified by (4–13), and measurable in $\rho$ for any fixed $u_1, u_2, m$. In addition we require the symmetry

$$\int \overline{K(u_1, u_2|\chi)}K_x(u_1, u_2|\chi)\,d\mu(u_1)\,d\mu(u_2)$$

$$= \int \overline{K(u_1, u_2|-\chi)}K_x(u_1, u_2|-\chi)\,d\mu(u_1)\,d\mu(u_2) \tag{4-51}$$

to hold for all $x \in \mathscr{C}^{\infty}$. Then we define a functional $p^*$ on $\mathscr{L}'$ by

$$(p^*, K_x) = \tfrac{1}{2}\int\limits_{C} d\rho \sum_{m=-\infty}^{+\infty} (m^2 + \rho^2)\int \overline{K(u_1, u_2|\chi)}$$

$$\times\, K_x(u_1, u_2|\chi)\,d\mu(u_1)\,d\mu(u_2) \tag{4-52}$$

assuming that the integrals and sums converge for all $K_x(\chi) \in \mathscr{Z}$ and the functional defined this way is continuous with respect to the convergence in $\mathscr{Z}$. We call such linear continuous functionals $p^*$ analytic functionals. This notation is sometimes also used for the distribution $p$ corresponding to $p^*$ by (4–49).

We recall that $K_x(\chi) \in \mathscr{Z}$ is entire in $\rho$. Therefore it is in many cases possible to deform the contour C in (4–52) leaving $p^*$ unaltered. Contours obtained from each other in this way are denoted equivalent. The contours obtained by deformation need no longer be symmetric. For the purpose of deforming a contour C an estimate of $K_x(\chi)$ is possibly of interest. For a number $N$ chosen such that $x(a) = 0$ if $|a| > N$, we have, for example,

$$|\rho^n K_x(u_1, u_2 \,|\, m, \rho)| \leq c_n N^{|\text{Im } \rho|}$$

where $c_n$ is independent of $\rho$, $u_1$, and $u_2$. For the proof of this assertion we refer to our remarks in Section 3–11.

If $p^*$ is an analytic functional such that C is the real axis and the distribution $p(a)$ corresponding to $p^*$ is regular, then we notice by a comparison of (4–50) and (4–52) that the function $K(\chi)$ on C determining the analytic functional is identical with the Fourier transform of $\overline{p(a)}$. This raises the question as to whether for other analytic functionals $p^*$ a function $p(a)$ can be found which is not necessarily square integrable, but such that

$$(p^*, K_x) = \int p(a)x(a)\, d\mu(a) \qquad (4\text{-}53)$$

is still a proper integral for all $x \in \mathscr{C}^\infty$. This is indeed possible in many cases.

The inverse Fourier transformation (4–30) suggests the following ansatz for $p(a)$

$$p(a) = \tfrac{1}{2} \int_C d\rho \sum_{m=-\infty}^{+\infty} (m^2 + \rho^2) \int \overline{K(u, u_a \,|\, \chi)} \alpha^\chi(u, a)\, d\mu(u) \qquad (4\text{-}54)$$

If we require that $K(u_1, u_2 \,|\, \chi)$ is continuous in $u_1$, $u_2$, and $\rho$ simultaneously, and that all integrals and the sum in (4–54) converge absolutely uniformly on any compact subset of SL(2, C), we can easily verify that (4–54) implies (4–53). In fact, from our premise follows that $p(a)$ is continuous and the integral

$$\int x(a)p(a)\, d\mu(a)$$

$$= \tfrac{1}{2} \int d\mu(a)x(a) \left\{ \int_C d\rho \sum_{m=-\infty}^{+\infty} (m^2 + \rho^2) \int \overline{K(u, u_a \,|\, \chi)} \alpha^\chi(u, a)\, d\mu(u) \right\} \qquad (4\text{-}55)$$

exists, even if all functions under the integral signs of the right-hand side are replaced by their moduli. The simultaneous integral exists therefore, too. The orders of integration and summation can then be changed arbitrarily in (4—55). As in the derivation of (4—12) we have in addition

$$\int \overline{K(u, u_a \,|\, \chi)} x(a) \alpha^x(u, a) \, d\mu(a) \, d\mu(u)$$

$$= \pi \int \overline{K(u_1, u_2 \,|\, \chi)} x(u_1^{-1} k u_2)$$

$$\times \lambda^{n_1 - 1} \overline{\lambda}^{n_2 - 1} \, d\mu_l(k) \, d\mu(u_1) \, d\mu(u_2)$$

$$= \int \overline{K(u_1, u_2 \,|\, \chi)} K_x(u_1, u_2 \,|\, \chi) \, d\mu(u_1) \, d\mu(u_2)$$

This completes the proof of (4—53).

We see that by means of Parseval's formula (4—49) and linear functionals it is possible to extend the notion of Fourier transformations to functions $p(a)$ which are no longer square integrable. If a function $p(a)$ can be expressed by a contour integral over a kernel $K(u_1, u_2 \,|\, \chi)$ as in (4—54), such that the conditions set up in the text are met, we can consider this well-defined kernel on the contour C, and not only an abstract functional $p^*$ over $\mathscr{L}$, as the Fourier transform of the function $p(a)$. To obtain this Fourier transform on its contour, we may always use a method which is based on analytic continuation of a distribution in a parameter. We shall study this technique and apply it to several kinds of functions $p(a)$ in Section 4—6.

## 4—5    THE REPRESENTATION FUNCTIONS OF THE SECOND KIND

Apart from the functions $d^\chi_{j_1 j_2 q}(\eta)$ introduced in Section 3—5, which we now denote functions of the first kind, there is a class of functions $e^\chi_{j_1 j_2 q}(\eta)$ of considerable importance. These functions are denoted functions of the second kind. Any function of the first kind can be expressed linearly by functions of the second kind and vice versa. Whereas the functions of the first kind are defined as certain matrix elements of the representations of SL(2, C), there is no direct group theoretical meaning of the functions of the second kind. The notation first and second kind has been chosen in analogy with Legendre functions (Section 6—4). As in that case it is possible to characterize the different kinds of functions as solutions of one system of differential equations which exhibit different analytic behavior at the singular points. We do not, however, want to discuss differential operators in this course. Instead we shall show by explicit construction how to derive functions of the second kind from those of the first kind.

The functions $e^{\chi}_{j_1 j_{2q}}(\eta)$ obtained in this fashion have the following properties. For $\eta > 0$ and $\rho = re^{i\varphi}$ we have:

(1) $\lim\limits_{r \to \infty} r^{\alpha} e^{(m; re^{i\varphi})}_{j_1 j_{2q}}(\eta) = 0$ for $0 < \varphi < \pi$ and arbitrary real $\alpha$.

$\lim\limits_{r \to \infty} r^{\alpha} e^{(m; \pm r)}_{j_1 j_{2q}}(\eta) = 0$ for $\alpha < 1$.

(2) $\lim\limits_{r \to \infty} r^{\alpha} [d^{(m; re^{i\varphi})}_{j_1 j_{2q}}(\eta) - e^{(m; re^{i\varphi})}_{j_1 j_{2q}}(\eta)] = 0$ for $\alpha < 1$.

for $\pi < \varphi < 2\pi$ and arbitrary real $\alpha$.

$\lim\limits_{r \to \infty} r^{\alpha} [d^{(m; \pm r)}_{j_1 j_{2q}}(\eta) - e^{(m; \pm r)}_{j_1 j_{2q}}(\eta)] = 0$ for $\alpha < 1$.

(3)  The function $e^{\chi}_{j_1 j_2 q}(\eta)$ is meromorphic in $\rho$ with a finite number of simple poles on the imaginary axis at the positions

$$\frac{i}{2} \rho = -j_1, -j_1 + 1, \ldots, j_2 - 1, j_2$$

which are independent of $\eta$.

Let us assume that we are given functions $e^{\chi}_{j_1 j_{2q}}(\eta)$ with the three properties listed. We write

$$d^{\chi}_{j_1 j_{2q}}(\eta) = e^{\chi}_{j_1 j_{2q}}(\eta) + f^{\chi}_{j_1 j_{2q}}(\eta)$$

Due to the symmetry relation (3–35) we get

$$e^{\chi}_{j_1 j_{2q}}(\eta) - (-1)^{j_1 - j_2} f^{-\chi}_{j_2 j_{1q}}(\eta) = (-1)^{j_1 - j_2} e^{-\chi}_{j_2 j_{1q}}(\eta) - f^{\chi}_{j_1 j_{2q}}(\eta)$$

The left-hand side of this equation tends to zero if $\rho$ goes to infinity in the upper half $\rho$-plane and on the real axis, whereas the right-hand side vanishes if $\rho$ tends to infinity in the lower half $\rho$-plane and on the real axis. Since $d^{\chi}_{j_1 j_{2q}}(\eta)$ is entire in $\rho$, the function $f^{\chi}_{j_1 j_{2q}}(\eta)$ possesses a finite number of first order poles which cancel the poles of $e^{\chi}_{j_1 j_{2q}}(\eta)$. Liouville's theorem implies therefore

$$f^{\chi}_{j_1 j_{2q}}(\eta) = (-1)^{j_1 - j_2} e^{-\chi}_{j_2 j_{1q}}(\eta) + \sum_{n=0}^{j_1 + j_2} \frac{a^{(n)}_{j_1 j_{2q}}(\eta)}{(i/2)\rho + j_1 - n}$$

where the residues $a^{(n)}_{j_1 j_2 q}(\eta)$ are independent of $\rho$. Since both $f^{\chi}_{j_1 j_2 q}(\eta)$ and $e^{-\chi}_{j_2 j_1 q}(\eta)$ go to zero faster than any power of $\rho$ in the lower half $\rho$-plane, we have necessarily

$$a^{(n)}_{j_1 j_2 q}(\eta) = 0$$

for all $n$. This yields

$$d^{\chi}_{j_1 j_2 q}(\eta) = e^{\chi}_{j_1 j_2 q}(\eta) + (-1)^{j_1 - j_2} e^{-\chi}_{j_2 j_1 q}(\eta) \tag{4-56}$$

Similarly we can prove the relation $[(3\text{–}35), (3\text{–}96)]$

$$e^{\chi}_{j_1 j_2 q}(\eta) = (-1)^{j_1 - j_2} (\beta^{j_1})^{-1} \beta^{j_2} e^{\chi}_{j_2 j_1 q}(\eta) \tag{4-57}$$

We turn now to an explicit calculation of $e^{\chi}_{j_1 j_2 q}(\eta)$. We start from the integral representation (3–32) for $d^{\chi}_{j_1 j_2 q}(\eta)$ and introduce a new parameter $\alpha$ by

$$e^{\alpha} = t e^{-\eta} + (1 - t) e^{\eta}$$

With the binomial expansion (2–30) for the functions $d^{j}_{(1/2)m, q}$ we obtain

$$
\begin{aligned}
d^{\chi}_{j_1 j_2 q}(\eta) \\
= &\left\{ (2j_1 + 1)(2j_2 + 1) \right. \\
&\left. \times \frac{(j_1 + (1/2)m)!(j_1 - (1/2)m)!(j_2 + (1/2)m)!(j_2 - (1/2)m)!}{(j_1 + q)!(j_1 - q)!(j_2 + q)!(j_2 - q)!} \right\}^{1/2} \\
&\times (2 \sinh \eta)^{-j_1 - j_2 - 1} \sum_{n_1, n_2} (-1)^{n_1 + n_2} \\
&\times \binom{j_1 + q}{n_1} \binom{j_1 - q}{n_1 - q + \frac{1}{2}m} \binom{j_2 + q}{n_2} \binom{j_2 - q}{n_2 - q + \frac{1}{2}m} \\
&\times \int_{-\eta}^{\eta} d\alpha \, e^{(i/2)\rho\alpha} (e^{\eta} - e^{\alpha})^{[2j_1 - 2n_1 + q - (1/2)m]/2} \\
&\times (e^{\alpha} - e^{-\eta})^{[2n_1 - q + (1/2)m]/2} (e^{\eta} - e^{-\alpha})^{[2n_2 - q + (1/2)m]/2} \\
&\times (e^{-\alpha} - e^{-\eta})^{[2j_2 - 2n_2 + q - (1/2)m]/2}
\end{aligned} \tag{4-58}
$$

where the sum over the nonnegative numbers $n_1$ and $n_2$ extends over that domain in which the binomial coefficients are finite nonzero.

We evaluate the integral (4–58) in two different fashions. First we set

$$v = \frac{t}{1-t} \qquad e^\alpha = e^\eta \frac{1 + ve^{-2\eta}}{1+v}$$

and get for the integral over $\alpha$

$$(2 \sinh \eta)^{j_1 + j_2 + 1} \exp \eta \left[ -2j_2 + 2n_2 - q + \tfrac{1}{2}m + \frac{i}{2}\rho - 1 \right]$$

$$\times \int_0^\infty dv v^{j_1 + j_2 - n_1 - n_2 + q - (1/2)m}$$

$$\times (1+v)^{(-i/2)\rho - j_1 - 1}(1 + ve^{-2\eta})^{+(i/2)\rho - j_2 - 1}$$

Then we expand the factor

$$(1 + ve^{-2\eta}) = (1+v)\left[ 1 - \frac{v}{1+v}(1 - e^{-2\eta}) \right]$$

into a binomial series and integrate each term separately. This yields for the integral over $v$

$$\frac{(j_1 + j_2 - n_1 - n_2 + q - (1/2)m)!\,(n_1 + n_2 - q + (1/2)m)!}{(j_1 + j_2 + 1)!}$$

$$\times {}_2F_1\left( j_1 + j_2 - n_1 - n_2 + q - \tfrac{1}{2}m + 1, j_2 - \frac{i}{2}\rho + 1, j_1 + j_2 + 2; 1 - e^{-2\eta} \right)$$

By means of a standard formula (GR 9.131.1) we get for the hypergeometric function instead

$$\exp \eta[+2j_2 - 2n_2 - 2n_1 + 2q - m - i\rho]$$

$$\times {}_2F_1\left( n_1 + n_2 - q + \tfrac{1}{2}m + 1, j_1 + \frac{i}{2}\rho + 1, j_1 + j_2 + 2; 1 - e^{-2\eta} \right);$$

and altogether

$$d^\chi_{j_1 j_2 q}(\eta) = \{\cdots\}^{1/2} \sum_{n_1 n_2} (-1)^{n_1 + n_2} \binom{j_1 + q}{n_1}\binom{j_1 - q}{n_1 - q + \tfrac{1}{2}m}$$

$$\times \binom{j_2 + q}{n_2}\binom{j_2 - q}{n_2 - q + \tfrac{1}{2}m}[(j_1 + j_2 + 1)!]^{-1}$$

$$\times (j_1 + j_2 - n_1 - n_2 + q - \tfrac{1}{2}m)!(n_1 + n_2 - q + \tfrac{1}{2}m)!$$

$$\times \exp \eta \left[ -2n_1 + q - \tfrac{1}{2}m - \frac{i}{2}\rho - 1 \right]$$

$$\times {}_2F_1\Bigg( n_1 + n_2 - q + \tfrac{1}{2}m + 1,$$

$$j_1 + \frac{i}{2}\rho + 1, j_1 + j_2 + 2; 1 - e^{-2\eta} \Bigg) \qquad (4\text{-}59)$$

The hypergeometric function involved in (4–59) is an infinite series, and it does not seem obvious that it is indeed an elementary function. If we attempt to express this function by hypergeometric polynomials, we are led to the functions $e^{\chi}_{j_1 j_2 q}(\eta)$.

The functions $e^{\chi}_{j_1 j_2 q}(\eta)$ can easily be obtained if we write (4–58) in abbreviated form

$$d^{\chi}_{j_1 j_2 q}(\eta) = \{\cdots\}^{1/2}(2 \sinh \eta)^{-j_1 - j_2 - 1} \int_{-\eta}^{\eta} d\alpha\, f(\alpha, \eta)$$

and take the primitive of the function $f(\alpha, \eta)$ in the unique form of a finite sum of exponentials

$$F(\alpha, \eta) = e^{(i/2)\rho\alpha} \sum_{\nu, \mu} \frac{c_{\nu\mu}}{(i/2)\rho + \nu}\, e^{\nu\alpha + \mu\eta}$$

where $\nu + j_2$ and $\mu + j_2$ are integers, and the $c_{\nu\mu}$ are certain rational constants. This gives

$$d^{\chi}_{j_1 j_2 q}(\eta) = \{\cdots\}^{1/2}(2 \sinh \eta)^{-j_1 - j_2 - 1}[F(\eta, \eta) - F(-\eta, \eta)]$$

$$= \{\cdots\}^{1/2}(2 \sinh \eta)^{-j_1 - j_2} \sum_{\nu, \mu} c_{\nu\mu} e^{\mu\eta} \frac{\sinh((i/2)\rho + \nu)\eta}{((i/2)\rho + \nu)\sinh \eta}$$

We set

$$e^{\chi}_{j_1 j_2 q}(\eta) = \{\cdots\}^{1/2}(2 \sinh \eta)^{-j_1 - j_2 - 1} F(\eta, \eta) \qquad (4\text{-}60)$$

and verify easily that this definition implies the properties (1) and (2) of our list and the existence of a finite number of first order poles. Since this suffices to prove (4–56) we have from (4–60)

$$e^{-\chi}_{j_2 j_1 q}(\eta) = (-1)^{j_1 - j_2 + 1}\{\cdots\}^{1/2}(2 \sinh \eta)^{-j_1 - j_2 - 1} F(-\eta, \eta)$$

We continue the computation of $F(-\eta, \eta)$. We have

$$-F(-\eta, \eta) = \sum_{n_1 n_2} \binom{j_1 + q}{n_1} \binom{j_1 - q}{n_1 - q + \frac{1}{2}m} \binom{j_2 + q}{n_2}$$

$$\times \binom{j_2 - q}{n_2 - q + \frac{1}{2}m} \sum_{r, s} (-1)^{r+s+q-(1/2)m+1} \left( \frac{i}{2}\rho - j_2 + r + s \right)^{-1}$$

$$\times \binom{j_1 + j_2 - n_1 - n_2 + q - \frac{1}{2}m}{r} \binom{n_1 + n_2 - q + \frac{1}{2}m}{s}$$

$$\times \exp \eta \left[ j_1 + j_2 - 2n_1 - 2r - \frac{i}{2}\rho + q - \frac{1}{2}m \right]$$

where the sum over $r$ and $s$ extends over that domain where the binomial coefficients do not vanish. In particular we have

$$\max(r + s) = j_1 + j_2$$

from which the rest of assertion (3) made at the beginning of this section follows. The sum over $s$ can be performed

$$\sum_s (-1)^s \left( \frac{i}{2}\rho - j_2 + r + s \right)^{-1} \binom{n_1 + n_2 - q + \frac{1}{2}m}{s}$$

$$= \frac{(n_1 + n_2 - q + (1/2)m)!\, \Gamma(r - j_2 + (i/2)\rho)}{\Gamma(r - j_2 + (i/2)\rho + n_1 + n_2 - q + (1/2)m + 1)}$$

The sum over $r$ can be expressed by a Gaussian hypergeometric polynomial

$$\sum_r \binom{j_1 + j_2 - n_1 - n_2 + q - \frac{1}{2}m}{r} (-e^{-2\eta})^r$$

$$\times \frac{\Gamma(r - j_2 + (i/2)\rho)}{\Gamma(r - j_2 + (i/2)\rho + n_1 + n_2 - q + (1/2)m + 1)}$$

$$= \frac{\Gamma(-j_2 + (i/2)\rho)}{\Gamma(-j_2 + (i/2)\rho + n_1 + n_2 - q + (1/2)m + 1)}$$

$$\times {}_2F_1\left( -j_1 - j_2 + n_1 + n_2 - q + \frac{1}{2}m, \; -j_2 + \frac{i}{2}\rho_1, \right.$$

$$\left. -j_2 + \frac{i}{2}\rho + 1 + n_1 + n_2 - q + \frac{1}{2}m; e^{-2\eta} \right)$$

Collecting all these results we get

$$e^{\chi}_{j_1 j_2 q}(\eta) = \{\cdots\}^{1/2}(2\sinh\eta)^{-j_1-j_2-1}(-1)^{j_1-j_2+q+(1/2)m+1}$$

$$\times \sum_{n_1 n_2} \binom{j_1+q}{n_1}\binom{j_1-q}{n_1-q-\tfrac{1}{2}m}\binom{j_2+q}{n_2}\binom{j_2-q}{n_2-q-\tfrac{1}{2}m}$$

$$\times \frac{(n_1+n_2-q-(1/2)m)!\,\Gamma(-j_1-(i/2)\rho)}{\Gamma(n_1+n_2-q-(1/2)m-(i/2)\rho-j_1+1)}$$

$$\times \exp\eta\left[j_1+j_2-2n_2+\frac{i}{2}\rho+q+\tfrac{1}{2}m\right] \qquad (4\text{-}61)$$

$$\times {}_2F_1\left(-j_1-j_2+n_1+n_2-q-\tfrac{1}{2}m,\ -j_1-\frac{i}{2}\rho,\right.$$

$$\left. n_1+n_2-q-\tfrac{1}{2}m-\frac{i}{2}\rho-j_1+1;\ e^{-2\eta}\right)$$

Instead of the hypergeometric polynomial (4–61) we can express the function $e^{\chi}_{j_1 j_2 q}(\eta)$ by an infinite series using *GR* 9.131.1 and the identity (*GR* 8.334.3)

$$\Gamma(z)\Gamma(1-z) = \frac{\pi}{\sin\pi z}$$

We get

$$e^{\chi}_{j_1 j_2 q}(\eta)$$

$$= \{\cdots\}^{1/2}\sum_{n_1 n_2}(-1)^{j_1-j_2+n_1+n_2}\binom{j_1+q}{n_1}\binom{j_1-q}{n_1-q-\tfrac{1}{2}m}$$

$$\times \binom{j_2+q}{n_2}\binom{j_2-q}{n_2-q-\tfrac{1}{2}m}$$

$$\times \frac{(n_1+n_2-q-(1/2)m)!\,\Gamma(-n_1-n_2+q+(1/2)m+(i/2)\rho+j_1)}{\Gamma(j_1+(i/2)\rho+1)}$$

$$\times \exp\eta\left[-2n_2+q+\tfrac{1}{2}m+\frac{i}{2}\rho-1\right]$$

$$\times {}_2F_1\left(n_1+n_2-q-\tfrac{1}{2}m+1,\ j_2-\frac{i}{2}\rho+1,\right.$$

$$\left. n_1+n_2-q-\tfrac{1}{2}m-\frac{i}{2}\rho-j_1+1;\ e^{-2\eta}\right)$$

If we compare this expression with *GR* 9.131.2 and (4–59) we get (4–56).

In order to identify the term $e^{\chi}_{j_1 j_2 q}(\eta)$ we replace the summation labels $n_1, n_2$ by $j_1 - n_1 + q, j_2 - n_2 + q$, respectively.

Let us now study the asymptotic behavior of $e^{\chi}_{j_1 j_2 q}(\eta)$ in $\eta$ at $\eta = \infty$ and $0$. We notice first that $F(0, 0)$ is finite nonzero. In fact, by means of the well-known formula (convergence presupposed)

$$_2F_1(a, b, c; 1) = \frac{\Gamma(c)\Gamma(c - a - b)}{\Gamma(c - a)\Gamma(c - b)}$$

we read off (4−61)

$$F(0, 0) = (-1)^{j_1 - j_2 + q + (1/2)m + 1} \binom{2j_1}{j_1 + \frac{1}{2}m} \binom{2j_2}{j_2 + \frac{1}{2}m}$$

$$\times \frac{\Gamma(-j_1 - (i/2)\rho)}{\Gamma(j_2 - (i/2)\rho + 1)}(j_1 + j_2)! \tag{4-62}$$

so that $e^{\chi}_{j_1 j_2 q}(\eta)$ develops a pole of order $j_1 + j_2 + 1$ in $\eta$ at $\eta = 0$. This must be compared with the behavior of $d^{\chi}_{j_1 j_2 q}(\eta)$, which possesses a zero in $\eta$ of order $|j_1 - j_2|$ at $\eta = 0$. To prove the last statement we make use of generators [see (3−31) and (3−81)]

$$\left(i\frac{\partial}{\partial\eta}\right)^n T_d^{\chi}\bigg|_{\eta=0} = (F_3)^n$$

$$\left(i\frac{\partial}{\partial\eta}\right)^n d^{\chi}_{j_1 j_2 q}(\eta)\bigg|_{\eta=0} = \langle j_1 q|(F_3)^n|j_2 q\rangle$$

Since $F_3$ is the third component of a vector operator with respect to the rotation group SU(2), we have

$$\langle j_1 q|(F_3)^n|j_2 q\rangle = \begin{cases} 0 & \text{if } n < |j_1 - j_2| \\ \text{finite in general for } n = |j_1 - j_2| \end{cases}$$

from which the assertion follows. In (4−56) the first $2\max(j_1, j_2) + 1$ terms in the Laurent expansions of $e^{\chi}_{j_1 j_2 q}(\eta)$ and $(-1)^{j_1 - j_2} e^{-\chi}_{j_2 j_1 q}(\eta)$ at $\eta = 0$ must cancel.

If we write (4−61) in the form

$$e^{\chi}_{j_1 j_2 q}(\eta) = \{\cdots\}^{1/2}(2\sinh\eta)^{-j_1 - j_2 - 1}$$

$$\times \sum_{\gamma=\min\gamma}^{\max\gamma} \exp\left\{\left(\frac{i}{2}\rho + \gamma\right)\eta\right\} \sum_{v+\mu=\gamma} \frac{c_{v\mu}}{(i/2)\rho + v} \tag{4-63}$$

the sum over $\gamma$ involves terms with

$$\gamma = \min \gamma + 2n \qquad n = 0, 1, 2, \dots$$

$$\max \gamma = -\min \gamma = j_1 + j_2 - |q + \tfrac{1}{2}m|$$

The asymptotic behavior of $e^\chi_{j_1 j_2 q}(\eta)$ at $\eta = \infty$ is determined by the term $\gamma = \max \gamma$ in (4–63). We have

$$\sum_{\nu + \mu = \max \gamma} \frac{c_{\nu\mu}}{(i/2)\rho + \nu} = (-1)^{j_2 - q} |q + \tfrac{1}{2}m|! \binom{j_1 + q}{q + \tfrac{1}{2}m}$$

$$\times \binom{j_2 + q}{q + \tfrac{1}{2}m} \frac{\Gamma(j_1 - (i/2)\rho + 1)\Gamma(-(1/2)m + (i/2)\rho)}{\Gamma(j_1 + (i/2)\rho + 1)\Gamma(q - (i/2)\rho + 1)}$$

$$(4\text{-}64)$$

if $q + \frac{1}{2}m \geqq 0$, and the expression (4–64) with $q$, $m$ replaced by $-q$, $-m$ if $q + \frac{1}{2}m \leqq 0$. From (4–63) we have in addition

$$e^\chi_{j_1 j_2 q}(\eta + i\pi) = \exp\left\{ i\pi\left( \frac{i}{2}\rho + \tfrac{1}{2}m + q + 1 \right) \right\} e^\chi_{j_1 j_2 q}(\eta)$$

This relation and (4–56) imply

$$2i \sin \pi\left( \tfrac{1}{2}m + \frac{i}{2}\rho \right) e^\chi_{j_1 j_2 q}(\eta)$$

$$= e^{-i\pi(q+1)} d^\chi_{j_1 j_2 q}(\eta + i\pi) - \exp\left\{ -i\pi\left( \tfrac{1}{2}m + \frac{i}{2}\rho \right) \right\} d^\chi_{j_1 j_2 q}(\eta) \qquad (4\text{-}65)$$

Relations (4–56) and (4–65) allow us to express functions of the first kind and functions of the second kind linearly by each other.

For the sake of later reference we note that full representation functions of the second kind can be introduced in analogy with (3–29)

$$E^\chi_{j_1 q_1 j_2 q_2}(a) = \sum_q D^{j_1}_{q_1 q}(u_1) D^{j_2}_{qq_2}(u_2) e^\chi_{j_1 j_2 q}(\eta)$$

$$a = u_1 \, du_2 \qquad \eta > 0 \qquad\qquad (4\text{-}66)$$

## 4–6   ELEMENTARY EXAMPLES OF DISTRIBUTIONS

In this section we study certain classes of distributions which are defined as continuous, polynomially bounded functions $p(a)$ over SL(2, C) (Section 3–4). We give their Fourier transform and the contour C corresponding to it (Section 4–4). We obtain the Fourier transform and the contour C by a method which is based on analytic continuation of the distributions in a para-meter $\sigma$. We say that a distribution $p_\sigma(a)$ depends analytically on the parameter $\sigma$ if

$$(p_\sigma, x) \quad x \in \mathscr{C}^\infty$$

is analytic in $\sigma$ in a domain which is common to all $x \in \mathscr{C}^\infty$. We assume that the domain of analyticity in $\sigma$ contains a subdomain in which $p_\sigma(a)$ is regular [i.e., $p_\sigma(a)$ is a square integrable function; see Section 4–4] and for which the con-tour C can be chosen as the real axis. We call this subdomain in $\sigma$ the regular subdomain of the distribution $p_\sigma(a)$. We find the Fourier transform of $p_\sigma(a)$ and the contour C by analytic continuation in $\sigma$ from the regular subdomain. We avoid the singularities moving against the real axis in the course of this continuation by a corresponding deformation of the contour.

All the distributions of this section are constant on two-sided cosets (or double cosets) of SU(2), namely,

$$p(u_1 \, au_2) = p(a)$$

for any $u_1, u_2 \in$ SU(2). Such distributions are of a particular simple structure which enables us to compute the Fourier transform in many cases explicitly. Distributions which are constant on double cosets present the most element-ary example of distributions which are covariant on double cosets. We denote such distributions "bicovariant," and we outline their formal theory in Section 4–7. Distributions which are constant on two-sided cosets of SU(2) depend in fact only on the variable $\eta$, as can be seen from (3–21), (3–22), and

$$p(a) = p(u_1 \, du_2) = p(d)$$

Their Fourier transform $K_P(u_1, u_2 | \chi)$ is independent of $u_1$ and $u_2$ and is nonzero only for $m = 0$. We can therefore write

$$K_p(u_1, u_2 \mid \chi) = K_p(\chi) = \delta_{m0} \, K_p(\rho)$$

If $p(a)$ is integrable and if $\chi$ is in the principal series, the operator $T_p^\chi$ annihilates the whole space $\mathscr{L}_m^2(U)$ if $\chi = (m, \rho)$ and $m \neq 0$, and annihilates all vectors of the canonical basis in the space $\mathscr{L}_0^2(U)$ whose spin $j$ is nonzero, so that

$$K_p(\rho) = \langle 00| \, T_p^{(0, \, \rho)} \, |00\rangle$$

From this relation follows

$$K_p(\rho) = K_p(-\rho)$$

From (4–21) we have

$$K_p(z_1, z_2 \,|\, \chi) = \pi^{-1}(1 + |z_1|^2)^{(i/2)\rho - 1}(1 + |z_2|^2)^{-(i/2)\rho - 1} \,\delta_{m0}\, K_p(\rho)$$

The inverse Fourier transform $p(a)$ of the functional $p^*$ determined by the function

$$K_{\bar{p}}(u_1, u_2 \,|\, \chi) = \delta_{m0}\, K_{\bar{p}}(\rho)$$

(provided it exists) is, according to (4–54),

$$p(a) = \tfrac{1}{2} \int\limits_C d\rho \rho^2 \overline{K_{\bar{p}}(\rho)} \int \alpha^\chi(u, a)\, d\mu(u)$$

$$\chi = (0, \rho)$$

The integral over $u$ yields

$$\int \alpha^{(0,\,\rho)}(u, a)\, d\mu(u) = \frac{2 \sin(1/2)\eta\rho}{\rho \sinh \eta}$$

Therefore we end up with

$$p(a) = \int\limits_C d\rho \rho \overline{K_{\bar{p}}(\rho)} \, \frac{\sin(1/2)\eta\rho}{\sinh \eta} \tag{4-67}$$

The function $\overline{K_{\bar{p}}(\rho)}$ in (4–67) is obtained by simultaneous analytic continuation in $\sigma$ off the regular subdomain and in $\rho$ off the real axis.

### a.   The Distribution $|a|^{2\sigma}$

The distribution

$$p_\sigma(a) = |a|^{2\sigma}$$

is not only analytic in $\sigma$ but also entire. This comes about because the norm squared, $|a|^2$, is bounded from below by 2. In the regular subdomain, Re $\sigma < -1$, the Fourier transform for $\chi$ in the principal series can be obtained by integrating [for simplicity we denote $K_p(\chi) = K_\sigma(\chi)$]

$$K_\sigma(\chi) = \tfrac{1}{2}(2\pi)^{-3} \int |u_1^{-1} k u_2|^{2\sigma} \lambda^{-(1/2)m + (i/2)\rho - 1}$$

$$\times\; \bar{\lambda}^{(1/2)m + (i/2)\rho - 1}\, D\lambda \, D\mu$$

$$= \frac{1}{16\pi} \delta_{m0} \int_0^\infty d\,|\lambda|^2\,|\lambda|^{i\rho-2}$$

$$\times \int_0^\infty d\,|\mu|^2 (|\lambda|^2 + |\lambda|^{-2} + |\mu|^2)^\sigma$$

Performing the elementary integrations (*GR* 3. 194.3) we get

$$K_\sigma(\rho) = \frac{\Gamma((i/4)\rho - (1/2)\sigma - 1/2)\Gamma(-(i/4)\rho - (1/2)\sigma - 1/2)}{32\pi\Gamma(-\sigma)} \tag{4-68}$$

In turn we have from (4-67) and $\overline{K_{\bar\sigma}(\bar\rho)} = K_\sigma(\rho)$

$$|a|^{2\sigma} = |2 \cosh \eta|^\sigma$$

$$= \frac{1}{32\pi\Gamma(-\sigma)} \int_{-\infty}^{+\infty} d\rho\rho\,\Gamma\left(\frac{i}{4}\rho - \frac{1}{2}\sigma - \frac{1}{2}\right)\Gamma\left(-\frac{i}{4}\rho - \frac{1}{2}\sigma - \frac{1}{2}\right)\frac{\sin(1/2)\eta\rho}{\sinh \eta}$$

.as can be verified easily by means of *GR* 6.422.3.

So far we have assumed Re $\sigma < -1$. Now we want to continue analytically in $\sigma$. In (4—67) we must then go away from the real $\rho$-axis. Since we do this by analytic continuation in $\rho$, and because on the real $\rho$-axis we have $\overline{K_{\bar\sigma}(\rho)} = K_\sigma(\rho)$, we may insert $K_\sigma(\rho)$ into (4—67) whatever the contour C is. The function $K_\sigma(\rho)$ given by (4—68) has poles at

$$\frac{i}{4}\rho - \tfrac{1}{2}\sigma - \tfrac{1}{2} = -n_1$$

$$-\frac{i}{4}\rho - \tfrac{1}{2}\sigma - \tfrac{1}{2} = -n_2, \qquad n_{1,2} = 0, 1, 2, \ldots$$

If Re $\sigma < -1$, those poles labeled $n_1$ lie in the upper half $\rho$-plane, and the other poles lie in the lower half $\rho$-plane. If we continue in $\sigma$ from the regular subdomain till the value $\sigma$ desired, and if Im $\sigma \neq 0$, it is easy to see how the contour can be deformed to evade the poles which move against the real axis from below or from above. This deformation can be done in particular such that two symmetric infinite intervals on the real axis remain unchanged and such that the contour remains symmetric. If Im $\sigma = 0$ but $\sigma$ is not an integer bigger than -2, we may continue in $\sigma$ along a path whose imaginary part is nonzero throughout except at the end point. If $\sigma$ is an integer not smaller than $-1$, the poles satisfying

$$n_1 + n_2 = \sigma + 1$$

coincide, and the contour is unavoidably pinched. We consider this case in more detail.

First we consider the case $\sigma = -1, n_1 = n_2 = 0$. There is only one pair of poles which pinches the contour at $\rho = 0$. The contribution of this pair vanishes because of the second order zero of the integrand of (4–67) at $\rho = 0$. This implies that the integral (4–67) may still be extended over the real axis. For $\sigma \geq 0$ integral, the function $\Gamma(-\sigma)$ in the denominator of (4–68) also possesses poles in $\sigma$, which cause all finite contributions of the integral (4–67) to vanish if $\sigma$ approaches such a pole. Only the pairs of poles pinching the contour at $\rho \neq 0$, which alone would contribute divergent terms, yield a finite contribution after the divergence has been cancelled against the pole of the function $\Gamma(-\sigma)$. In this case the integration contour reduces to a finite number of circles around points on the imaginary axis. Defining analytic delta functionals (in the sense of analytic functionals discussed in Section 4–4) which operate on entire functions $f(w)$,

$$f(w') = \int f(w) \, \delta(w - w') \, dw$$

$$= \frac{1}{2\pi i} \int_{C_{w'}} \frac{f(w)}{w - w'} \, dw \qquad (4\text{-}69)$$

where $C_{w'}$ encircles $w'$ once in the positive sense, we can express the Fourier transform of $|a|^{2\sigma}$, $\sigma \geq 0$ integral, as a finite sum of such analytic delta functionals

$$K_\sigma(\rho) = -\frac{1}{4(\sigma + 1)} \sum_{n=0}^{\sigma+1} \binom{\sigma + 1}{n} \delta(\rho - 2i(\sigma + 1 - 2n)) \qquad (4\text{-}70)$$

Inserting (4–70) into (4–67) yields in turn

$$\frac{1}{2} \sum_{n=0}^{\sigma+1} \binom{\sigma + 1}{n} \frac{\sigma - 2n + 1}{\sigma + 1} \frac{\sinh \eta(\sigma - 2n + 1)}{\sinh \eta}$$

$$= \frac{1}{4 \sinh \eta} \sum_{n=0}^{\sigma} \binom{\sigma}{n} [e^{\eta(\sigma - 2n + 1)}$$

$$- e^{-\eta(\sigma - 2n + 1)} - e^{\eta(\sigma - 2n - 1)} + e^{-\eta(\sigma - 2n - 1)}]$$

$$= [2 \cosh \eta]^\sigma$$

In these cases it was easy to verify directly that the inverse Fourier transformation (4–67) reproduces the original function. We can achieve this check, however, with another technique also, which is worth mentioning because of its importance in applications. This method is based on the residue theorem.

The function $d_{000}^{(0;\rho)}(\eta)$ (3–36) can be split into two functions of the second kind (4–56)

$$d_{000}^{(0;\rho)}(\eta) = \frac{2\sin(1/2)\eta\rho}{\rho\sinh\eta} = e_{000}^{(0,\rho)}(\eta) + e_{000}^{(0,-\rho)}(\eta)$$

From (4–61) we find

$$e_{000}^{(0,\rho)}(\eta) = \frac{e^{(i/2)\eta\rho}}{i\rho\sinh\eta}$$

For a symmetric contour C and because of the symmetry of $K_\sigma(\rho)$ in $\rho$, we can rewrite the integral (4–67) with $K_\sigma(\rho)$ given by (4–68) as

$$\frac{1}{32\pi i\Gamma(-\sigma)}\int_C d\rho\rho\,\Gamma\left(\frac{i}{4}\rho - \sigma - \frac{1}{2}\right)\Gamma\left(-\frac{i}{4}\rho - \frac{1}{2}\sigma - \frac{1}{2}\right)\frac{e^{(i/2)\eta\rho}}{\sinh\eta}$$

The right asymptotic behavior of the function of the second kind allows us to evaluate this integral by shifting the contour to $+i\infty$. Each pole of the function $K_\sigma(\rho)$ crossed gives the residue ($n = 0, 1, 2, \ldots$)

$$\frac{(-1)^n}{2n!\,\Gamma(-\sigma)}\Gamma(n - \sigma - 1)(2n - \sigma - 1)\frac{e^{-\eta(2n-\sigma-1)}}{\sinh\eta}$$

These residues can be summed up and yield

$$\frac{\Gamma(-\sigma - 1)}{2\Gamma(-\sigma)\sinh\eta}\left(-\frac{\partial}{\partial\eta}\right)\sum_{n=0}^{\infty}\binom{\sigma + 1}{n}e^{-\eta(2n-\sigma-1)} = [2\cosh\eta]^\sigma$$

Since this is already the result we want, we need not prove that the contributions of the infinite parts of the contour vanish, though the proof would in fact be simple.

b.    *The Distribution $(|a|^2 - \alpha)^\sigma$*

We study now in brief the more general class of distributions

$$p(a) = (|a|^2 - \alpha)^\sigma$$

defined for all values in the complex $\alpha$-plane which we cut along

$$2 \leq \alpha < \infty$$

If $\alpha$ is on this cut, the distribution is analytic but not entire; if $\alpha$ is any other complex number, the distribution is entire in $\sigma$. If we put $\sigma = -1$ and approach the cut in $\alpha$ from below, we obtain the denominator in the dispersion integral

$$p(a) = \int\limits_{2}^{\infty} \frac{\kappa(\alpha)}{\alpha - |a|^2 - i0+} \, d\alpha$$

We shall come back to this special case later.

For Re $\sigma < -1$ and all $\alpha$ in the interior of the cut $\alpha$-plane we obtain the Fourier transform $K_p(\rho)$ by evaluating the integrals

$$K_p(\rho) = \frac{1}{16\pi} \int\limits_{0}^{\infty} d|\lambda|^2 |\lambda|^{i\rho-2} \int\limits_{0}^{\infty} d|\mu|^2 (|\lambda|^2 + |\lambda|^{-2} + |\mu|^2 - \alpha)^\sigma$$

$$= \frac{-1}{16\pi(\sigma + 1)} \int\limits_{0}^{\infty} (1 - \alpha t + t^2)^{\sigma+1} t^{(i/2)\rho - \sigma - 2} \, dt$$

When we write

$$1 - \alpha t + t^2 = (1 + t)^2 - (\alpha + 2)t$$

and expand the integrand in powers of $\alpha + 2$, we obtain a uniformly convergent series in the circle $|\alpha + 2| < 4$, which may be integrated termwise

$$K_p(\rho) = \frac{1}{8\pi\Gamma(-2\sigma - 1)} \Gamma\left(\frac{i}{2}\rho - \sigma - 1\right)\Gamma\left(-\frac{i}{2}\rho - \sigma - 1\right)$$

$$\times {}_2F_1\left(\frac{i}{2}\rho - \sigma - 1, \ -\frac{i}{2}\rho - \sigma - 1, \ -\sigma - \frac{1}{2}; \frac{1}{4}(\alpha + 2)\right) \quad (4\text{-}71)$$

However, we may also set

$$1 - \alpha t + t^2 = (1 + t^2) - \alpha t$$

and expand the integrand in powers of $\alpha$. This series converges uniformly in the circle $|\alpha| < 2$ and can then be put in the form

$$
K_p(\rho) = \frac{1}{32\pi\Gamma(-\sigma)}\left\{\Gamma\left(\frac{i}{4}\rho - \frac{1}{2}\sigma - \frac{1}{2}\right)\Gamma\left(-\frac{i}{4}\rho - \frac{1}{2}\sigma - \frac{1}{2}\right)\right.
$$

$$
\times\, {}_2F_1\left(\frac{i}{4}\rho - \frac{1}{2}\sigma - \frac{1}{2},\, -\frac{i}{4}\rho - \frac{1}{2}\sigma - \frac{1}{2}; \frac{1}{2}; \frac{1}{4}\alpha^2\right)
$$

$$
+ \alpha\Gamma\left(\frac{i}{4}\rho - \frac{1}{2}\sigma\right)\Gamma\left(-\frac{i}{4}\rho - \frac{1}{2}\sigma\right)
$$

$$
\left.\times\, {}_2F_1\left(\frac{i}{4}\rho - \frac{1}{2}\sigma,\, -\frac{i}{4}\rho - \frac{1}{2}\sigma, \frac{3}{2}; \frac{1}{4}\alpha^2\right)\right\} \tag{4-72}
$$

Both results (4–71), (4–72) are known to be identical (*GR* 9.136.2). They exhibit a cut in $\alpha$ from $\alpha = 2$ to $\infty$ as desired. From (4–72) we regain (4–68) if we set $\alpha = 0$.

As in Section 4–6a we can reach the values of $\sigma$ with $\mathrm{Re}\,\sigma \geq -1$ by an analytic continuation. The contour can be deformed in a manner similar to the former case yielding an analytic functional in all cases ($\alpha$ in the interior of the cut $\alpha$-plane). A pinch occurs when $\sigma \geq -1$ is an integer or a half-integer. In the case of no pinch the inverse Fourier transformation (4–67) has the form [$\overline{K_{\bar{p}}(\rho)}$ is replaced by $K_p(\rho)$ as in Section 4–6a]

$$
\frac{1}{8\pi\Gamma(-2\sigma - 1)}\int_C d\rho\rho\,\Gamma\left(\frac{i}{2}\rho - \sigma - 1\right)\Gamma\left(-\frac{i}{2}\rho - \sigma - 1\right)
$$

$$
\times\, {}_2F_1\left(\frac{i}{2}\rho - \sigma - 1,\, -\frac{i}{2}\rho - \sigma - 1,\, -\sigma - \frac{1}{2}; \frac{1}{4}(\alpha + 2)\right)\frac{\sin(1/2)\eta\rho}{\sinh\eta}
$$

Again making use of the symmetry in $\rho$ to replace the function of the first kind by a function of the second kind, the integral can be expanded into a series of residues ($n = 0, 1, 2, \ldots$)

$$
(-1)^n \frac{(n - \sigma - 1)\Gamma(n - 2\sigma - 2)}{n!\,\Gamma(-2\sigma - 1)}\frac{e^{-\eta(n-\sigma-1)}}{\sinh\eta}
$$

$$
\times\, {}_2F_1(-n, n - 2\sigma - 2, -\sigma - \tfrac{1}{2}; \tfrac{1}{4}(\alpha + 2))
$$

each of which belongs to a pole in the upper half $\rho$-plane. These residue terms can be simplified if we use Gegenbauer polynomials (*GR* 8.93); we get

$$
\frac{n - \sigma - 1}{-2\sigma - 2}\frac{e^{-\eta(n-\sigma-1)}}{\sinh\eta}\, C_n^{-\sigma-1}\left(\frac{1}{2}\alpha\right)
$$

Such Gegenbauer polynomials can be defined by a generating function

$$(1 - \alpha t + t^2)^\beta = \sum_{n=0}^{\infty} C_n^{-\beta}(\tfrac{1}{2}\alpha)t^n \qquad (4\text{-}73)$$

As (4–73) shows, the series of residue terms reduces to a finite series

$$0 \leqq n \leqq 2\sigma + 2$$

if $\sigma \geqq$ -1 is an integer. In this case the Fourier transform can be expressed as a finite sum of delta functionals just as in the case treated in Section 4–6a. If $\sigma \geqq$ -1 is half-integral, there is still an infinite series of terms left. Those up to $n = 2\sigma + 2$ stem from the pinching pairs of poles, but the residues with $n > 2\sigma + 2$ remain finite nonzero also. If we use the generating function (4–73) to sum up the series of residues, we get

$$\sum_{n=0}^{\infty} \frac{n - \sigma - 1}{-2\sigma - 2} \frac{e^{-\eta(n-\sigma-1)}}{\sinh \eta} C_n^{-\sigma-1}(\tfrac{1}{2}\alpha)$$

$$= \frac{1}{(2\sigma + 2)\sinh \eta} \frac{\partial}{\partial \eta} \{e^{\eta(\sigma+1)}(1 - \alpha e^{-\eta} + e^{-2\eta})^{\sigma+1}\}$$

$$= [2 \cosh \eta - \alpha]^\sigma$$

By specialization of (4–71) we can treat the dispersion integral

$$p(a) = \int_2^\infty \frac{\kappa(\alpha)}{\alpha - |a|^2 - i0+} \, d\alpha$$

which might be of interest in physical applications (the forward dispersion relation for elastic scattering of spinless particles is of this structure). For simplicity we shall assume that the weight function $\kappa(\alpha)$ is integrable and square integrable

$$\int_2^\infty |\kappa(\alpha)| \, d\alpha < \infty \qquad \int_2^\infty |\kappa(\alpha)|^2 \, d\alpha < \infty$$

As inverse Fourier transformation we get the integral representation

$$p(a) = \frac{1}{4} \int_{-\infty}^{+\infty} d\rho \, \frac{\sin(1/2)\eta\rho}{\sinh \eta \, \sinh(\pi/2)\rho} \kappa^*(\rho) \qquad (4\text{-}74)$$

with the weight function $\kappa^*(\rho)$ defined by

$$\kappa^*(\rho) = -2 \int_0^\infty \kappa(2 \cosh \tau)\cos \tfrac{1}{2}\rho(\tau + i\pi)\sinh \tau \, d\tau \qquad (4\text{-}75)$$

We emphasize that the relation (4–75) connecting $\kappa(\rho)$ and $\kappa^*(\rho)$ is a classical Fourier transformation.

## 4–7   SU(2) BICOVARIANT DISTRIBUTIONS

SU(2) bicovariant functions (distributions) are defined as a $(2j_1 + 1) \times (2j_2 + 1)$ array of functions $x_{j_1 q_1 j_2 q_2}(a)$ [distributions $p_{j_1 q_1 j_2 q_2}(a)$] which are covariant on double cosets of SU(2) in the sense

$$x_{j_1 q_1 j_2 q_2}(u_1 a u_2) = \sum_{qq'} D_{q_1 q}^{j_1}(u_1) \, D_{q' q_2}^{j_2}(u_2) x_{j_1 q j_2 q'}(a) \qquad (4\text{-}76)$$

respectively,

$$p_{j_1 q_1 j_2 q_2}(u_1 a u_2) = \sum_{qq'} D_{q_1 q}^{j_1}(u_1) \, D_{q' q_2}^{j_2}(u_2) p_{j_1 q j_2 q'}(a) \qquad (4\text{-}77)$$

Equation (4–76) and the orthogonality of the functions $D_{q_1 q_2}^j(u)$ imply, on the other hand,

$$x_{j_1 q j_2 q'}(a) = (2j_1 + 1)(2j_2 + 1) \int x_{j_1 q_1 j_2 q_2}(u_1^{-1} a u_2^{-1})$$

$$\times D_{qq_1}^{j_1}(u_1) \, D_{q_2 q'}^{j_2}(u_2) \, d\mu(u_1) \, d\mu(u_2)$$

with arbitrary $q_1$ and $q_2$, so that one function suffices to reconstruct the whole array. We show next that any function of $\mathscr{C}^\infty$ (distribution of $\mathscr{C}^{\infty\prime}$) can be expanded into a series of bicovariant functions (distributions) which converges in the sense of $\mathscr{C}^\infty$ ($\mathscr{C}^{\infty\prime}$).

For any element $x \in \mathscr{C}^\infty$ we define the functions

$$x_{j_1 q_1 j_2 q_2}^{q_3 q_4}(a) = \int x(u_1^{-1} a u_2^{-1}) \, D_{q_1 q_3}^{j_1}(u_1) \, D_{q_4 q_2}^{j_2}(u_2) \, d\mu(u_1) \, d\mu(u_2) \qquad (4\text{-}78)$$

which are all in $\mathscr{C}^\infty$, have a uniformly bounded support for fixed $x$, and are bicovariant as required by (4–76) (the labels $q_3$ and $q_4$ are fixed). Given an element $x \in \mathscr{C}^\infty$, (4–78) supplies us with a $(2j_1 + 1) \times (2j_2 + 1)$ array of bicovariant functions of $\mathscr{C}^\infty$ for any fixed $j_1, j_2, q_3, q_4$. With the Plancherel theorem for SU(2) (Sections 3–3 and 3–11) we get the series

$$x(u_1^{-1}\, a\, u_2^{-1}) = \sum_{\substack{j_1 j_2 \\ q_1 q_2 q_3 q_4}} (2j_1 + 1)(2j_2 + 1)\, \overline{D_{q_1 q_3}^{j_1}(u_1)}\; \overline{D_{q_4 q_2}^{j_2}(u_2)} x_{j_1 q_1 j_2 q_2}^{q_3 q_4}(a)$$

which converges in the sense of $\mathscr{C}^\infty$. In particular we set $u_1 = u_2 = e$ and have

$$x(a) = \sum_{\substack{j_1 j_2 \\ q_1 q_2}} (2j_1 + 1)(2j_2 + 1)x_{j_1 q_1 j_2 q_2}^{q_1 q_2}(a) \tag{4-79}$$

This series (4–79) picks out one function of each array.

  To derive the corresponding results for the distributions we proceed in a different fashion. We define the distributions $p_{j_1 q_1 j_2 q_2}^{q_3 q_4}(a)$ by the requirement

$$\int p_{j_1 q_1 j_2 q_2}^{q_3 q_4}(a)\overline{x(a)}\; d\mu(a) = \int p(a)\overline{x_{j_1 q_3 j_2 q_4}^{q_1 q_2}(a)}\; d\mu(a) \tag{4-80}$$

for all $x \in \mathscr{C}^\infty$. Covariance (4–77) follows from this definition and

$$x'(a) = x(u_1^{-1} a u_2^{-1})$$

$$\overline{x'^{q_1 q_2}_{j_1 q_3 j_2 q_4}(a)} = \sum_{qq'} D_{q_1 q}^{j_1}(u_1)\, D_{q' q_2}^{j_2}(u_2)\overline{x_{j_1 q_3 j_2 q_4}^{qq'}(a)}$$

The continuity of this functional is obvious from (4–80). The continuity of $p$ implies

$$\int p(a)\overline{x(a)}\; d\mu(a) = \sum_{\substack{j_1 j_2 \\ q_1 q_2}} (2j_1 + 1)(2j_2 + 1) \int p(a)\overline{x_{j_1 q_1 j_2 q_2}^{q_1 q_2}(a)}\; d\mu(a)$$

$$= \sum_{\substack{j_1 j_2 \\ q_1 q_2}} (2j_1 + 1)(2j_2 + 1) \int p_{j_1 q_1 j_2 q_2}^{q_1 q_2}(a)\overline{x(a)}\; d\mu(a)$$

which gives the asserted expansion of $p(a)$ in analogy with (4–79).

  The definition (4–76) or (4–77) of bicovariant functions or distributions has some elementary but important implications. If we insert $u_1 = u_2 = e_- = -e$, we obtain as a necessary condition

$$(-1)^{2(j_1 - j_2)} = +1$$

If we insert for $a \in SL(2, C)$ its diagonalized form $a = u_1 d u_2$ (3–21), (3–22), we obtain from (4–77)

$$p_{j_1 q_1 j_2 q_2}(a) = \sum_q D_{q_1 q}^{j_1}(u_1)\, D_{q q_2}^{j_2}(u_2) p_{j_1 j_2 q}(\eta) \tag{4-81}$$

where we made use of

$$p_{j_1 q_1 j_2 q_2}(d) = \delta_{q_1 q_2} p_{j_1 j_2 q_1}(\eta)$$

We assume first that the distributions $p_{j_1 q_1 j_2 q_2}(a)$ are all regular. For convenience we compute the Fourier transform of the complex conjugate distribution $\overline{p_{j_1 q_1 j_2 q_2}(a)}$. In general we denote the Fourier transform of the conjugate of a bicovariant function $x_{j_1 q_1 j_2 q_2}(a)$ or distribution $p_{j_1 q_1 j_2 q_2}(a)$ by $K^*_{j_1 q_1 j_2 q_2}(u_1, u_2 | \chi)$. We shall also make use of "reduced kernels" $K^*_{j_1 j_2 q}(\chi)$ and $K_{j_1 j_2 q}(\chi)$. These are to be regarded as analytic functions in $\rho$ if possible. On the real $\rho$-axis ($\chi$ in the principal series) they are related by

$$K^*_{j_1 j_2 q}(\chi) = \overline{K_{j_1 j_2 q}(\bar\chi)} \qquad \bar\chi = (m, -\rho) \tag{4-82}$$

From (4–12) we have

$$K^*_{j_1 q_1 j_2 q_2}(u_1, u_2 | \chi) = \pi \int \overline{p_{j_1 q_1 j_2 q_2}(u_1^{-1} k u_2)} \lambda^{n_1 - 1} \bar\lambda^{n_2 - 1} \, d\mu_l(k)$$

$$= \pi \sum_{q_1' q_2'} D^{j_1}_{q_1' q_1}(u_1) \, \overline{D^{j_2}_{q_2' q_2}(u_2)}$$

$$\times \int \overline{p_{j_1 q_1' j_2 q_2'}(k)} \lambda^{n_1 - 1} \bar\lambda^{n_2 - 1} \, d\mu_l(k)$$

It is easy to see that the integral vanishes if $q_1'$ or $q_2'$ differs from $\frac{1}{2}m$. We have therefore from (3–20)

$$K^*_{j_1 q_1 j_2 q_2}(u_1, u_2 | \chi) = (2j_1 + 1)^{-1/2}(2j_2 + 1)^{-1/2} \varphi^{j_1}_{q_1}(u_1) \overline{\varphi^{j_2}_{q_2}(u_2)}$$

$$\times \pi \int \overline{p_{j_1(1/2)m j_2(1/2)m}(k)} \lambda^{n_1 - 1} \bar\lambda^{n_2 - 1} \, d\mu_l(k) \tag{4-83}$$

As (4–83) shows, the Fourier transform of $\overline{p_{j_1 q_1 j_2 q_2}(a)}$ is an operator in $\mathscr{L}_m^2$ (U), which maps the one-dimensional subspace spanned by $\varphi^{j_2}_{q_2}(u)$ into the one-dimensional subspace spanned by $\varphi^{j_1}_{q_1}(u)$ and annihilates all vectors which are orthogonal to $\varphi^{j_2}_{q_2}(u)$. It is nonzero for at most those $m$ which satisfy

$$0 \leq |\tfrac{1}{2}m| \leq j \qquad 2j \cong m \bmod 2 \qquad j = \min(j_1, j_2)$$

We continue the discussion of the integral in (4–83). We simplify it by decomposing $k$ into three matrices

$$k = u_1 d u_2^{-1}$$

such that (3–17)

$$u_2 = (u_1)_d \qquad \lambda = \lambda(u_1, d)$$

Regarding $u_1$ and $d$ as variables and fixing the phase of $u_2$ by any requirement, say by $(u_2)_{22} \geq 0$, yields (Appendix A–1d)

$$d\mu_l(k) = (2\pi)^{-2} d\mu(u_1) \sinh^2 \eta \, d\eta$$

Inserting this into (4–83) we get

$$\pi \int \overline{p_{j_1(1/2)m j_2(1/2)m}(k)} \lambda^{n_1 - 1} \bar{\lambda}^{n_2 - 1} \, d\mu_l(k)$$

$$= \frac{1}{4\pi} (2j_1 + 1)^{-1/2} (2j_2 + 1)^{-1/2} \sum_q \int_0^\infty d\eta \, \sinh^2 \eta \, p_{j_1 j_2 q}(\eta)$$

$$\times \int d\mu(u) \overline{\varphi_q^{j_1}(u)} \alpha^\chi(u, d) \varphi_q^{j_2}(u_d)$$

$$= \frac{1}{4\pi} (2j_1 + 1)^{-1/2} (2j_2 + 1)^{-1/2} \sum_q \int_0^\infty d\eta \, \sinh^2 \eta \, \overline{p_{j_1 j_2 q}(\eta)} \, d_{j_1 j_2 q}^\chi(\eta)$$

With the new notation

$$K_{j_1 j_2 q}^*(\chi) = \frac{1}{4\pi} (2j_1 + 1)^{-1} (2j_2 + 1)^{-1}$$

$$\times \int_0^\infty d\eta \, \sinh^2 \eta \, \overline{p_{j_1 j_2 q}(\eta)} \, d_{j_1 j_2 q}^\chi(\eta) \tag{4-84}$$

we get finally

$$K_{j_1 q_1 j_2 q_2}^*(u_1, u_2 \mid \chi) = \varphi_{q_1}^{j_1}(u_1) \overline{\varphi_{q_2}^{j_2}(u_2)} \sum_q K_{j_1 j_2 q}^*(\chi) \tag{4-85}$$

For the inverse Fourier transformation we refer to the general formula (4–30) or (4–54) which together with (4–82) give

$$p_{j_1 q_1 j_2 q_2}(a) = \frac{1}{2} \int_{-\infty}^{+\infty} d\rho \sum_{m=-j}^{j} (m^2 + \rho^2) \int \overline{K_{j_1 q_1 j_2 q_2}^*(u, u_a \mid \chi)} \alpha^\chi(u, a) \, d\mu(u)$$

$$= \frac{1}{2} \int_{-\infty}^{+\infty} d\rho \sum_{m=-j}^{j} (m^2 + \rho^2) \, D_{j_1 q_1 j_2 q_2}^\chi(a) \sum_q K_{j_1 j_2 q}(\bar{\chi})$$

Since the functions $p_{j_1 j_2 q}(\eta)$ are independent as functions of $\eta$, and because of (4–81), (3–29), (3–31) we read off (4–85)

$$p_{j_1 j_2 q}(\eta) = \tfrac{1}{2} \int_{-\infty}^{+\infty} d\rho \sum_{m=-j}^{j} (m^2 + \rho^2)\, d^{\chi}_{j_1 j_2 q}(\eta) K_{j_1 j_2 q}(\bar{\chi}) \qquad (4\text{–}86)$$

and the orthogonality relation for $q \neq q'$ in the sense of distributions

$$\int_{-\infty}^{+\infty} d\rho \sum_{m=-j}^{j} (m^2 + \rho^2)\, d^{\chi}_{j_1 j_2 q}(\eta)\, d^{\bar{\chi}}_{j_1 j_2 q'}(\eta') = 0 \qquad j = \min(j_1, j_2) \quad (4\text{–}87)$$

The inverse Fourier transformation can also be formulated by means of functions of the second kind. Because of (4–56) and (4–57)

$$d^{\chi}_{j_1 j_2 q}(\eta) = e^{\chi}_{j_1 j_2 q}(\eta) + (\beta^{j_1})^{-1}\beta^{j_2} e^{-\chi}_{j_1 j_2 q}(\eta) \qquad \eta > 0 \qquad (4\text{–}88)$$

and the symmetry relations (3–96), (4–84)

$$K^{*}_{j_1 j_2 q}(\chi) = (\beta^{j_1})^{-1}\beta^{j_2} K^{*}_{j_1 j_2 q}(-\chi)$$
$$K_{j_1 j_2 q}(\chi) = (\beta^{j_1})^{-1}\beta^{j_2} K_{j_1 j_2 q}(-\chi) \qquad (4\text{–}89)$$

we get

$$p_{j_1 j_2 q}(\eta) = \int_{-\infty}^{\infty} d\rho \sum_{m=-j}^{j} (m^2 + \rho^2) e^{\chi}_{j_1 j_2 q}(\eta) K_{j_1 j_2 q}(\bar{\chi}) \qquad (4\text{–}90)$$

$$\eta > 0 \qquad j = \min(j_1, j_2)$$

provided the integral exists. The existence of the integral (4–90) is not questionable because of the boundaries $\pm\infty$ in $\rho$. The corresponding problem for the integral (4–86) is settled by the Plancherel theorem, which asserts that these limits exist in the $\mathscr{L}^2$ sense. It is easy to see that this carries over to the integral (4–90). However, the functions of the second kind exhibit first order poles in $\rho$ (Section 4–5), one of which falls on the real axis at $\rho = 0$, if $j_1$ and $j_2$ are integers. In this case we need only cut out a symmetric interval of length $\varepsilon$ around $\rho = 0$ in the original integral (4–86), then replace the functions of the first kind by functions of the second kind, and let $\varepsilon$ tend to zero. In other words, the inverse Fourier transformation (4–90) involving functions of the second kind has to be understood as a principal value integral if $j_1$ and $j_2$ are integers.

The functions of the second kind decrease exponentially in $\rho$ in the upper half-plane if $\eta > 0$. When $K_{j_1 j_2 q}(\overline{\chi})$ is meromorphic in the upper half $\rho$-plane for all $m$ in the interval $-j \leqq m \leqq j$, $j = \min(j_1, j_2)$ with a square integrable boundary value on the real axis, it may be possible to shift the contour to $+i\infty$. This yields a representation of $p_{j_1 j_2 q}(\eta)$ as a series of residues of poles. Of course, the poles of the functions of the second kind (e-functions) at

$$\rho = 2ij_1, \, 2i(j_1 - 1), \ldots \qquad \text{Im } \rho \geqq 0$$

(Section 4–5) must in general also be taken into account. In the case that $j_1$ and $j_2$ are integers, the pole at $\rho = 0$ contributes only half its residue because the principal value has to be taken. However, not all poles of the e-functions in the upper half-plane yield nonzero residue terms. In fact, (4–89) implies that $K_{j_1 j_2 q}(\overline{\chi})$ is meromorphic also in the lower half-plane if it is meromorphic in the upper half-plane for all $m$, and that it is continuous on the real axis if it possesses a square integrable limit there. Let $j_1$ be bigger than $j_2$ and let $K_{j_1 j_2 q}(\overline{\chi})$ be holomorphic at the points

$$\rho = \pm 2ij_1, \, \pm 2i(j_1 - 1), \ldots, \pm 2i(j_2 + 1)$$

Due to the constraint (4–89), $K_{j_1 j_2 q}(\overline{\chi})$ then has first order zeros at the positions

$$\rho = 2ij_1, \, 2i(j_1 - 1), \ldots, 2i(j_2 + 1)$$

which cancel the poles of the e-functions. For a big class of functions it suffices therefore to take account of the poles of the e-functions only in the interval

$$0 \leqq \text{Im } \rho \leqq 2j \qquad j = \min(j_1, j_2)$$

With a different argument we can show easily that the sum of all contributions of the poles of the e-functions in the upper half-plane vanishes if $K_{j_1 j_2 q}(\overline{\chi})$ is the Fourier transform of a distribution $p_{j_1 q_1 j_2 q_2}(a)$ with compact support (but not necessarily a function of $\mathscr{C}^\infty$). We need only prove that

$$\sum_{m=-j}^{j} \sum_{\text{Im } \rho \geqq 0} \text{Res}\{(m^2 + \rho^2) \, d^{\overline{\chi}}_{j_1 j_2 q}(\eta) e^{\chi}_{j_1 j_2 q}(\eta')\} = 0$$

In fact, it can be seen from (4–84) and (4–90) that the function

$$f_\varepsilon(\eta, \eta') = [4\pi(2j_1 + 1)(2j_2 + 1)]^{-1} \sum_{m=-j}^{j} \int_{-\infty}^{+\infty} d\rho$$

$$\times (m^2 + \rho^2) \, d^{\overline{\chi}}_{j_1 j_2 q}(\eta) e^{\chi}_{j_1 j_2 q}(\eta')(1 - i\varepsilon\rho)^{-2}$$

possesses the distribution limit

$$\lim_{\varepsilon \to 0+} f_\varepsilon(\eta, \eta') = (\sinh \eta)^{-2} \, \delta(\eta - \eta')$$

On the other hand, we can evaluate $f_\varepsilon(\eta, \eta')$ for $\varepsilon > 0$ and $\eta' > \eta$ by shifting the integration contour to $+i\infty$. This leaves only the contribution of the poles of the $e$-function. In this fashion we obtain the desired result for $\eta' > \eta$, but it follows immediately for all $\eta'$ by analytic continuation.

If we abandon the premise that $p_{j_1 q_1 j_2 q_2}(a)$ are regular distributions [which led us to (4–90)], and if we permit instead functions which are measurable and essentially (i.e., apart from a set of measure zero) polynomially bounded functions, we can always find a procedure by which the final distributions are obtained from regular distributions by analytic continuation in a parameter. One possibility is to multiply the distributions with $|a|^{2\sigma}$ and continue from a sufficiently low negative value $\sigma$ till $\sigma = 0$. But this is not necessarily the most convenient device. If the inverse Fourier transformation involves functions of the first kind, in the course of the continuation in $\sigma$ singularities may move from above and below against the real $\rho$-axis. If we use functions of the second kind, the same thing happens, but there are in addition the fixed poles of these functions. In the first case we can try to avoid the singularities by essentially arbitrary deformations of the contour. In the second case the fixed poles may cause additional pinches. The distances which the singularities move in the complex $\rho$-plane can always be made finite for polynomially bounded functions.

As we shall see in Chapter 6, it is most convenient in dealing with harmonic analysis on the group SU(1, 1) to use functions of the second kind to do the Fourier transformations and functions of the first kind for the inverse transformations. It is therefore worth mentioning that Fourier transforms on SL(2, C) involving functions of the second kind can in general not be defined (e.g., they do not exist for the whole space $\mathscr{C}^\infty$). The inverse Fourier transformation (4–86) shows, namely, that a sufficiently regular bicovariant function exhibits a zero at $\eta = 0$ of the same order as does a representation function of the first kind, that is, a zero of order $|j_1 - j_2|$. This zero does not in general match the pole of the $e$-function of the order $j_1 + j_2 + 1$ (Section 4–5).

## 4–8    REMARKS

In the presentation of the Fourier transformation we follow Naimark [24] and Gelfand *et al.* [12 (Volume 5)]. Our definition of functions of the second kind is in agreement with Sciarrino and Toller [30]. For the analytic properties of the functions of both the second and the first kinds in $z = \cosh \eta$ see Akyeampong [2]. The concept of constructing Fourier transforms of polynomially bounded functions by continuing the distribution

$$p_\sigma(a) = |a|^{2\sigma} x(a)$$

analytically in $\sigma$ from its regularity domain till $\sigma = 0$, which we sketched
briefly in Section 4–7, leads to the general problem of giving the Fourier
transform of a product

$$x(a) = x_1(a) x_2(a)$$

in terms of the Fourier transforms of its factors $x_{1,2}(a)$. This problem is
intimately connected with the task of decomposing tensor products of
representations of the principal series of $SL(2, C)$ into irreducible components.
Neither of these problems is treated in this course. For a discussion of the
relation between these problems and details of the continuation procedure
see Rühl [29].

# Chapter 5
# Representations of the Group SL(2, R)

The representations of the group SL(2, R) can, to a large extent, be
treated with the same methods which we used for the group SL(2, C).
In order not to repeat things too often, we will be rather short whenever
arguments can directly be carried over from SL(2, C) to SL(2, R). A novel
feature of the group SL(2, R) is the appearance of a class of reducible
representations, called analytic representations, which after reduction give
the discrete series of unitary representations. They require a careful study.

In Sections 5–1 to 5–3 we introduce spaces of homogeneous functions,
define group operations in them, and use the maximal compact subgroup of
SL(2, R) to construct a canonical basis in such space. Similarly, as in the case
of the group SL(2, C), we define a principal series of representations by
inducing from one-dimensional unitary representations of the subgroup K
of triangular matrices. Among the class of unitary representations of SL(2, R)
obtained in this fashion, there is one strange candidate, which is reducible
into two representations of the discrete series, whereas all the other representations are irreducible. This reducible representation reappears later among the
analytic representations, and we exclude it therefore from the principal series.

In Section 5–4 we give the representation of the generators as differential operators and methods to diagonalize them. In this context we take the
opportunity to sketch briefly but rather completely how to realize nonunitary
representations of SL(2, R) in Banach spaces with invariant norm, and how
to decompose these spaces into one-dimensional (improper) subspaces, each
of which carries a unitary irreducible representation of a noncompact one-
dimensional subgroup R(1) of SL(2, R). This is interesting in several respects.
First it shows that apart from Hilbert spaces carrying the principal series or
other series of unitary representations, there exist more general linear spaces
with invariant norm. Second it turns out that it is not necessary to extend
harmonic analysis on a group [in this case R(1)] from an $\mathscr{L}^2$-space directly
to a space of distributions, but that it is also possible and may be useful to
stop the extension at the much smaller class of Banach spaces. Third we are

135

interested in the general problem of decomposing a representation of a Lie group into irreducible representations of a noncompact subgroup [here the groups $SL(2, R) \supset R(1)$], and devote the full Chapter 7 [there the groups $SL(2, C) \supset SL(2, R)$] to this issue. We emphasize that the method of analytic functionals applied in Chapter 7 is not the only one possible.

In Sections 5–5 to 5–7 and 5–9 we study bilinear invariant functionals and interwining operators quite the same way as we did in Chapter 3. In Sections 5–8 and 5–10 we study the discrete series of representations in more detail.

## 5–1   ELEMENTARY PROPERTIES OF THE GROUP SL(2, R)

Any element $a \in SL(2, R)$ can be decomposed into the products

$$a = k_1 u \qquad a = k_2 \zeta \qquad \text{if} \quad a_{22} \neq 0$$

just as in the case of the group $SL(2, C)$. In the first case $u$ and $e_- u$ are simultaneous solutions, $e$  is the central element

$$e_- = -e = \begin{pmatrix} -1 & 0 \\ 0 & -1 \end{pmatrix}$$

The elements $k, u,$ and $\zeta$ of $SL(2, R)$ are parametrized throughout as

$$k = \begin{pmatrix} \lambda^{-1} & \mu \\ 0 & \lambda \end{pmatrix} \qquad \lambda, \mu \text{ real}$$

$$u = u(\psi) = \exp\left\{ -\frac{i}{2} \psi \sigma_2 \right\}$$

$$= \begin{pmatrix} \cos \tfrac{1}{2}\psi & -\sin \tfrac{1}{2}\psi \\ \sin \tfrac{1}{2}\psi & \cos \tfrac{1}{2}\psi \end{pmatrix}$$

$$\zeta = \begin{pmatrix} 1 & 0 \\ z & 1 \end{pmatrix} \qquad z \text{ real}$$

If the matrices $u$ and $\zeta$ are connected by $u = k\zeta$, we have

$$z = \tan \tfrac{1}{2}\psi \qquad \lambda = \cos \tfrac{1}{2}\psi$$

Similarly we can decompose any element of $SL(2, R)$ as

$$a = u_1 d u_2 \tag{5-1}$$

such that

$$d = \begin{pmatrix} e^{1/2\eta} & 0 \\ 0 & e^{-1/2\eta} \end{pmatrix} \qquad \eta \geq 0 \qquad\qquad (5\text{-}2)$$

It is perhaps worthwhile to give the arguments leading to (5–1) in the case of real matrices $a$. We get

$$a = a_s u(\psi) = \begin{pmatrix} a_{11} & a_{12} \\ a_{21} & a_{22} \end{pmatrix}$$

such that $a_s$ is symmetric, if we choose the angle $\psi$ as

$$\tan\frac{1}{2}\psi = \frac{a_{21} - a_{12}}{a_{11} + a_{22}}$$

Diagonalizing the real symmetric matrix $a_s (|\mathrm{Tr}(!)a_s| \geq 2)$ gives

$$a_s = u_1\, d u_1^{-1} \quad \text{or} \quad a_s = u_1\, de_-u_1^{-1}$$

which implies (5–1) with

$$u_2 = u_1^{-1}u(\psi) \quad \text{or} \quad u_2 = e_- u_1^{-1}u(\psi)$$

The matrices $d$ and $de_-$ form an Abelian subgroup whose elements are denoted $\delta$ in general. The matrices $u$ form an Abelian subgroup as well. Both subgroups are maximal Abelian in SL(2, R) and have only the center in common. This center consists of the two matrices $e_{\pm} = \pm e$ as for the group SL(2, C).

When two elements $a_1$, $a_2$ are called equivalent if and only if an element $s \in$ SL(2, R) exists such that $a_1 = s^{-1}a_2 s$, the group SL(2, R) decomposes into classes of equivalent elements. For the elements

$$a = s\, \delta s^{-1}$$

we find

$$\mathrm{Tr}\, a = \pm 2 \cosh \tfrac{1}{2}\eta \qquad |\mathrm{Tr}\, a| \geq 2$$

The trace of

$$a = sus^{-1}$$

is

$$\mathrm{Tr}\, a = 2 \cos \tfrac{1}{2}\psi \qquad |\mathrm{Tr}\, a| \leq 2$$

On the other hand if

$$|\mathrm{Tr}\, a| > 2$$

there exists an element $s \in SL(2, R)$ such that

$$a = s \, \delta s^{-1}$$

as can be seen from the characteristic equation for $a$. There exist therefore two sets of equivalence classes, each class characterized by the trace, if the trace is bigger than 2 or smaller than $-2$. Similarly, each trace in the interval

$$-2 < \mathrm{Tr}\, a < 2$$

characterizes one equivalence class which consists of elements

$$a = sus^{-1}$$

Finally, if

$$|\mathrm{Tr}\, a| = 2$$

$a$ may be equal to one of the two central elements $e_\pm$, each of which forms a class, or belong to one of four exceptional classes, each of which contains one and only one of the four elements

$$\pm \begin{pmatrix} 1 & 1 \\ 0 & 1 \end{pmatrix} \qquad \pm \begin{pmatrix} 1 & -1 \\ 0 & 1 \end{pmatrix}$$

On $SL(2, R)$ the matrix norm $|a|$ may be defined by

$$|a|^2 = \mathrm{Tr}(aa^T) = \sum_{i,\, j = 1,\, 2} a_{ij}^2$$

which implies

$$|a|^2 \geq 2$$

The decomposition (5–1) yields

$$|a|^2 = 2 \cosh \eta$$

To close this section we recall the standard isomorphism (1–19) which maps $SL(2, R)$ onto $SU(1, 1)$ and vice versa, such that

$$\underline{SL(2, R)} \qquad\qquad \underline{SU(1, 1)}$$

$$a \leftrightarrow \tfrac{1}{2}(e + i\sigma_1)a(e - i\sigma_1)$$

$$\exp\left\{ -\frac{i}{2}\,\psi\sigma_2 \right\} \leftrightarrow \exp\left\{ +\frac{i}{2}\,\psi\sigma_3 \right\} \qquad\qquad (5\text{-}3)$$

$$\exp\{ +\tfrac{1}{2}\eta\sigma_3 \} \leftrightarrow \exp\{ +\tfrac{1}{2}\eta\sigma_2 \}$$

## 5–2 LINEAR SPACES OF HOMOGENEOUS FUNCTIONS

We proceed in complete analogy to our discussion of the group SL(2, C). We consider homogeneous functions of two real variables $z_1$, $z_2$. A function $F(z_1, z_2)$ is called homogeneous of degree $\sigma$ and of parity $\varepsilon$, where $\sigma$ may be any complex number and $\varepsilon$ takes on the values 0 and 1, if

$$F(\alpha z_1, \alpha z_2) = \alpha^\sigma F(z_1, z_2) \qquad \text{for all} \quad \alpha > 0 \tag{5-4}$$

and

$$F(-z_1, -z_2) = (-1)^\varepsilon F(z_1, z_2) \tag{5-5}$$

We construct topological spaces $\mathcal{D}_\chi$, $\chi = \{\sigma, \varepsilon\}$, of such homogeneous functions with the defining properties:

(1) The space $\mathcal{D}_\chi$ consists of homogeneous functions $F$ of degree $\sigma - 1$ and parity $\varepsilon$, and is a linear vector space.

(2) These functions possess derivatives of all orders at every point $z_1$, $z_2$ with possible exception of the point $z_1 = z_2 = 0$.

(3) The space $\mathcal{D}_\chi$ possesses a topology analogous to that in the case of the group SL(2, C) (Section 3–1).

We shall sometimes use the notation $-\chi = \{-\sigma, \varepsilon\}$ if $\chi = \{\sigma, \varepsilon\}$.

Proceeding as in the case of SL(2, C), it is possible to find other realizations of the spaces $\mathcal{D}_\chi$. We may define functions $f(z)$ by

$$f(z) = F(z, 1)$$

The relation between $f$ and $F$ can be inverted

$$F(z_1, z_2) = |z_2|^{\sigma - 1}(\text{sign } z_2)^\varepsilon f\left(\frac{z_1}{z_2}\right) \tag{5-6}$$

Such functions $f(z) \in \mathcal{D}_\chi$ allow for a simultaneous asymptotic expansion at $\pm\infty$,

$$f(z) \cong |z|^{\sigma - 1}(\text{sign } z)^\varepsilon \sum_{j=0}^\infty a_j z^{-j} \tag{5-7}$$

A third realization of $\mathcal{D}_\chi$ can be obtained with the definition

$$\varphi(u) = \varphi(u(\psi)) \equiv \varphi(\psi) = F(\sin \tfrac{1}{2}\psi, \cos \tfrac{1}{2}\psi)$$

which can be inverted as

$$F(z_1, z_2) = (z_1{}^2 + z_2{}^2)^{(\sigma-1)/2} \varphi(\psi)$$

$$\psi = 2 \arctan \frac{z_1}{z_2} \tag{5-8}$$

Replacing $\psi$ by $\psi + 2\pi$ leads to a covariance constraint on $\varphi(\psi)$ or $\varphi(u)$

$$\varphi(\psi + 2\pi) = (-1)^\varepsilon \varphi(\psi) \qquad \varphi(e_- u) = (-1)^\varepsilon \varphi(u) \tag{5-9}$$

In the transition from $F(z_1, z_2)$ to $f(z)$ or $\varphi(u)$ we carry the topology of $\mathscr{D}_\chi$ over to the new spaces.

In the space $\mathscr{D}_\chi$ we define an operator $T_a{}^\chi$ by

$$T_a{}^\chi F(z_1, z_2) = F(z_1', z_2')$$

$$(z_1', z_2') = (z_1, z_2) \begin{pmatrix} a_{11} & a_{12} \\ a_{21} & a_{22} \end{pmatrix} \tag{5-10}$$

$$= (a_{11} z_1 + a_{21} z_2, a_{12} z_1 + a_{22} z_2)$$

This operator is continuous in $F$ and $a$ and satisfies the group law

$$T_{a_1}^\chi T_{a_2}^\chi = T_{a_1 a_2}^\chi$$

The boundedness of this operator $T_a{}^\chi$ in a specific norm is proved in the subsequent section. Using the realizations of $\mathscr{D}_\chi$ in terms of spaces of functions $f(z)$ or $(u)$, we get instead of (5–10)

$$T_a{}^\chi f(z) = \alpha(z, a) f(z_a)$$

$$\alpha(z, a) = |\lambda(z, a)|^{\sigma-1} (\operatorname{sign} \lambda(z, a))^\varepsilon$$

$$\lambda(z, a) = a_{12} z + a_{22} \tag{5-11}$$

$$z_a = (a_{11} z + a_{21}) \lambda(z, a)^{-1}$$

and

$$T_a{}^\chi \varphi(u) = \alpha(u, a) \varphi(u_a)$$

$$\alpha(u, a) = |\lambda(u, a)|^{\sigma-1} (\operatorname{sign} \lambda(u, a))^\varepsilon$$

$$\lambda(u, a) = \frac{a_{12} u_{21} + a_{22} u_{22}}{(u_a)_{22}} \tag{5-12}$$

$$(u_a)_{21} = \sin \tfrac{1}{2}\psi_a \qquad (u_a)_{22} = \cos \tfrac{1}{2}\psi_a$$

$$\tan \tfrac{1}{2}\psi_a = \frac{a_{11} \tan(1/2)\psi + a_{21}}{a_{12} \tan(1/2)\psi + a_{22}}$$

The matrix $u_a$ is determined up to a factor $e_-$.

It is easy to see that $z$, respectively $\psi$, $0 \leq \psi < 2\pi$, can be considered as parameters describing the right cosets of the subgroup K of triangular matrices $k$,

$$k = \begin{pmatrix} \lambda^{-1} & \mu \\ 0 & \lambda \end{pmatrix}$$

In fact, we have

$$\begin{pmatrix} 1 & 0 \\ z & 1 \end{pmatrix}\begin{pmatrix} a_{11} & a_{12} \\ a_{21} & a_{22} \end{pmatrix} = \begin{pmatrix} \lambda(z, a)^{-1} & \mu \\ 0 & \lambda(z, a) \end{pmatrix}\begin{pmatrix} 1 & 0 \\ z_a & 1 \end{pmatrix}$$

with $\lambda(z, a)$ as in (5−11), and

$$\begin{pmatrix} u_{11} & u_{12} \\ u_{21} & u_{22} \end{pmatrix}\begin{pmatrix} a_{11} & a_{12} \\ a_{21} & a_{22} \end{pmatrix} = \begin{pmatrix} \lambda(u, a)^{-1} & \mu \\ 0 & \lambda(u, a) \end{pmatrix}u_a$$

with $\lambda(u, a)$ as in (5−12). The function $\beta(k)$ corresponding to the multipliers $\alpha(z, a)$ and $\alpha(u, a)$ (Section 1−4a) is

$$\beta(k) = |\lambda|^\sigma (\text{sign } \lambda)^\varepsilon$$

If $\sigma$ is purely imaginary, this function $\beta(k)$ presents a one-dimensional unitary representation of the group K of triangular matrices $k$. By induction we may in turn obtain unitary representations of SL(2, R) from these representations of K. We call these representations of SL(2, R) representations of the principal series of SL(2, R) except for the case in which $\sigma = 0$ and $\varepsilon = 1$. We learn later (Sections 5−6, 5−7) that this particular representation $X = \{0, 1\}$ belongs to the class of analytic representations which are reducible representations, whereas the representations of the principal series will turn out to be irreducible (Section 5−9).

We take $dz$ as the quasi-invariant measure on SL(2, C)/K. The representations of the principle series are then carried by the Hilbert space $\mathcal{L}^2 (Z)$ of measurable functions on the real line whose norm $\|f\| = (f, f)^{1/2}$ is finite, where $(f_1, f_2)$ denotes the scalar product

$$(f_1, f_2) = \int_{-\infty}^{+\infty} \overline{f_1(z)} f_2(z) \, dz \qquad (5\text{-}13)$$

The group transformations in the space $\mathcal{L}^2(Z)$ are obtained from (5−11) by extension.

## 5–3    THE CANONICAL BASIS AND THE CANONICAL NORM

In the space $\mathscr{D}_\chi$ of functions $\varphi(\psi)$ with period $4\pi$ we may introduce a basis

$$\varphi_q(\psi) = e^{iq\psi} \qquad (5\text{--}14)$$

where $q$ runs over integers and half-integers. Because of the constraint (5–9) the functions $\varphi_q$ with integral $q$ only are already complete in $\mathscr{D}_\chi$ when $\varepsilon = 0$, whereas for $\varepsilon = 1$ we need only consider the functions $\varphi_q$ with half-integral $q$. This basis in $\mathscr{D}_\chi$ is denoted the canonical basis. With the help of the norm

$$\|\varphi\|^2 = \frac{1}{4\pi} \int_0^{4\pi} |\varphi(\psi)|^2 \, d\psi \qquad (5\text{--}15)$$

which we call the canonical norm, we complete the spaces $\mathscr{D}_\chi$. We obtain Hilbert spaces $\mathscr{L}_\varepsilon^2(U)$ with the scalar product

$$(\varphi_1, \varphi_2) = \frac{1}{4\pi} \int_0^{4\pi} d\psi \, \overline{\varphi_1(\psi)} \varphi_2(\psi)$$

in which the canonical basis is complete and orthonormal. We compute the operator norm of $T_a^\chi$ with respect to the vector norm (5–15).

For an operator $T_u^\chi$, $u = u(\psi)$, we have

$$T_u^\chi \varphi(\psi_1) = \varphi(\psi_1 + \psi)$$

which implies

$$\|T_u^\chi\| = 1 \qquad (5\text{--}16)$$

Since any element $a \in SL(2, R)$ can be decomposed as in (5–1), (5–2) we have

$$\|T_a^\chi \varphi\|^2 = \frac{1}{4\pi} \int_0^{4\pi} |T_{\delta u_2}^\chi \varphi(\psi)|^2 \, d\psi$$

$$= \frac{1}{4\pi} \int_0^{4\pi} |T_a^\chi \varphi'(\psi)|^2 \, d\psi$$

$$= \frac{1}{4\pi} \int_0^{4\pi} |\varphi'(\psi_d)|^2 \left| e^{-(1/2)\eta} \frac{\cos(1/2)\psi}{\cos(1/2)\psi_d} \right|^{2\,\mathrm{Re}\,\sigma - 2} d\psi \qquad (5\text{--}17)$$

where we used the notation

$$T_{u_2}^\chi \varphi = \varphi' \qquad \|\varphi'\| = \|\varphi\|$$

The transformed angle $\psi_d$ can be read off (5–12)

$$\tan \tfrac{1}{2}\psi_d = e^\eta \tan \tfrac{1}{2}\psi$$

$$\frac{d\psi}{d\psi_d} = e^{-\eta} \frac{\cos^2(1/2)\psi}{\cos^2(1/2)\psi_d} = \frac{1 + \tan^2(1/2)\psi_d}{e^\eta + e^{-\eta}\tan^2(1/2)\psi_d} \tag{5-18}$$

Inserting (5–18) into (5–17) we get

$$\|T_a^\chi \varphi\|^2 = \frac{1}{4\pi} \int\limits_0^{4\pi} d\psi \, |\varphi'(\psi)|^2 \left(\frac{1 + \tan^2(1/2)\psi}{e^\eta + e^{-\eta}\tan^2(1/2)\psi}\right)^{\mathrm{Re}\,\sigma}$$

which implies immediately

$$\|T_a^\chi\| = \sup_\varphi \frac{\|T_a^\chi \varphi\|}{\|\varphi\|} = \exp\{\tfrac{1}{2}\eta \, |\mathrm{Re}\,\sigma|\} \tag{5-19}$$

The operators $T_a^\chi$ can therefore be extended onto the Hilbert space $\mathscr{L}_\varepsilon^2(U)$; the operators $T_u^\chi$ are unitary. In this fashion we obtain representations of SL(2, R) in a Hilbert space for each $\chi$. If $\sigma$ is purely imaginary, we see from (5–19) and the fact that for each $T_a^\chi$ an inverse is given by $T_{a^{-1}}^\chi$, that this representation in $\mathscr{L}_\varepsilon^2(U)$ is unitary. These unitary representations belong to the principal series (the case $\sigma = 0$, $\varepsilon = 1$ is again excluded) which we obtained by induction in Section 5–2. The Hilbert spaces $\mathscr{L}_\varepsilon^2(U)$ and $\mathscr{L}^2(Z)$ are related as follows.

In order to map the Hilbert spaces $\mathscr{L}_\varepsilon^2(U)$ on spaces of functions $f(z)$ we use (5–6) and (5–8). With the additional normalizing factor $\pi^{1/2}$ and sign $z_2 = +1$ we get

$$\varphi(\psi) = \pi^{1/2}(1 + z^2)^{-(\sigma-1)/2}f(z)$$

$$\psi = 2 \arctan z \tag{5-20}$$

and

$$\frac{1}{2\pi} \int\limits_{-\pi}^{+\pi} |\varphi(\psi)|^2 \, d\psi = \int\limits_{-\infty}^{+\infty} |f(z)|^2(1 + z^2)^{-\mathrm{Re}\,\sigma} \, dz$$

We define therefore Hilbert spaces $\mathscr{L}_{\sigma_1}^2(Z)$, $\sigma_1 = -\mathrm{Re}\,\sigma$, $[\mathscr{L}_0^2(Z) \equiv \mathscr{L}^2(Z)]$, consisting of measurable functions on the real line, which have finite norm

$\|f\| = (f, f)^{1/2}$, where $(f_1, f_2)$ denotes the scalar product

$$(f_1, f_2) = \int_{-\infty}^{+\infty} \overline{f_1(z)} f_2(z)(1 + z^2)^{\sigma_1} \, dz$$

The point transformation (5–20) establishes in this way an isomorphism between $\mathcal{L}_{\varepsilon_1}^2(U)$ and $\mathcal{L}_{\sigma_1}^2(Z)$. The orthonormal canonical basis in $\mathcal{L}_{\varepsilon}^2(Z)$ has the elements

$$f_q(z) = \pi^{-1/2}(1 + z^2)^{(\sigma - 1)/2} \left( \frac{1 + iz}{1 - iz} \right)^q \tag{5-21}$$

$$-\infty < q < \infty \qquad 2q \cong \varepsilon \bmod 2$$

Now we compute the matrix elements of the operator $T_a{}^\chi$ with respect to the canonical basis. For an element $u = u(\psi)$ of SL(2, R) we get

$$T_u{}^\chi \varphi_q(\psi_1) = e^{iq\psi} \varphi_q(\psi_1) \tag{5-22}$$

This property of being eigenvectors for the operators $T_a{}^\chi$ in $\mathcal{L}_{\varepsilon}^2(U)$ characterizes the elements of the canonical basis up to a constant phase factor. With (5–1), (5–2)

$$a = u_1 \, du_2 \qquad u_{1,2} = u(\psi_{1,2})$$

we get for a matrix element of $T_a{}^\chi$

$$(\varphi_{q_1}, T_a{}^\chi \varphi_{q_2}) \equiv \langle q_1 | T_a{}^\chi | q_2 \rangle \equiv C_{q_1 q_2}^\chi(a)$$
$$= \exp\{iq_1\psi_1 + iq_2\psi_2\}\langle q_1 | T_d{}^\chi | q_2 \rangle$$

and

$$\langle q_1 | T_d{}^\chi | q_2 \rangle \equiv c_{q_1 q_2}^\chi(\eta)$$

$$= \frac{1}{2\pi} \int_0^{2\pi} d\psi \, \exp\{-iq_1\psi + iq_2\psi_d\} \left| e^{-\eta/2} \frac{\cos(1/2)\psi}{\cos(1/2)\psi_d} \right|^{\sigma - 1} \tag{5-23}$$

since we can always make

$$\text{sign}\left( \frac{\cos(1/2)\psi}{\cos(1/2)\psi_d} \right)$$

constant equal to 1 [recall (5–18)]. Inserting the relations

$$e^{-\eta} \frac{\cos^2(1/2)\psi}{\cos^2(1/2)\psi_d} = e^{-\eta} \cos^2 \tfrac{1}{2}\psi + e^{\eta} \sin^2 \tfrac{1}{2}\psi$$

$$= (e^{i\psi} \cosh \tfrac{1}{2}\eta - \sinh \tfrac{1}{2}\eta)(e^{-i\psi} \cosh \tfrac{1}{2}\eta - \sinh \tfrac{1}{2}\eta)$$

and

$$\exp\{i\psi_d\} = \frac{1 + ie^{\eta} \tan(1/2)\psi}{1 - ie^{\eta} \tan(1/2)\psi}$$

$$= e^{-i\psi}(e^{i\psi} \cosh \tfrac{1}{2}\eta - \sinh \tfrac{1}{2}\eta)$$

$$\times (e^{-i\psi} \cosh \tfrac{1}{2}\eta - \sinh \tfrac{1}{2}\eta)^{-1}$$

into the integral (5–23) we obtain

$$c^{\chi}_{q_1 q_2}(\eta) = \frac{1}{2\pi} \int_0^{2\pi} d\psi \, \exp\{-i(q_1 + q_2)\psi\}$$

$$\times [e^{i\psi} \cosh \tfrac{1}{2}\eta - \sinh \tfrac{1}{2}\eta]^{(\sigma-1)/2 + q_2}$$

$$\times [e^{-i\psi} \cosh \tfrac{1}{2}\eta - \sinh \tfrac{1}{2}\eta]^{(\sigma-1)/2 - q_2} \tag{5-24}$$

where both square brackets are to be taken on the principal sheet at $\psi = 0$. The integral representation (5–24) for the function $c^{\chi}_{q_1 q_2}(\eta)$ can be expanded into a binomial series

$$c^{\chi}_{q_1 q_2}(\eta) = (-1)^{q_1 - q_2}(\cosh \tfrac{1}{2}\eta)^{\sigma - 1}$$

$$\times \sum_{n = \max(0, \, q_2 - q_1)}^{\infty} \binom{\tfrac{1}{2}(\sigma - 1) + q_2}{n}$$

$$\times \binom{\tfrac{1}{2}(\sigma - 1) - q_2}{n + q_1 - q_2}(\tanh \tfrac{1}{2}\eta)^{2n + q_1 - q_2} \tag{5-25}$$

which converges for all real $\eta$. The series (5–25) is equivalent with a Gaussian hypergeometric series

$$c^{\chi}_{q_1 q_2}(\eta) = (-1)^{q_1 - q_2}(\cosh \tfrac{1}{2}\eta)^{\sigma - 1}(\tanh \tfrac{1}{2}\eta)^{q_1 - q_2} \binom{\tfrac{1}{2}(\sigma - 1) - q_2}{q_1 - q_2}$$

$$\times \, {}_2F_1(-\tfrac{1}{2}(\sigma - 1) - q_2, \, -\tfrac{1}{2}(\sigma - 1)$$

$$+ q_1, q_1 - q_2 + 1; \tanh^2 \tfrac{1}{2}\eta) \tag{5-26}$$

for $q_1 \geqq q_2$, and

$$c^\chi_{q_1 q_2}(\eta) = (-1)^{q_2 - q_1}(\cosh \tfrac{1}{2}\eta)^{\sigma - 1}(\tanh \tfrac{1}{2}\eta)^{q_2 - q_1}\begin{pmatrix} \tfrac{1}{2}(\sigma - 1) + q_2 \\ q_2 - q_1 \end{pmatrix}$$

$$\times\, _2F_1(-\tfrac{1}{2}(\sigma - 1) + q_2,\, -\tfrac{1}{2}(\sigma - 1)$$

$$- q_1,\, q_2 - q_1 + 1;\, \tanh^2 \tfrac{1}{2}\eta) \tag{5-27}$$

for $q_2 \geqq q_1$. We shall come back to these functions in Chapter 6.

## 5–4   GENERATORS AND THEIR DIAGONALIZATIONS

Since we do not intend to study covariant operators of SL(2, R) as in the case of SL(2, C), we prefer to discuss the generators prior to the invariant bilinear functionals. We expand an infinitesimal group element as

$$a = e + i\left[\varepsilon_2 \frac{1}{2}\sigma_2 + \eta_1 \frac{i}{2}\sigma_1 + \eta_3 \frac{i}{2}\sigma_3\right] + O(\varepsilon^2, \eta^2)$$

and the operator $T_a^\chi$ on $\mathcal{D}_\chi$ correspondingly as

$$T_a^\chi = E + i[\varepsilon_2 H_2 + \eta_1 F_1 + \eta_3 F_3] + O(\varepsilon^2, \eta^2) \tag{5-28}$$

We can give $H_2, F_1, F_3$ as differential operators on the space $\mathcal{D}_\chi$ realized by functions $f(z)$

$$H_2 = +\frac{i}{2}(1 + z^2)\frac{d}{dz} - \frac{i}{2}(\sigma - 1)z$$

$$F_1 = +\frac{i}{2}(1 - z^2)\frac{d}{dz} + \frac{i}{2}(\sigma - 1)z \tag{5-29}$$

$$F_3 = +iz\frac{d}{dz} - \frac{i}{2}(\sigma - 1)$$

We define the Casimir operator of SL(2, R) by

$$I = H_2{}^2 - F_1{}^2 - F_3{}^2 \tag{5-30}$$

Inserting the differential operators (5–29) into (5–30) gives

$$I = \tfrac{1}{4}(\sigma^2 - 1)E$$

which can also be written

$$I = J(J + 1)E$$

if we introduce the parameter $J$ by

$$J = \tfrac{1}{2}(\sigma - 1) \qquad \sigma = 2J + 1 \tag{5-31}$$

In Chapters 6 and 7 the parameter $J$ replaces $\sigma$ throughout.

We verify easily that $H_2$ is diagonal on the basis functions (5–21)

$$H_2 f_q(z) = -q f_q(z) \tag{5-32}$$

as it should be, because of (5–22) and

$$T^\chi_{u(\psi)} f_q(z) = \exp\{-i\psi H_2\} f_q(z) \tag{5-33}$$

The diagonalization of the generator $F_3$ is more complicated but also more interesting. This generator creates the one-parameter noncompact subgroup R(1) of dilatations $T_a{}^\chi$ (here we let $\eta$ vary over the whole real axis). First therefore we construct the eigendistributions of these operators. We consider distributions $p(z)$ acting on spaces $\mathscr{C}^\infty$ of functions $f(z) \in \mathscr{D}_{-\chi}$ with compact support. Transformations of these distributions can be defined by

$$\int_{-\infty}^{+\infty} (T_a{}^\chi p(z)) f(z)\, dz = \int_{-\infty}^{+\infty} p(z)(T_{a^{-1}}^{-\chi} f(z))\, dz \tag{5-34}$$

for all $f(z) \in \mathscr{C}^\infty$ and those $a \in$ SL(2, R) which make $T_{a^{-1}}^{-\chi} f$ again a function of compact support. In fact, the definition (5–34) anticipates the bilinear invariant functional (5–53). The eigendistributions of the dilatation operators $T_a{}^\chi$ are the homogeneous distributions

$$p_{\beta,\gamma}(z) = c(\text{sign } z)^\beta |z|^{(\sigma-1)/2 + i\gamma}$$
$$\beta = 0, 1 \qquad \tfrac{1}{2}(\sigma - 1) + i\gamma \neq -1, -2, \ldots \tag{5-35}$$

respectively,

$$p_{\beta,\gamma}(z) = c(z + i0 \pm)^{(\sigma-1)/2 + i\gamma}$$
$$\beta = 0, 1 \qquad (-1)^\beta = \pm \tag{5-36}$$
$$\tfrac{1}{2}(\sigma - 1) + i\gamma = -1, -2, \ldots$$

They satisfy

$$T_a{}^\chi p_{\beta,\gamma}(z) = e^{i\gamma\eta} p_{\beta,\gamma}(z) \tag{5-37}$$

If $\sigma$ is purely imaginary, the distributions

$$p^\chi_{\beta,\gamma}(z) = (4\pi)^{-1/2} (\text{sign } z)^\beta |z|^{(\sigma-1)/2 + i\gamma} \tag{5-38}$$

for arbitrary real $\gamma$, fixed $\sigma$, and both $\beta = 0$ and $1$, are orthogonal and complete in the sense

$$\sum_{\beta = 0, 1} \int_{-\infty}^{+\infty} d\gamma \, \overline{p_{\beta, \gamma}^{\chi}(z)} p_{\beta, \gamma}^{\chi}(z') = \delta(z - z') \tag{5-39}$$

Similarly are the distributions $\varphi_{\beta, \gamma}^{\chi}(\psi)$ that correspond to $p_{\beta, \gamma}^{\chi}(z)$ through (5–6) and (5–8) orthogonal and complete

$$\varphi_{\beta, \gamma}^{\chi}(\psi) = (4\pi)^{-1/2} [\text{sign}(\sin \tfrac{1}{2}\psi)]^{\beta} [\text{sign}(\cos \tfrac{1}{2}\psi)]^{\beta + \varepsilon}$$
$$\times |\sin \tfrac{1}{2}\psi|^{(\sigma - 1)/2 + i\gamma} |\cos \tfrac{1}{2}\psi|^{(\sigma - 1)/2 - i\gamma} \tag{5-40}$$

$$\sum_{\beta = 0, 1} \int_{-\infty}^{+\infty} d\gamma \, \overline{\varphi_{\beta, \gamma}^{\chi}(\psi)} \varphi_{\beta, \gamma}^{\chi}(\psi') = \tfrac{1}{2}[\delta(\psi - \psi') + (-1)^{\varepsilon} \, \delta(\psi - \psi' \pm 2\pi)] \tag{5-41}$$

In the space $\mathscr{L}_{\varepsilon}^{2}(U)$ of the principal series of unitary representations of SL(2, R) we can therefore perform a harmonic analysis based on the unitary irreducible representations of the one-dimensional subgroup R(1) of dilatations. For $\varphi(\psi)$ in a space $\mathscr{L}_{\varepsilon}^{2}(U)$ which carries the representation $\chi = \{\sigma, \varepsilon\}$ of the principal series, we define a Fourier transform by

$$\Phi^{\chi}(\beta, \gamma) = \int_{0}^{4\pi} \varphi_{\beta, -\gamma}^{-\chi}(\psi) \varphi(\psi) \, d\psi \tag{5-42}$$

[note that $\varphi_{\beta, -\gamma}^{-\chi}(\psi) = \overline{\varphi_{\beta, \gamma}^{\chi}(\psi)}$ for the principal series]. A dilatation operates on the Fourier transform as

$$T_d^{\chi} \Phi^{\chi}(\beta, \gamma) = \int_{0}^{4\pi} \varphi_{\beta, -\gamma}^{-\chi}(\psi) (T_d^{\chi} \varphi(\psi)) \, d\psi$$

$$= \int_{0}^{4\pi} (T_{d^{-1}}^{-\chi} \varphi_{\beta, -\gamma}^{-\chi}(\psi)) \varphi(\psi) \, d\psi$$

$$= e^{in\gamma} \Phi^{\chi}(\beta, \gamma)$$

The Fourier transformation (5–42) establishes in this way a decomposition of $\mathscr{L}_{\varepsilon}^{2}(U)$ into a direct integral of Hilbert spaces $\mathscr{H}_{\beta, \gamma}$,

$$\mathscr{L}_{\varepsilon}^{2}(U) = \sum_{\beta = 0, 1}^{\oplus} \int_{-\infty}^{+\infty \oplus} \mathscr{H}_{\beta, \gamma} \, d\gamma$$

each of which carries an irreducible (one-dimensional) unitary representation of the group of dilatations. This harmonic analysis can be generalized to non-unitary representations of SL(2, R) in the following characteristic fashion.

We define Banach spaces $\mathscr{L}_\varepsilon^q(U)$, $1 < q < \infty$, consisting of measurable functions $\varphi(\psi)$, $0 \leq \psi < 4\pi$, which are covariant as

$$\varphi(\psi + 2\pi) = (-1)^\varepsilon \varphi(\psi)$$

and have the finite $q$-norm

$$\|\varphi\|_q = \left[ \int_0^{4\pi} |\varphi(\psi)|^q \, d\psi \right]^{1/q}$$

Two such spaces $\mathscr{L}_\varepsilon^q(U)$ and $\mathscr{L}_\varepsilon^p(U)$ with

$$\frac{1}{q} + \frac{1}{p} = 1$$

are dual to each other, that is, any function of one space creates a linear continuous functional on the other space, and any continuous linear functional on one space can be generated by a set of functions of the other space, which are equal almost everywhere. The operators $T_a{}^\chi$ have the $q$-norm

$$\|T_a{}^\chi\|_q = \exp\left\{ \frac{1}{2} \, \eta \left| \operatorname{Re} \sigma - 1 + \frac{2}{q} \right| \right\}$$

and can therefore be extended from $\mathscr{D}_\chi$ to all spaces $\mathscr{L}_\varepsilon^q(U)$. If

$$-1 < \operatorname{Re} \sigma < 1$$

and if we set

$$q = q(\sigma) = \frac{2}{1 - \operatorname{Re} \sigma} \qquad \frac{1}{q(\sigma)} + \frac{1}{q(-\sigma)} = 1$$

the $q(\sigma)$-norm of $T_a{}^\chi$ is equal to 1. For any pair $\varphi_1 \in \mathscr{L}_\varepsilon^{q(\sigma)}(U), \varphi_2 \in \mathscr{L}_\varepsilon^{q(-\sigma)}(U)$, the bilinear form

$$[\varphi_1, \varphi_2] = \int_0^{4\pi} \varphi_1(\psi)\varphi_2(\psi) \, d\psi$$

is invariant,

$$[\varphi_1, \varphi_2] = [T_a{}^\chi \varphi_1, T_a{}^{-\chi} \varphi_2]$$

for all $a \in SL(2, R)$.

If

$$-1 < \mathrm{Re}\ \sigma \leq 0$$

it is possible to base a harmonic analysis of the space $\mathscr{L}_\varepsilon^{q(\sigma)}(U)$ on the functions (5-40). We define an analog of a Fourier transform by

$$\Phi^\chi(\beta, \gamma) = \int_0^{4\pi} \varphi_{\beta, -\gamma}^{-\chi}(\psi)\varphi(\psi)\,d\psi \qquad \varphi \in \mathscr{L}_\varepsilon^{q(\sigma)}(U) \tag{5-43}$$

and its inversion by

$$\varphi(\psi) = \sum_{\beta=0, 1} \int_{-\infty}^{+\infty} \Phi^\chi(\beta, \gamma)\varphi_{\beta, \gamma}^\chi(\psi)\,d\gamma \tag{5-44}$$

These integrals must be interpreted as certain limits yet to be specified. To prove (5-43) and (5-44) it suffices for us to show that for measurable functions $\varphi(\psi)$, $0 \leq \psi \leq \pi$, with finite norm,

$$\int_0^\pi |\varphi(\psi)|^{q(\sigma)}\,d\psi < \infty$$

the integral

$$\Phi^\chi(\gamma) = \int_0^\pi \varphi_{-\gamma}^{-\chi}(\psi)\varphi(\psi)\,d\psi \tag{5-45}$$

exists in the $q(-\sigma)$ -mean and can be inverted. Here we use the notation

$$\varphi_\gamma^\chi(\psi) = (4\pi)^{-1/2}\,|\sin\tfrac{1}{2}\psi|^{(\sigma-1)/2+i\gamma}\,|\cos\tfrac{1}{2}\psi|^{(\sigma-1)/2-i\gamma}$$

similarly as in (5-40).

With the parameter $t$ introduced by

$$e^t = \tan\tfrac{1}{2}\psi \qquad \frac{d\psi}{dt} = \sin\psi = \frac{1}{\cosh t}$$

we obtain

$$\int_0^\pi |\varphi(\psi)|^q\,d\psi = \int_{-\infty}^{+\infty} |\varphi(2\arctan e^t)(\cosh t)^{-1/q}|^q\,dt < \infty$$

that is, if $\varphi(\psi)$ is in $\mathscr{L}^q(0, \pi)$ as a function of $\psi$, then

$$f(t) = \varphi(2 \arctan e^t)(\cosh t)^{-1/q}$$

is in $\mathscr{L}^q(-\infty, \infty)$ as a function of $t$. Further we have from (5–45)

$$\Phi^x(\gamma) = \pi^{-1/2} \int_{-\infty}^{+\infty} dt \, e^{-i\gamma t}(2 \cosh t)^{(\sigma-1)/2 + 1/q}$$

$$\times [\varphi(2 \arctan e^t)(2 \cosh t)^{-1/q}]$$

If we set

$$\mathrm{Re}\left\{\frac{1}{2}(\sigma - 1) + \frac{1}{q}\right\} = 0 \quad \text{viz.} \quad q = q(\sigma)$$

we have reduced our problem to the issue of Fourier transforming a space $\mathscr{L}^q(-\infty, \infty)$ in the standard sense, where $q$ lies in the interval $1 < q \leq 2$. The solution of this problem is contained in the textbooks. It tells us that the integral (5–45) converges in the $q(-\sigma)$-mean towards an element of the space $\mathscr{L}^{q(-\sigma)}(-\infty, \infty)$ such that the inequality

$$\left[\int_{-\infty}^{\infty} |\Phi^x(\gamma)|^{q(-\sigma)} \, d\gamma\right]^{1/\{q(-\sigma)\}} \leq (4\pi)^{(1/2) - 1/q(\sigma)} \left[\int_{0}^{\pi} |\varphi(\psi)|^{q(\sigma)} \, d\psi\right]^{1/\{q(\sigma)\}} \tag{5-46}$$

holds. This integral transformation can moreover be inverted by

$$\varphi(\psi) = (\sin \psi)^{(\sigma + 1)/2} \frac{d}{d\psi} \int_{-\infty}^{+\infty} d\gamma \, \Phi^x(\gamma)$$

$$\times \int_{(1/2)\pi}^{\psi} d\psi' \, (\sin \psi')^{-(\sigma + 1)/2} \varphi_\gamma{}^x(\psi') \tag{5-47}$$

where the equality is true almost everywhere.

An analogous procedure allows us to analyze certain Banach spaces $\mathscr{L}^q$, $1 < q \leq 2$, by means of unitary representations of the groups SL(2, C) and SL(2, R) instead of R(1). In this context the particular problem arises of decomposing a nonunitary representation of SL(2, C) realized in a Banach space with invariant norm into unitary irreducible representations of SL(2, R) or SU(1, 1). We attack a similar problem in Chapter 7, however, using the more general method of analytic functionals, which includes also nonunitary representations of SU(1, 1) in the decomposition. The price we have to pay for the unitarity of the representations appearing in the decomposition is that the Banach spaces analyzed are rather small compared with the spaces of distributions treated in Chapters 4, 6, and 7. For the physical applications we have in mind, these spaces are too small in general.

## 5–5    BILINEAR INVARIANT FUNCTIONALS

In order to study the irreducibility, equivalence, and unitarity of the representations on $\mathcal{D}_\chi$, we proceed exactly as in the case of the group SL(2, C) and investigate the bilinear invariant functionals on the spaces $\mathcal{D}_\chi$. The difference between the group structures of SL(2, C) and SL(2, R) comes into the discussion through the fact that for real variables more homogeneous distributions exist than for complex variables. Indeed, for the homogeneities $s = -1, -2, \ldots$, there exist not only the derivatives of the delta-function $\delta^{(-s-1)}(z)$ as in the complex case but also the principal value distributions $z^s$. These two kinds of distributions may be combined into the linear combinations $(z + i0\pm)^s$. The latter distributions give rise to the so-called analytic representations.

Bilinear invariant functionals are defined verbatim as in Section 3–6. In order to verify invariance of a functional, we make use of the fact that any element of SL(2, R), just as the elements of SL(2, C), can be decomposed into a product of the elements $\zeta, \delta, \varepsilon$, where

$$\zeta = \begin{pmatrix} 1 & 0 \\ z & 1 \end{pmatrix}$$

is a translation,

$$\delta = \begin{pmatrix} \lambda^{-1} & 0 \\ 0 & \lambda \end{pmatrix} \qquad \lambda > 0 \quad \text{or} \quad < 0$$

is a dilatation, and

$$\varepsilon = \begin{pmatrix} 0 & 1 \\ -1 & 0 \end{pmatrix}$$

is an inversion. It suffices to check invariance against these three types of transformations separately. In applying the theory of distributions to spaces of functions $f(z)$ we make use of a space of test functions $\mathscr{C}^\infty$ which consists of functions possessing derivatives of all orders and a compact support. The convergence in $\mathscr{C}^\infty$ is as usual.

Let

$$\chi_1 = \{\sigma_1, \varepsilon_1\} \qquad \chi_2 = \{\sigma_2, \varepsilon_2\}$$

be two representations, let $f(z)$ transform as an element of $\mathscr{D}_{\chi_1}$ and $h(z)$ as an element of $\mathscr{D}_{\chi_2}$, and let $f(z)$ and $h(z)$ be in $\mathscr{C}^\infty$. Taking into account invariance against translations, we make the following ansatz for the bilinear invariant functional

$$B(h, f) = \int_{-\infty}^{+\infty} dz_2\, M(z_2) \int_{-\infty}^{+\infty} dz_1 h(z_1 + z_2) f(z_1) \qquad (5\text{-}48)$$

where $M(z)$ is any distribution. One can show that this ansatz (5—48) is the most general one that is in agreement with the requirements of bilinearity, continuity, and invariance against translations. Imposing on (5—48) the requirement of dilatational invariance, we obtain

$$B(T_\delta^{\chi_2} h,\, T_\delta^{\chi_1} f) = |\lambda|^{\sigma_1 + \sigma_2 - 2}(\text{sign }\lambda)^{\varepsilon_1 + \varepsilon_2}$$

$$\times \int_{-\infty}^{+\infty} dz_2\, M(z_2) \int_{-\infty}^{+\infty} dz_1\, h(\lambda^{-2}(z_1 + z_2)) f(\lambda^{-2} z_1)$$

$$= |\lambda|^{\sigma_1 + \sigma_2}(\text{sign }\lambda)^{\varepsilon_1 + \varepsilon_2}$$

$$\times \int_{-\infty}^{+\infty} dz_2\, M(z_2) \int_{-\infty}^{+\infty} dz_1\, h(\lambda^{-2} z_2 + z_1) f(z_1)$$

$$= B(h, f)$$

If we set $\delta = e_-$, it follows that

$$\varepsilon_1 = \varepsilon_2 = \varepsilon \qquad (5\text{-}49)$$

Moreover this constraint implies that $M(z)$ is a homogeneous distribution of degree

$$s = -\tfrac{1}{2}(\sigma_1 + \sigma_2) - 1$$

Such distributions are of the form

$$c|z|^s \quad \text{or} \quad c|z|^s \text{ sign } z \qquad (5\text{-}50)$$

if $s \neq -1, -2, -3, \ldots$, and equal to

$$c\delta^{(-s-1)}(z) \quad \text{or the principal value} \quad cz^s \qquad (5\text{-}51)$$

if $s = -1, -2, -3, \ldots$. In case (5—51) the linear combinations

$$(z + i\,0\pm)^s$$

which have no definite parity, present another possibility of writing the homogeneous distribution $M(z)$. We call the case (5–50) the regular alternative, and case (5–51) the singular alternative.

In the regular alternative the functional is a linear combination of the two possible forms

$$\int |z_2 - z_1|^{-(\sigma_1+\sigma_2)/2 - 1}[\text{sign}(z_2 - z_1)]^\beta h(z_2) f(z_1)\, dz_1 dz_2$$

with $\beta = 0$ or 1. If we apply an inversion, the functional becomes

$$
\begin{aligned}
B(T_\varepsilon^{\chi_2} h,\, T_\varepsilon^{\chi_1} f) = \sum_\beta c_\beta \int |z_2 - z_1|^{-(\sigma_1+\sigma_2)/2 - 1} \\
\times |z_1|^{\sigma_1 - 1}|z_2|^{\sigma_2 - 1}[\text{sign}(z_2 - z_1)]^\beta \\
\times [\text{sign}(z_1 z_2)]^\varepsilon h(-z_2^{-1}) f(-z_1^{-1})\, dz_1\, dz_2
\end{aligned}
$$

Provided $f$ and $h$ vanish in a neighborhood of $z = 0$, we may make the substitutions

$$z_1 \rightarrow -z_1^{-1} \qquad z_2 \rightarrow -z_2^{-1}$$

and get the requirement

$$
\begin{aligned}
B(h,\, f) = \sum_\beta c_\beta \int |z_2 - z_1|^{-(\sigma_1+\sigma_2)/2 - 1} |z_1|^{(\sigma_2-\sigma_1)/2} \\
\times |z_2|^{(\sigma_1-\sigma_2)/2}[\text{sign}(z_2 - z_1)]^\beta [\text{sign}(z_1 z_2)]^{\beta+\varepsilon} \\
\times h(z_2) f(z_1)\, dz_1\, dz_2
\end{aligned}
$$

Invariance for general functions $f$ and $h$ is obviously possible only if

$$c_\beta = 0 \qquad \text{for} \quad \beta \neq \varepsilon \quad \text{and} \quad \sigma_1 = \sigma_2$$

In this fashion we arrive at a necessary condition on the existence of a bilinear invariant functional for the regular alternative, which we may formulate as follows. If $\frac{1}{2}(\sigma_1 + \sigma_2)$ is neither positive integral nor zero, a bilinear invariant functional exists only if

$$\chi_1 = \chi_2 = \{\sigma, \varepsilon\}$$

It is of the form

$$
\begin{aligned}
B(h,\, f) = \int |z_2 - z_1|^{-\sigma - 1}[\text{sign}(z_2 - z_1)]^\varepsilon \\
\times h(z_2) f(z_1)\, dz_1\, dz_2
\end{aligned}
\tag{5-52}
$$

if the functions $f$ and $h$ are as required in our derivation. On the other hand, it can be shown that the integral (5–52) does indeed define a bilinear invariant functional on the whole space $\mathscr{D}_\chi$, that is, our condition is also sufficient, provided we understand the integral (5–52) to be regularized if $\mathrm{Re}\,\sigma \geqq 0$.

Now we consider the singular alternative (5–51). The functional is a linear combination of the form

$$B_1(h, f) = \int h(z)\left(\frac{d}{dz}\right)^{(\sigma_1 + \sigma_2)/2} f(z)\, dz$$

and the principal value integral

$$B_2(h, f) = \int (z_2 - z_1)^{-(\sigma_1 + \sigma_2)/2 - 1} h(z_2) f(z_1)\, dz_1\, dz_2$$

Applying an inversion to the functional $B_1(h, f)$ gives

$$\int h(-z^{-1}) |z|^{\sigma_2 - 1}\left(\frac{d}{dz}\right)^{(\sigma_1 + \sigma_2)/2} |z|^{\sigma_1 - 1} f(-z^{-1})\, dz$$

$$= \int h(z) |z|^{-\sigma_2 - 1}\left(z^2 \frac{d}{dz}\right)^{(\sigma_1 + \sigma_2)/2}$$

$$\times\ |z|^{-\sigma_1 + 1} f(z)\, dz$$

provided $f$ and $h$ vanish in a neighborhood of $z = 0$. From the formulas (3–50), (3–52) we see that the functional $B_1(h, f)$ is separately invariant if either

$$\sigma_1 = -\sigma_2$$

in which case it reduces to

$$B_1(h, f) = \int_{-\infty}^{+\infty} h(z) f(z)\, dz$$

or if

$$\sigma_1 = \sigma_2 = \sigma$$

in which case we get

$$B_1(h, f) = \int_{-\infty}^{+\infty} h(z)\left(\frac{d}{dz}\right)^{\sigma} f(z)\, dz$$

The principal value integral $B_2(h, f)$ goes under inversion into

$$\int (z_2 - z_1)^{-(\sigma_1 + \sigma_2)/2 - 1} |z_1|^{\sigma_1 - 1} |z_2|^{\sigma_2 - 1}$$

$$\times [\text{sign}(z_1 z_2)]^\varepsilon h(-z_2^{-1}) f(-z_1^{-1}) \, dz_1 \, dz_2$$

Provided $f$ and $h$ vanish in a neighborhood of $z = 0$, this expression can be transformed as

$$B_2(h, f) = \int (z_2 - z_1)^{-(\sigma_1 + \sigma_2)/2 - 1} |z_1|^{(\sigma_2 - \sigma_1)/2} |z_2|^{(\sigma_1 - \sigma_2)/2}$$

$$\times [\text{sign}(z_1 z_2)]^{\varepsilon + 1 + (\sigma_1 + \sigma_2)/2} h(z_2) f(z_1) \, dz_1 \, dz_2$$

This constraint requires

$$\sigma_1 = \sigma_2 \quad \text{and} \quad \varepsilon + 1 \cong \tfrac{1}{2}(\sigma_1 + \sigma_2) \bmod 2$$

It is now easy to prove that invariance of the linear combination of the two functionals $B_1(h, f)$ and $B_2(h, f)$ can be reduced to invariance of either functional. We may, for example, choose $f$ and $h$ such that the intersection of their supports is void. Then the whole functional consists only of the principal value integral. The requirement of invariance of the principal value integral for such particular functions $f$ and $h$ suffices to draw the same conclusions as before on the necessary form of $B_2(h, f)$. One functional in a linear combination of two being invariant, the other must also be invariant separately.

The necessary condition on the existence of the bilinear invariant functional for the singular alternative arrived at can be formulated as follows. If $\tfrac{1}{2}(\sigma_1 + \sigma_2)$ is positive integral or zero, a bilinear invariant functional may exist only if (5–49) holds and in either of the cases

(1)   $\sigma_1 = -\sigma_2$, $\sigma_1$ arbitrary complex

$$B(h, f) = \int_{-\infty}^{+\infty} h(z) f(z) \, dz \tag{5-53}$$

(2)   $\sigma_1 = \sigma_2 = \sigma$, $\sigma$ positive integral or zero,
   (a)   $\varepsilon \cong \sigma \bmod 2$,

$$B(h, f) = \int h(z) \left( \frac{d}{dz} \right)^\sigma f(z) \, dz \tag{5-54}$$

(b)   $\varepsilon + 1 \cong \sigma \bmod 2$,

$$B(h, f) = c_1 \int h(z)\left(\frac{d}{dz}\right)^{\sigma} f(z)\, dz$$

$$+ c_2 \int (z_2 - z_1)^{-\sigma-1} h(z_2) f(z_1)\, dz_1\, dz_2 \qquad (5\text{-}55)$$

with arbitrary constants $c_1$ and $c_2$.

The bilinear invariant functionals in the explicit form of the integrals (5–53), (5–54), (5–55), which were shown to exist for particular sets of functions $f$ and $h$, can in fact be extended onto all of $\mathscr{D}_{\chi_1}$, and $\mathscr{D}_{\chi_2}$. The principal value integrals must be regularized of course; the other integrals exist in the proper sense.

The representations $\chi = \{\sigma, \varepsilon\}$ with integral $\sigma$ and $\varepsilon + 1 \cong \sigma \bmod 2$, which are denoted "analytic representations," play a particular role. Their multipliers (5–11) are rational functions of $z$,

$$\alpha(z, a) = (\lambda(z, a))^{\sigma - 1}$$

Analytic representations with $\sigma \geqq 0$ possess two independent bilinear invariant functionals, as we have just seen. The representation $\sigma = 0$, $\varepsilon = 1$ (Section 5–2) is analytic in this terminology.


## 5–6   INVARIANT SUBSPACES

As in the case of the group SL(2, C) a closed invariant subspace of $\mathscr{D}_{\chi_1}$ is obtained if we consider all those elements $f \in \mathscr{D}_{\chi_1}$ for which $B(h, f) = 0$ for all $h \in \mathscr{D}_{\chi_2}$. If $B(h, f)$ is not unique, we may consider any choice. Such annihilation of the bilinear invariant functional can obviously occur only for analytic representations $\chi_1$.

If $\chi_1$ is an analytic representation with $\sigma_1 \geqq 0$ and $\chi_2 = \chi_1$, the invariant bilinear functional (5–55) can also be obtained as a combination of ($\sigma = \sigma_1 = \sigma_2$, $\chi = \chi_1 = \chi_2$)

$$B_+(h, f) = c_+ \int (z_1 - z_2 - i0+)^{-\sigma-1} h(z_2) f(z_1)\, dz_2\, dz_1 \qquad (5\text{-}56)$$

$$B_-(h, f) = c_- \int (z_1 - z_2 - i0-)^{-\sigma-1} h(z_2) f(z_1)\, dz_2\, dz_1 \qquad (5\text{-}57)$$

These integrals may be regularized by formal partial integrations, which yield

$$B_+(h, f) = \frac{1}{2\pi i} \int\limits_{-\infty}^{+\infty} dz_2 \, h(z_2) \int\limits_{-\infty}^{+\infty} dz_1$$

$$\times (z_1 - z_2 - i0+)^{-1} \left(\frac{d}{dz_1}\right)^\sigma f(z_1) \tag{5-58}$$

$$B_-(h, f) - \frac{-1}{2\pi i} \int\limits_{-\infty}^{+\infty} dz_2 \, h(z_2) \int\limits_{-\infty}^{+\infty} dz_1$$

$$\times (z_1 - z_2 - i0-)^{-1} \left(\frac{d}{dz_1}\right)^\sigma f(z_1) \tag{5-59}$$

where we gave the constants $c_\pm$ a definite value. The integrals $(5-58), (5-59)$ exist for all $f$ and $h \in \mathscr{D}_\chi$ in the proper sense if we integrate over $z_1$ first as indicated. In fact, the asymptotic series $(5-7)$ for $f(z)$ and $h(z)$ starts with the power $z^{\sigma-1}$ The series for

$$\left(\frac{d}{dz_1}\right)^\sigma f(z_1)$$

begins correspondingly with $z_1^{-\sigma-1}$. This implies that the integral over $z_1$ exists. The integral over $z_1$ itself has an asymptotic expansion with the highest power $z_2^{-\sigma-1}$, so that the final integration over $z_2$ converges, too.

For a square integrable function $f(z)$ the Hilbert transforms

$$f_\pm(z) = \frac{\pm 1}{2\pi i} \int\limits_{-\infty}^{+\infty} \frac{f(z_1)}{z_1 - z - i0\pm} \, dz_1 \tag{5-60}$$

are themselves square integrable, and each is the mean square limit of a holomorphic function in one half-plane, $f_+(z)$ from the upper and $f_-(z)$ from the lower half-plane. Their sum is

$$f_+(z) + f_-(z) = f(z) \tag{5-61}$$

Moreover we have

$$(f_+)_+(z) = f_+(z) \qquad (f_-)_-(z) = f_-(z) \tag{5-62}$$

If $f(z)$ is square integrable and in $\mathscr{D}_\chi$, the Hilbert transforms $f_\pm(z)$ are also in $\mathscr{D}_\chi$ and satisfy

$$\left(\frac{d}{dz}\right)^n f_\pm(z) = \left[\left(\frac{d}{dz}\right)^n f\right]_\pm(z)$$

Together with all their derivatives they are proper limits of holomorphic functions. Applying the Hilbert transformations (5–60) to

$$\left(\frac{d}{dz}\right)^\sigma f(z) \qquad f \in \mathscr{D}_\chi$$

where $f$ is not necessarily square integrable, is equivalent to the following procedure. We split $f(z)$ into two parts

$$f(z) = q_{\sigma-1}(z) + f^*(z) \tag{5-63}$$

where $q_{\sigma-1}(z)$ is a polynomial of maximal degree $\sigma-1$ and is identical to the first $\sigma$ terms in the asymptotic series (5–7), so that $f^*(z)$ is square integrable and in $\mathscr{D}_\chi$. With the Hilbert transforms $f_\pm^*(z)$ of $f^*(z)$ we get

$$\left[\left(\frac{d}{dz}\right)^\sigma f(z)\right]_\pm = \left(\frac{d}{dz}\right)^\sigma f_\pm^*(z) \tag{5-64}$$

Functions $f(z)$ of $\mathscr{D}_\chi$ which are boundary values of functions that are holomorphic in the upper half-plane constitute a closed invariant subspace $\mathscr{D}_\chi^+$ of $\mathscr{D}_\chi$. In fact,

$$B_-(h, f) = \int h(z) \left[\left(\frac{d}{dz}\right)^\sigma f(z)\right]_- dz$$

$$= \int h(z) \left(\frac{d}{dz}\right)^\sigma f_-^*(z)\, dz = 0$$

for all $h \in \mathscr{D}_\chi$ if and only if

$$f_-^*(z) = 0 \qquad \text{viz.} \quad f(z) = q_{\sigma-1}(z) + f_+^*(z)$$

Similarly, the functions of $\mathscr{D}_\chi$ which are boundary values of holomorphic functions in the lower half-plane constitute a closed invariant subspace $\mathscr{D}_\chi^-$ of $\mathscr{D}_\chi$ on which $B_+$ vanishes. Since each element of $\mathscr{D}_\chi^\pm$ can be written

$$f(z) = q_{\sigma-1}(z) + f_\pm^*(z) \tag{5-65}$$

the intersection of $\mathscr{D}_\chi^+$ and $\mathscr{D}_\chi^-$ is a space of polynomials of maximal degree $\sigma - 1$. We denote this space $\mathscr{E}_\chi$. In the case $\sigma > 0$ the space $\mathscr{D}_\chi$ of an analytic representation has therefore three proper closed invariant subspaces

$$\mathscr{D}_\chi^+ \qquad \mathscr{D}_\chi^- \qquad \mathscr{E}_\chi$$

where

$$\mathscr{E}_\chi = \mathscr{D}_\chi^+ \cap \mathscr{D}_\chi^- \qquad \mathscr{D}_\chi = \mathscr{D}_\chi^+ + \mathscr{D}_\chi^- \qquad \text{(direct sum)}$$

For $\sigma = 0$ the space $\mathscr{E}_\chi$ contains only the null vector.

In the case of a negative integral $\sigma$ with analytic representations, the unique bilinear functional is given by (5–52)

$$B(h, f) = \int (z_2 - z_1)^{-\sigma-1} h(z_2) f(z_1) \, dz_1 \, dz_2$$

It vanishes on the subspace $\mathscr{F}_\chi$ consisting of such functions $f(z)$ whose moments

$$\int\limits_{-\infty}^{+\infty} z^k f(z) \, dz \qquad k = 0, 1, 2, \ldots, -\sigma - 1$$

vanish. This space is infinitely dimensional and the quotient space $\mathscr{D}_\chi / \mathscr{F}_\chi$ has dimension $-\sigma$. The spaces $\mathscr{F}_\chi$ can be decomposed into the direct sum of two closed subspaces $\mathscr{D}_\chi^+$ and $\mathscr{D}_\chi^-$, which consist of boundary values of analytic functions holomorphic in one half-plane. We shall show in a moment that the spaces $\mathscr{D}_\chi^\pm$ are separately invariant. The proof of the invariance of $\mathscr{D}_\chi^\pm$ is a by-product of an investigation of the elements of the canonical basis. This is of course not the most elegant proof, but it serves also to prepare some later considerations.

For analytic representations (the sign of $\sigma$ is arbitrary now) we may write the functions of the canonical basis (5–21) as

$$f_q(z) = \pi^{-1/2}(1 + iz)^{(\sigma-1)/2+q}(1 - iz)^{(\sigma-1)/2-q}$$

where both powers $\frac{1}{2}\sigma - \frac{1}{2} \pm q$ are integers. The basis functions are therefore rational in $z$. Those functions $f_q(z)$ with a label

$$q \geqq -\tfrac{1}{2}(\sigma - 1)$$

are holomorphic in the upper half-plane, those with

$$q \leqq \tfrac{1}{2}(\sigma - 1)$$

are holomorphic in the lower half-plane. We consider the subspace $\mathscr{D}_\chi^{+\prime}$ of $\mathscr{D}_\chi$ which is obtained by closing the linear space spanned by the functions $f_q(z)$ with $q \geq -\frac{1}{2}\sigma + \frac{1}{2}$. Similarly the basis functions $f_q(z)$ with $q \leq \frac{1}{2}\sigma - \frac{1}{2}$ yield a closed subspace $\mathscr{D}_\chi^{-\prime}$ of $\mathscr{D}_\chi$. If $\sigma > 0$, the intersection of $\mathscr{D}_\chi^{+\prime}$ and $\mathscr{D}_\chi^{-\prime}$ has a basis of functions $f_q(z)$ with

$$-\tfrac{1}{2}(\sigma - 1) \leq q \leq \tfrac{1}{2}(\sigma - 1)$$

This intersection is seen to be identical with the space of polynomials $\mathscr{E}_\chi$. Our aim is to show that the spaces $\mathscr{D}_\chi^{\pm\prime}$ coincide with $\mathscr{D}_\chi^\pm$, and that the spaces $\mathscr{D}_\chi^\pm$ are invariant.

We change the realization of the spaces $\mathscr{D}_\chi$. We recall (5–6) and (5–8)

$$\varphi(\psi) = (1 + z(\psi)^2)^{-(\sigma - 1)/2}[\text{sign}(\cos \tfrac{1}{2}\psi)]^\varepsilon f(z(\psi))$$
$$z(\psi) = \tan \tfrac{1}{2}\psi \tag{5-66}$$

For analytic representations we introduce the new functions

$$\Phi(e^{i\psi}) = e^{(i/2)(\sigma - 1)\psi}\varphi(\psi) \tag{5-67}$$

which are periodic with period $2\pi$ in $\psi$ (instead of $4\pi$). They possess derivatives of all orders in $\psi$ and represent the elements of $\mathscr{D}_\chi$ uniquely. In the sequel we limit the angle $\psi$ to the interval $-\pi \leq \psi \leq \pi$. A null sequence in $\mathscr{D}_\chi$ corresponds to a sequence of functions $\Phi(e^{i\psi})$ which together with all the derivatives with respect to $\psi$ converges to zero uniformly. These functions $\Phi(e^{i\psi})$ transform as [see (5–12), (5–67)]

$$T_a^\chi\Phi(e^{i\psi}) = \alpha(u(\psi), a)\exp\left\{\frac{i}{2}(\sigma - 1)(\psi - \psi_a)\right\}\Phi(e^{i\psi_a})$$
$$= (-\bar{\beta}e^{i\psi} + \alpha)^{\sigma - 1}\Phi(e^{i\psi_a}) \tag{5-68}$$

where

$$\alpha = \tfrac{1}{2}(a_{11} + a_{22} + ia_{12} - ia_{21})$$
$$\beta = \tfrac{1}{2}(a_{11} - a_{22} - ia_{12} - ia_{21}) \tag{5-69}$$

$$\exp\{i\psi_a\} = \frac{\bar{\alpha}e^{i\psi} - \beta}{-\bar{\beta}e^{i\psi} + \alpha} \tag{5-70}$$

The transformation law (5–68) is considerably simpler than (5–12). This is of particular importance when we want to continue analytically in $z$ or $e^{i\psi}$.

The conformal mapping

$$w = \frac{1 + iz}{1 - iz} \qquad z = -i\frac{w - 1}{w + 1}$$

$$1 + iz = \frac{2w}{1 + w} \qquad 1 - iz = \frac{2}{1 + w} \tag{5-71}$$

maps the upper half $z$-plane onto the interior of the unit circle in $w$. A function $f(z)$ of $\mathscr{D}_\chi$ with a holomorphic continuation in the upper half-plane goes into a function $\Phi(e^{i\psi})$ by (5–66), (5–67), which can be continued analytically into the interior of the unit circle and is holomorphic there due to

$$\Phi(w) = w^{(\sigma-1)/2}\varphi(\psi(w))$$

$$= [\tfrac{1}{2}(1 + w)]^{\sigma-1}f(z(w)) \tag{5-72}$$

The basis functions $f_q(z)$ (5–21) go correspondingly into

$$\Phi_q(w) = \pi^{-1/2}w^{(\sigma-1)/2+q} \tag{5-73}$$

Now to the proof announced earlier.

The invariance of the spaces $\mathscr{D}_\chi^\pm$ can be seen directly from the transformation law (5–68), (5–70). Further we can expand any $\Phi(e^{i\psi}) \in \mathscr{D}_\chi^+$ into a Fourier series

$$\Phi(e^{i\psi}) = \sum_n a_n e^{in\psi}$$

where $n$ is integral. This series converges in the topology of $\mathscr{D}_\chi$. Because it possesses a holomorphic continuation inside the unit circle, we must have

$$a_n = 0 \qquad \text{for} \quad n < 0$$

This implies that $\Phi(e^{i\psi})$ can be expanded into a series of the basis elements $\Phi_q(e^{i\psi})$, with $q \geq -\tfrac{1}{2}(\sigma-1)$, which converges in the sense of $\mathscr{D}_\chi$, that is,

$$\mathscr{D}_\chi^+ \subset \mathscr{D}_\chi^{+\prime}$$

If $\Phi(e^{i\psi})$ is in $\mathscr{D}_\chi^{+\prime}$, it possesses a series expansion

$$\Phi(e^{i\psi}) = \sum_{q \geq -(\sigma-1)/2} a'_q \Phi_q(e^{i\psi})$$

which converges in the sense of $\mathscr{D}_\chi$. Since each basis element appearing in this series possesses a holomorphic continuation inside the unit circle, the maximum principle guarantees that the series of functions $\Phi_q(w)$ converges towards a holomorphic function $\Phi(w)$, which is the analytic continuation of $\Phi(e^{i\psi})$, that is,

$$\mathscr{D}_\chi^{+\prime} \subset \mathscr{D}_{\bar{\chi}}^{\pm}$$

Therefore $\mathscr{D}_\chi^+$ and $\mathscr{D}_\chi^{+\prime}$ are identical. Similarly we can prove the identity of $\mathscr{D}_\chi^-$ and $\mathscr{D}_{\bar{\chi}}^{-\prime}$ by means of antiholomorphic continuations inside the unit circle.

## 5–7   UNITARY REPRESENTATIONS

Any Hermitian invariant form in a linear vector space is a bilinear invariant functional, we need only put

$$(h, f) = B(\bar{h}, f) \tag{5-74}$$

If $f$ and $h$ are in $\mathscr{D}_\chi$, then $\bar{h}$ is in $\mathscr{D}_\chi'$, where

$$\chi = \{\sigma, \varepsilon\} \quad \text{and} \quad \chi' = \{\bar{\sigma}, \varepsilon\}$$

The results of Section 5–5 allow only the following cases:

   (i)  $\sigma = -\bar{\sigma}$, $\sigma$ is purely imaginary,
   (ii)  $\sigma = \bar{\sigma}$, $\sigma$ is real

In case (i) the Hermitian form (5–53) is

$$(h, f) = \int_{-\infty}^{+\infty} \overline{h(z)} f(z)\, dz \tag{5-75}$$

The form (5–75) is obviously positive definite, and the space $\mathscr{D}_\chi$ can be completed to yield the Hilbert space $\mathscr{L}^2(Z)$. We have rediscovered in this fashion the principle series of unitary representations together with the analytic representation $\chi = \{0, 1\}$, which we found by induction in Section 5–2.

   In case (ii) the Hermitian form is (5–52)

$$(h, f) = \int |z_2 - z_1|^{-\sigma-1}[\text{sign}(z_2 - z_1)]^\varepsilon$$

$$\times \overline{h(z_2)} f(z_1)\, dz_1\, dz_2 \tag{5-76}$$

if $\sigma \neq 0, 1, 2, \ldots$ ; and (5–54)

$$(h, f) = \int_{-\infty}^{+\infty} \overline{h(z)} \left(\frac{d}{dz}\right)^{\sigma} f(z) \, dz \tag{5-77}$$

for $\sigma = 1, 2, 3, \ldots$ , and nonanalytic representations [$\varepsilon \cong \sigma \bmod 2$; the case $\sigma = 0$, $\varepsilon = 0$ may be neglected because it is a principal series representation and is included already in (5–75)] ; and (5–58), (5–59)

$$(h, f) = c_+ \int_{-\infty}^{+\infty} \overline{h(z)} \left[\left(\frac{d}{dz}\right)^{\sigma} f(z)\right]_+ dz + c_- \int_{-\infty}^{+\infty} \overline{h(z)} \left[\left(\frac{d}{dz}\right)^{\sigma} f(z)\right]_- dz \tag{5-78}$$

with arbitrary constants $c_\pm$ for $\varepsilon = 0, 1, 2, \ldots$ , and analytic representations ($\varepsilon + 1 \cong \sigma \bmod 2$). We start our discussion with the case (5–76) and $\sigma < 0$.

The integral (5–76) need not be regularized for $\sigma < 0$. It can most conveniently be studied in a matrix form with respect to the canonical basis, where

$$f(z) = \sum_q \alpha_q f_q(z) \qquad h(z) = \sum_q \beta_q f_q(z) \tag{5-79}$$

and

$$(h, f) = \sum_{q_1 q_2} \tau^\varepsilon_{q_1 q_2} \bar\beta_{q_1} \alpha_{q_2} \tag{5-80}$$

To perform the integrations we replace $z$ by $\psi$ as usual,

$$z_i = \tan \tfrac{1}{2}\psi_i \qquad -\pi \leq \psi_i \leq \pi \qquad i = 1, 2$$

and get from (5–21), (5–76)

$$\tau^\varepsilon_{q_2 q_1} = \frac{(-1)^\varepsilon}{4\pi} \int_{-\pi}^{\pi} d\psi_1 \int_{-\pi}^{\pi} d\psi_2 \, \exp\{i(q_1\psi_1 - q_2\psi_2)\}$$

$$\times |\sin \tfrac{1}{2}(\psi_1 - \psi_2)|^{-\sigma-1}[\mathrm{sign}(\sin \tfrac{1}{2}(\psi_1 - \psi_2))]^\varepsilon$$

which simplifies to

$$\tau^\varepsilon_{q_2 q_1} = \delta_{q_1 q_2}(-1)^\varepsilon \tfrac{1}{2} \int_{-\pi}^{+\pi} d\psi \, |\sin \psi|^{-\sigma-1}[\mathrm{sign}(\sin \psi)]^\varepsilon \exp\{2iq_1\psi\}$$

For $\varepsilon = 0$ this integral gives (*GR* 3.631.8)

$$\tau_{qq}^0 = \int_0^\pi d\psi (\sin \psi)^{-\sigma-1} \cos 2q\psi$$

$$= \frac{2^{\sigma+1}\pi(-1)^q\Gamma(-\sigma)}{\Gamma((1/2) - (1/2)\sigma + q)\Gamma((1/2) - (1/2)\sigma - q)} \tag{5-81}$$

whereas $\varepsilon = 1$ leads to (*GR* 3.631.1)

$$\tau_{qq}^1 = -i \int_0^\pi d\psi (\sin \psi)^{-\sigma-1} \sin 2q\psi$$

$$= \frac{2^{\sigma+1}\pi i(-1)^{q+1/2}\Gamma(-\sigma)}{\Gamma((1/2) - (1/2)\sigma + q)\Gamma((1/2) - (1/2)\sigma - q)} \tag{5-82}$$

For $\varepsilon = 0$ and $\sigma \neq -1, -3, -5, \ldots$, namely, for nonanalytic representations, $\tau_{qq}^0$ is always different from zero. Moreover

$$\tau_{qq}^0 = \tau_{-q,-q}^0$$

From

$$\tau_{qq}^0 = 2^{\sigma+1}\Gamma(-\sigma)\sin \frac{1}{2}\pi(\sigma+1)\frac{\Gamma((1/2) + (1/2)\sigma + q)}{\Gamma((1/2) - (1/2)\sigma + q)}$$

we see that all $\tau_{qq}^0$ are positive if $-1 < \sigma < 0$. For these $\sigma$ the Hermitian form (5–76) is definite. We can complete the space $\mathscr{D}_\chi$ with respect to this norm and obtain a Hilbert space $\mathscr{H}_\sigma$. Because of the invariance of the Hermitian form, the new norm of the operators $T_a{}^\chi$ is 1, and these operators can be extended onto the Hilbert space $\mathscr{H}_\sigma$. The series of unitary representations obtained in this fashion is denoted the supplementary series. For $\varepsilon = 0$ and other $\sigma < 0$ the Hermitian form (5–76) is indefinite. For $\varepsilon = 1$ and $\sigma \neq -2, -4, -6, \ldots$, the Hermitian form (5–76) is indefinite throughout since $\tau_{qq}^1 = -\tau_{-q-q}^1$.

As we shall show in Section 5–9, two representations $\chi_1 = \{\sigma, \varepsilon\}, \chi_2 = \{-\sigma, \varepsilon\}$ which are nonanalytic are equivalent. We can therefore spare ourselves the investigation of nonanalytic representations with $\sigma > 0$. It remains for us to study the analytic representations. We start with $\sigma \geq 0$. For $f(z)$ and $h(z)$ in $\mathscr{D}_\chi^+$ and with the decomposition (5–65), the Hermitian form (5–78) reduces to

$$(h, f) = c_+ \int_{-\infty}^{+\infty} \overline{h_+^*(z)}\left(\frac{d}{dz}\right)^\sigma f_+^*(z)\, dz \tag{5-83}$$

After a Fourier transformation

$$f_+^*(z) = \int_{-\infty}^{+\infty} g_f(y)e^{iyz}\,dy$$

$$(5\text{-}84)$$

$$h_+^*(z) = \int_{-\infty}^{+\infty} g_h(y)e^{iyz}\,dy$$

for which

$$g_f(y) = g_h(y) = 0 \qquad \text{if} \quad y < 0$$

we get

$$(h,\,f) = 2\pi c_+(i)^\sigma \int_0^\infty \overline{g_h(y)}g_f(y)y^\sigma\,dy \qquad (5\text{-}85)$$

Definite Hermitian forms can thus be constructed on the quotient spaces $\mathscr{D}_\chi^+/\mathscr{E}_\chi$, and on the space $\mathscr{D}_\chi^+$ itself if $\chi = \{0, 1\}$. The same can be shown for the spaces $\mathscr{D}_\chi^-/\mathscr{E}_\chi$. The representations induced on these quotient spaces are unitary in this metric and give rise to the so-called discrete series. The representations on the spaces $\mathscr{D}_\chi^+/\mathscr{E}_\chi$ are denoted the positive discrete series, those on $\mathscr{D}_\chi^-/\mathscr{E}_\chi$ the negative discrete series. We show in Section 5–9 that the spaces $\mathscr{D}_\chi^\pm$ carry representations which are equivalent to those on the quotient spaces $\mathscr{D}_\chi^\pm/\mathscr{E}_\chi$ if $\sigma > 0$. A scalar product can therefore also be defined on these spaces; note, however, that (5–76) vanishes on $\mathscr{D}_\chi^\pm$. The only spaces left whose Hermitian invariant forms have not yet been studied are the finite-dimensional spaces $\mathscr{E}_\chi$. It is easy to verify that invariant Hermitian forms exist on these spaces, though they are indefinite except in the trivial one-dimensional case.

Let us consider in somewhat more detail the analytic representation $\chi = \{0, 1\}$, which we obtained first by induction together with the principal series in Section 5–2. For this representation the invariant polynomial subspace $\mathscr{E}_\chi$ is the null space, and in agreement with what we said earlier, the spaces $\mathscr{D}_\chi^\pm$ themselves carry unitary representations. In fact, any element $f(z) \in \mathscr{D}_\chi$ is square integrable and decomposes into the sum of two Hilbert transforms (as for any representation of the principal series)

$$f(z) = f_+(z) + f_-(z)$$

but contrary to the representations of the principal series, these Hilbert transforms span invariant subspaces $\mathscr{D}_\chi^\pm$. The scalar product (5–13) or (5–75) splits always as

$$\int\limits_{-\infty}^{+\infty} \overline{f_1(z)} f_2(z)\, dz = \int\limits_{-\infty}^{+\infty} \overline{(f_1)_+(z)}(f_2)_+(z)\, dz + \int\limits_{-\infty}^{+\infty} \overline{(f_1)_-(z)}(f_2)_-(z)\, dz$$

but each part is separately invariant only if $\sigma = 0$ and $\varepsilon = 1$.

The form (5–85) of the scalar product for the discrete series is not very convenient. A simpler expression can be obtained if we realize the representation spaces of the discrete series by means of the holomorphic functions $\Phi(w)$ (Section 5–6). Though we could derive it from (5–83) or (5–85) directly, we prefer to find it by an independent method, namely, the method of induced representations.

## 5–8   REPRESENTATIONS OF THE DISCRETE SERIES AS INDUCED REPRESENTATIONS

We consider right cosets of the subgroup of matrices $u(\psi)$ in SL(2, R). Since any element of SL(2, R) can be decomposed into the product

$$a = us$$

[see the proof of (5–1)], such that $s$ is symmetric and positive, the right cosets are unambiguously described by such symmetric matrices $s$. We set

$$s = s(w) = (1 - w_1{}^2 - w_2{}^2)^{-1/2}\begin{pmatrix} 1 - w_1 & w_2 \\ w_2 & 1 + w_1 \end{pmatrix}$$

$$w = w_1 + iw_2 \qquad |w|^2 < 1 \tag{5-86}$$

Points in the interior of the unit circle in the complex $w$-plane can therefore be used to parametrize the right cosets as well.

The transformation formula for the number $w$ representing the right coset follows from the definition (1–25)

$$s(w)a = u(\psi(w, a))s(w_a) \tag{5-87}$$

After some algebra we get [compare (5–70)]

$$w_a = \frac{\bar{\alpha}w - \beta}{-\bar{\beta}w + \alpha} \tag{5-88}$$

where $\alpha$ and $\beta$ were defined in (5–69). Moreover we have

$$\exp\{i\psi(w, a)\} = \frac{1 - |w|^2}{1 - |w_a|^2}(-\bar{\beta}w + \alpha)^{-2} \tag{5-89}$$

The easiest way to find these results is to map $SL(2, R)$ on $SU(1, 1)$ by the standard isomorphism $(1-19)$ or $(5-3)$

$$SL(2, R) \qquad\qquad SU(1, 1)$$

$$a \leftrightarrow \begin{pmatrix} \bar{\alpha} & -i\bar{\beta} \\ i\beta & \alpha \end{pmatrix} \tag{5-90}$$

$$s(w) \leftrightarrow (1 - |w|^2)^{-1/2} \begin{pmatrix} 1 & i\bar{w} \\ -iw & 1 \end{pmatrix} \tag{5-91}$$

From $(5-88)$ and $(5-89)$ we obtain in addition

$$|-\bar{\beta}w + \alpha|^2 = \frac{1 - |w|^2}{1 - |w_a|^2} \tag{5-92}$$

and

$$\left| \frac{dw_a}{dw} \right| = |-\bar{\beta}w + \alpha|^{-2} = \frac{1 - |w_a|^2}{1 - |w|^2} \tag{5-93}$$

We define a Hilbert space $\mathcal{H}^*$ of functions $\Phi^*(w)$ which are measurable in $|w| < 1$ and have finite norm $\|\Phi^*\| = (\Phi^*, \Phi^*)^{1/2}$, where $(\Phi_1^*, \Phi_2^*)$ denotes the scalar product

$$(\Phi_1^*, \Phi_2^*) = \int\limits_{|w| < 1} \overline{\Phi_1^*(w)} \Phi_2^*(w) \, Dw \tag{5-94}$$

If $e^{ik\psi}$, $k$ integral or half-integral, is a unitary one-dimensional representation of the group of matrices $u(\psi)$, a unitary representation of $SL(2, R)$ in $\mathcal{H}^*$ is obtained by induction (Section $1-4b$)

$$T_a \Phi^*(w) = \exp\{ik\psi(w, a)\} \left| \frac{dw_a}{dw} \right| \Phi^*(w_a) \tag{5-95}$$

The multiplier can be rewritten as

$$\exp\{ik\psi(w, a)\} \left| \frac{dw_a}{dw} \right| = \left[ \frac{1 - |w|^2}{1 - |w_a|^2} \right]^{k-1} (-\bar{\beta}w + \alpha)^{-2k}$$

which suggests the following redefinitions

$$\Phi(w) = (1 - |w|^2)^{-k+1} \Phi^*(w)$$

$$T_a \Phi(w) = (-\bar{\beta}w + \alpha)^{-2k} \Phi(w_a) \tag{5-96}$$

$$\|\Phi\|^2 = \int_{|w| < 1} |\Phi(w)|^2 (1 - |w|^2)^{2k-2} \, Dw \qquad (5\text{-}97)$$

So far $\Phi(w)$ has been an arbitrary measurable function, square integrable in the sense (5—97). With such functions, however, the representation defined by (5—96) is reducible. In fact, since

$$|\alpha| > |\beta|$$

$T_a$ (5—96) maps a function which is holomorphic inside the unit circle onto a function with the same property. Functions which are holomorphic inside the unit circle span an invariant subspace (which is not necessarily closed a priori, see below). If these holomorphic functions are moreover differentiable to all orders on the boundary of the unit circle, we may identify them with elements $\Phi(w)$ of $\mathscr{D}_\chi^+$, as defined by (5—72). In order to fit (5—96) to (5—68) we set

$$-2k = \sigma - 1$$

We remember that the spaces $\mathscr{D}_\chi^+$ possess nontrivial invariant subspaces $\mathscr{E}_\chi$ if $\sigma > 0$ (Section 5—6). The same fact is reflected by the integral (5—97), which converges with holomorphic functions in the proper sense only for $\sigma < 0$. We can, however, extend the definition of this scalar product to $\sigma = 0$ if we multiply the integral with

$$-\frac{\sigma}{\pi} = \frac{2k - 1}{\pi}$$

and approach $\sigma = 0$ from below. From now on we use this normalization of (5—97) throughout.

In this manner we have accomplished the construction of a scalar product in the spaces $\mathscr{D}_\chi^+$ with $\sigma \leq 0$. Completing these spaces with respect to the norm (5—97) yields unitary representations of the positive discrete series, which we label $(k, +)$, realized in Hilbert spaces $\mathscr{H}(k, +)$.

With the same method we can arrive at a Hilbert space $\mathscr{H}(k, -)$ carrying a representation of the negative discrete series $(k, -)$, where $k$ runs over the positive integers and half-integers. In the first step we reduce the space $\mathscr{H}^*$ of square integrable functions $\Phi^*(w)$ to an invariant but not necessarily closed subspace of antiholomorphic functions inside the unit circle, using the relation

$$\exp\{ik\psi(w, a)\} \left| \frac{dw_a}{dw} \right| = \left[ \frac{1 - |w|^2}{1 - |w_a|^2} \right]^{-k-1} (-\beta \bar{w} + \bar{\alpha})^{2k}$$

In the second step we change the sign of $k$ for convenience and complete the space of antiholomorphic functions with the norm (5–97).

Let us denote the Hilbert space of measurable functions $\Phi'(w)$ with finite norm (5–97) by $\mathcal{H}'$, such that $\mathcal{H}'$ and $\mathcal{H}^*$ are isomorphically related by the point transformation

$$\Phi'(w) = (1 - |w|^2)^{-k+1}\Phi^*(w)$$

Of course, $\mathcal{H}(k, +)$ is a Hilbert subspace of $\mathcal{H}'$, that is, there is an orthogonal projection $P$ such that

$$\mathcal{H}(k, +) = P\mathcal{H}'$$

We want now to gain a deeper insight into the nature of this projection operator $P$.

For the positive discrete series $(k, +)$ with $k = 1/2, 1, 3/2, \ldots$, the set of functions

$$\Phi_q(w) = N_q{}^k w^{q-k} \qquad q = k, k+1, k+2, \ldots \tag{5-98}$$

which differ from the functions (5–73) only by the normalization constants $N_q{}^k$, presents an orthonormal basis in $\mathcal{H}(k, +)$ if

$$(N_q^k)^2 = \binom{k+q-1}{q-k} = (-1)^{q-k}\binom{-2k}{q-k} \tag{5-99}$$

In fact, with (5–99) we get

$$\|\Phi_q\|^2 = \frac{2k-1}{\pi}(N_q^k)^2 \int\limits_{|w|<1} |w|^{2q-2k}(1 - |w|^2)^{2k-2}\, Dw$$

$$= 1$$

We consider the sum over the basis elements (5–98)

$$G(w_1, w_2 \,|\, k, +) = \sum_{q=k}^{\infty} \Phi_q(w_1)\overline{\Phi_q(w_2)}$$

$$= (1 - w_1\overline{w}_2)^{-2k} \tag{5-100}$$

which obviously converges if $|w_1\overline{w}_2| < 1$. We call this function $G(w_1, w_2 | k, +)$ the Cauchy kernel for holomorphic functions in the unit circle. It has the property

$$\Phi(w_1) = \frac{2k-1}{\pi} \int\limits_{|w_2|<1} G(w_1, w_2 \,|\, k, +)\Phi(w_2)(1 - |w_2|^2)^{2k-2}\, Dw_2 \tag{5-101}$$

for any function $\Phi(w)$ which is holomorphic inside the unit circle and continuous on the boundary. For an arbitrary measurable function $\Phi'(w)$ which is square integrable in the sense (5–97), namely, $\Phi'(w) \in \mathscr{H}'$, the integral

$$\Phi(w_1) = \frac{2k-1}{\pi} \int\limits_{|w_2| < 1} G(w_1, w_2 \,|\, k, +)\Phi'(w_2)$$

$$\times (1 - |w_2|^2)^{2k-2} \, Dw_2 \tag{5-102}$$

defines a function $\Phi(w)$ which is holomorphic in the interior of the unit circle and square integrable at the boundary such that

$$\|\Phi\|^2 \leq \|\Phi'\|^2 \tag{5-103}$$

We prove the inequality (5–103).

We define regularized functions $\Phi_\varepsilon(w), \Phi'_\varepsilon(w)$ by

$$\Phi'_\varepsilon(w) = \theta(1 - \varepsilon - |w|)\Phi'(w)$$

and $\Phi_\varepsilon(w)$ by $\Phi'_\varepsilon(w)$ through (5–102). The function $\Phi_\varepsilon(w)$ is holomorphic in the circle

$$|w| < (1 - \varepsilon)^{-1}$$

and the norm $\|\Phi_\varepsilon\|^2$ is therefore finite. We have

$$\|\Phi_\varepsilon\|^2 = \left(\frac{2k-1}{\pi}\right)^2 \sum_{q=k}^{\infty} \left| \int \Phi'_\varepsilon(w)\overline{\Phi_q(w)} \right.$$

$$\times (1 - |w|^2)^{2k-2} \, Dw \Big|^2$$

$$= \left(\frac{2k-1}{\pi}\right)^2 \int \overline{\Phi'_\varepsilon(w_1)}\Phi'_\varepsilon(w_2)G(w_1, w_2 \,|\, k, +)$$

$$\times (1 - |w_1|^2)^{2k-2}(1 - |w_2|^2)^{2k-2} \, Dw_1 \, Dw_2$$

because the sum and the integrals converge absolutely and uniformly. Performing the integration over $w_2$ gives

$$\|\Phi_\varepsilon\|^2 = \frac{2k-1}{\pi} \int \overline{\Phi'_\varepsilon(w)}\Phi_\varepsilon(w)(1 - |w|^2)^{2k-2} \, Dw$$

$$\leq \|\Phi_\varepsilon\| \, \|\Phi'_\varepsilon\|$$

by means of Schwartz's inequality. Finally we have

$$\|\Phi_\varepsilon\| \le \|\Phi_\varepsilon'\|$$

In the limit $\varepsilon \to 0+$ the right-hand side of the inequality tends to a finite limit. Since

$$\lim_{\varepsilon \to 0+} \Phi_\varepsilon(w) = \Phi(w)$$

uniformly on any compact set in the open circle, it results that $\Phi(w)$ is holomorphic inside the unit circle and square integrable as required by (5–103). In addition $\Phi(w)$ is an element of $\mathscr{H}(k, +)$. As a corollary we find that these Cauchy kernels generate the orthogonal projection operators $P$, which map $\mathscr{H}'$ onto the Hilbert spaces $\mathscr{H}(k, +)$, and that $\mathscr{H}(k, +)$ consists of functions which are holomorphic in the interior of the unit circle and square integrable at the boundary.

We note that the Cauchy kernels for antiholomorphic functions in the unit circle can be defined by

$$G(w_1, w_2 \,|\, k, -) = G(w_2, w_1 \,|\, k, +) \tag{5-104}$$

## 5–9  EQUIVALENCE AND IRREDUCIBILITY

As in the case of the group SL(2, C) we study the problems of equivalence and irreducibility of representations by means of the device of an intertwining operator $A$ that maps a space $\mathscr{D}_\chi$ continuously into another space $\mathscr{D}_{\chi'}$ and intertwines the representations $\chi$ and $\chi'$ in the sense

$$T_a^{\chi'} A = A T_a^\chi$$

for all $a \in$ SL(2, C). If this mapping is one-to-one and bicontinuous, we call the representations $\chi$ and $\chi'$ equivalent. If the representation $\chi = \chi'$ is unitary, and if the only bounded operator commuting with the representation on its Hilbert space is proportional to the unit operator, then the argument presented in Section 3–8 shows that this unitary representation is irreducible.

Our study of bilinear invariant functionals in Section 5–5 gave the result that for any pair $\chi = \{\sigma, \varepsilon\}$ and $-\chi = \{-\sigma, \varepsilon\}$ of representations a bilinear invariant functional (5–53)

$$[h, f] = \int_{-\infty}^{+\infty} h(z) f(z) \, dz \qquad f \in \mathscr{D}_\chi \qquad h \in \mathscr{D}_{-\chi} \tag{5-105}$$

exists. With the help of the definition

$$B(h, f) = [h, Af] \qquad f \in \mathcal{D}_\chi \qquad h \in \mathcal{D}_{-\chi'} \tag{5-106}$$

we can reduce the problem of finding an intertwining operator for two representations $\chi$ and $\chi'$ to that of constructing a bilinear invariant functional for the representations $\chi$ and $-\chi'$. The results on the bilinear invariant functionals imply that an intertwining operator $A$ can exist only if $\varepsilon = \varepsilon'$ and

(i)   $\sigma = \sigma'$,
(ii)   $\sigma = -\sigma'$.

In the first case we get a trivial intertwining operator from (5–53)

$$Af(z) = cf(z) \tag{5-107}$$

In other words, any continuous operator intertwining a representation on a space $\mathcal{D}_\chi$ with itself is proportional to the unit operator. Since this assertion still holds if the representation is unitary and the intertwining operator is allowed to map $\mathcal{D}_\chi$ into its Hilbert space completion, the irreducibility of the representations of the principal and supplementary series follows. The discrete series of representations is realized only on part of the space $\mathcal{D}_\chi$ and its completion, and we cannot be sure a priori whether the operators commuting with the representation on $\mathcal{D}_\chi$ exhaust all commuting operators on the invariant subspaces. However, the following direct arguments prove irreducibility for the representations of the discrete series. We refer to realizations on Hilbert spaces of holomorphic (antiholomorphic) functions in the unit circle for which the powers $w^n(\overline{w}^n)$, $n = 0, 1, 2, \ldots$, are a basis (Section 5–8). A projection operator $P$ which commutes with all operators $T_a^{(k,\pm)}$ commutes in particular with the operators $T_u^{(k,\pm)}P$ is therefore diagonal on the powers $w^n(\overline{w}^n)$. We define a label $n_0$ by

$$n_0 = \min\{n \mid Pw^n = w^n(P\overline{w}^n = \overline{w}^n)\}$$

Applying $T_a^{(k,\pm)}$ to $w^{n_0}(\overline{w}^{n_0})$ (5–96) and expanding in a Taylor series around $w = 0$ shows that $n_0 = 0$ if it exists at all. The same comes out for $E - P$. Consequently $P$ is either the null or the unit operator, and the representations of both discrete series are irreducible.

In  case (ii) the intertwining operator is (5–52)

$$Af(z) = c \int\limits_{-\infty}^{+\infty} |z - z_1|^{-\sigma-1} [\text{sign}(z - z_1)]^\varepsilon f(z_1)\, dz_1$$

provided $\sigma \neq 0, 1, 2, \ldots$, and (5-54)

$$Af(z) = c\left(\frac{d}{dz}\right)^{\sigma} f(z)$$

for $\sigma = 0, 1, 2, \ldots$, but nonanalytic representations. The analytic representations are to be studied later in this section. With a standard formula of distribution theory we may unite these two expressions into one

$$Af(z) = \frac{1}{\Gamma(-\sigma)} \int\limits_{-\infty}^{+\infty} |z - z_1|^{-\sigma-1}[\text{sign}(z - z_1)]^{\epsilon} f(z_1)\, dz_1 \qquad (5\text{-}108)$$

by an appropriate normalization. The analogous intertwining operator which maps $\mathscr{D}_{-\chi}$ into $\mathscr{D}_{\chi}$ is

$$A'f(z) = \frac{1}{\Gamma(\sigma)} \int\limits_{-\infty}^{+\infty} |z - z_1|^{\sigma-1}[\text{sign}(z - z_1)]^{\epsilon} f(z_1)\, dz_1 \qquad (5\text{-}109)$$

The product $A'A$ intertwines $\chi$ with itself and is, according to (5-107), proportional to the unit operator

$$A'A = \gamma E \qquad (5\text{-}110)$$

To obtain the number $\gamma$ we compute the matrix elements of $A$ and $A'$ in the canonical basis.

Because $A$ intertwines the generator $H_2$ (Section 5-4), we have

$$Af_q^{\chi}(z) = \beta^q(\chi) f_q^{-\chi}(z) \qquad (5\text{-}111)$$

It is remarkable that $\beta^q(\chi)$ depends indeed on $q$ (see the proof for the independence of $\beta^j(\chi)$ of $q$ in Section 3-9, which was based on the existence of the operator $H_+$). With the elements $f_q(z)$ of the canonical basis (5-21) we can express the coefficient $\beta^q(\chi)$ as the matrix element in $\mathscr{L}^2_{\text{Re}\sigma}(Z)$

$$\beta^q(\chi) = \int\limits_{-\infty}^{+\infty} \overline{f_q^{-\chi}(z)} Af_q^{\chi}(z)(1 + z^2)^{\text{Re}\,\sigma}\, dz$$

Comparing this expression with (5-79), (5-80) we get for real $\sigma$

$$\beta^q(\chi) = [\Gamma(-\sigma)]^{-1} \tau_{qq}^{\epsilon}(\chi)$$

Using the analytic expressions (5–81), (5–82) for $\tau^\varepsilon_{qq}(\chi)$, this result can be continued analytically into the whole complex $\sigma$-plane. For $\varepsilon = 0$ (5–81) yields

$$\beta^q(\sigma, 0) = 2^{\sigma+1} \cos \frac{1}{2} \pi\sigma \, \frac{\Gamma((1/2) + (1/2)\sigma + q)}{\Gamma((1/2) - (1/2)\sigma + q)} \tag{5-112}$$

from which follows

$$\gamma = \beta^q(\sigma, 0)\beta^q(-\sigma, 0) = 4 \cos^2 \tfrac{1}{2}\pi\sigma$$

For $\varepsilon = 1$ we obtain from (5–82)

$$\beta^q(\sigma, 1) = 2^{\sigma+1}i \sin \frac{1}{2} \pi\sigma \, \frac{\Gamma((1/2) + (1/2)\sigma + q)}{\Gamma((1/2) - (1/2)\sigma + q)} \tag{5-113}$$

which leads to

$$\gamma = \beta^q(\sigma, 1)\beta^q(-\sigma, 1) = 4 \sin^2 \tfrac{1}{2}\pi\sigma$$

For nonanalytic representations the number $\gamma$ is different from zero, and the intertwining operator $A$ is one-to-one and bicontinuous. Consequently the representations $\chi$ and $-\chi$ are equivalent. Because of the uniqueness of Hermitian invariant functionals, we can conclude as in Section 3–7 that for unitary representations of the principal and supplementary series $A$ is isometric up to a constant factor, and that $\chi$ and $-\chi$ are equivalent in the standard sense. From the eigenvalues of the operators $T_u^{(k,\pm)}$ we see that the discrete series does not contain a pair of equivalent representations and no representation which is equivalent to any unitary representation of the principal or supplementary series.

To complete our discussion of intertwining operators we have still to study analytic representations in the case (ii). If the space $\mathscr{D}_\chi$ with $\sigma > 0$ carries an analytic representation, there exist two intertwining operators $A_\pm$ which map $\mathscr{D}_\chi$ into the space $\mathscr{D}_{-\chi}$. They can be read off (5–58) and (5–59)

$$A_\pm f(z) = \left[\left(\frac{d}{dz}\right)^\sigma f(z)\right]_\pm \tag{5-114}$$

Their kernels are $\mathscr{D}_\chi^\mp$ and their images $\mathscr{D}_{-\chi}^\pm$, respectively. The representations induced on the quotient spaces

$$\mathscr{D}_\chi/\mathscr{D}_\chi^\mp = \mathscr{D}_\chi^\pm/\mathscr{E}_\chi$$

are therefore equivalent to the representations on the spaces $\mathscr{D}_{-\chi}^\pm$. In particular the representations induced on

$$\mathscr{D}_\chi / \mathscr{E}_\chi = \mathscr{D}_\chi / \mathscr{D}_\chi^+ + \mathscr{D}_\chi / \mathscr{D}_\chi^- \qquad \text{(direct sum)}$$

are equivalent to the representations on the invariant subspaces $\mathscr{F}_{-\chi}$. These assertions, namely, the one-to-one and bicontinuous nature of the mappings of the quotient spaces onto the invariant subspaces, can easily be proved if we take into account that $A_\pm$ can be inverted by multiple integration

$$f_\pm^*(z) = \int\limits_{-\infty}^{z} dt_1 \int\limits_{-\infty}^{t_1} dt_2 \cdots \int\limits_{-\infty}^{t_{\sigma-1}} dt_\sigma \left[ \left( \frac{d}{dt_\sigma} \right)^\sigma f(t_\sigma) \right]_\pm$$

The functions $f_\pm^*(z)$ are defined in (5–65).

There exists also an intertwining operator mapping $\mathscr{D}_{-\chi}$ into $\mathscr{D}_\chi$, where $\chi$ is again analytic with $\sigma > 0$. It has the form (5–52)

$$A'f(z) = \int\limits_{-\infty}^{+\infty} (z - z_1)^{\sigma - 1} f(z_1) \, dz_1 \qquad (5\text{-}115)$$

and has the kernel $\mathscr{F}_{-\chi}$ and the image $\mathscr{E}_\chi$. Therefore we also find equivalence of the representation induced on the quotient space $\mathscr{D}_{-\chi} / \mathscr{F}_{-\chi}$ with the finite-dimensional representation on $\mathscr{E}_\chi$.

The matrix elements of the intertwining operators $A_\pm$ (5–114) and $A'$ (5–115) in the canonical basis (5–21) can be found easily by direct comp-utation. For analytic representations $\chi = \{\sigma, \varepsilon\}$ with $\sigma > 0$ we get

$$A_+ f_q^\chi(z) = \delta_+{}^q(\chi) f_q^{-\chi}(z) \qquad (5\text{-}116)$$

$$A_- f_q^\chi(z) = \delta_-{}^q(\chi) f_q^{-\chi}(z) \qquad (5\text{-}117)$$

$$A' f_q^{-\chi}(z) = \delta'^q(\chi) f_q^\chi(z) \qquad (5\text{-}118)$$

where

$$\delta_+{}^q(\chi) = (2i)^\sigma \sigma! \binom{\frac{1}{2}(\sigma - 1) + q}{\sigma} \qquad (5\text{-}119)$$

$$\delta_-^q(\chi) = (-2i)^\sigma \sigma! \binom{\frac{1}{2}(\sigma - 1) - q}{\sigma} \qquad (5\text{-}120)$$

$$\delta'^q(\chi) = (2i)^{-\sigma+1} (-1)^{(\sigma-1)/2 - q} \pi \binom{\sigma - 1}{\frac{1}{2}(\sigma - 1) - q} \qquad (5\text{-}121)$$

We verify easily that the coefficients (5–119), (5–120), (5–121) vanish on the invariant subspaces $\mathscr{D}_\chi^-$, $\mathscr{D}_\chi^+$, $\mathscr{F}_{-\chi}$, respectively.

## 5–10   MATRIX ELEMENTS OF THE GROUP OPERATORS IN THE DISCRETE SERIES

We consider the Hilbert space $\mathcal{H}(k, +)$ of functions $\Phi(w)$ which are holomorphic in the interior of the unit circle and square integrable at the boundary in the sense (5–97). We span the Hilbert space $\mathcal{H}(k, +)$ by the orthonormal basis (5–98). The operator $T_u^{(k,+)}$ is diagonal on this basis. Its eigenvalues can be obtained from (5–96) inserting (5–69)

$$\alpha = e^{(-i/2)\psi} \qquad \beta = 0$$

We get

$$T_{u(\psi)}^{(k,+)}\Phi_q(w) = e^{iq\psi}\Phi_q(w) \tag{5-122}$$

The matrix elements for a general group element $a \in SL(2, R)$, decomposed as in (5–1),

$$a = u(\psi_1)\, du(\psi_2)$$

are therefore obtained in this basis from

$$C_{q_1 q_2}^{(k,+)}(a) = \exp\{i(q_1\psi_1 + q_2\psi_2)\}c_{q_1 q_2}^{(k,+)}(\eta) \tag{5-123}$$

and

$$c_{q_1 q_2}^{(k,+)}(\eta) = \frac{2k-1}{\pi} \int\limits_{|w| < 1} \overline{\Phi_{q_1}(w)}T_d^{(k,+)}\Phi_{q_2}(w)$$

$$\times (1 - |w|^2)^{2k-2}\, Dw$$

With a little algebra we derive in this way the integral representation

$$c_{q_1 q_2}^{(k,+)}(\eta) = \left[\frac{(q_1 - k)!(k + q_2 - 1)!}{(q_2 - k)!(k + q_1 - 1)!}\right]^{1/2} \frac{1}{2\pi} \int\limits_0^{2\pi} d\psi \, \exp\{-i(q_1 + q_2)\psi\}$$

$$\times [e^{i\psi}\cosh \tfrac{1}{2}\eta - \sinh \tfrac{1}{2}\eta]^{-k+q_2}$$

$$\times [e^{-i\psi}\cosh \tfrac{1}{2}\eta - \sinh \tfrac{1}{2}\eta]^{-k-q_2} \tag{5-124}$$

It differs from the matrix element of $T_a^\chi$ in the canonical basis (5–24) only by the square root factor in front and by the fact that $\sigma$ has been replaced by $-2k + 1$. These functions $c_{q_1 q_2}^{(k,+)}(\eta)$ are real.

The representations $(k, -)$ are realized in Hilbert spaces $\mathcal{H}(k, -)$ of antiholomorphic functions $\Phi(\overline{w})$ in the interior of the unit circle which have finite norm (5—97). These functions transform as

$$T_a^{(k,-)}\Phi(\overline{w}) = (-\beta\overline{w} + \overline{\alpha})^{-2k}\Phi(\overline{w}_a) \qquad (5\text{-}125)$$

The functions

$$\Phi_q(\overline{w}) = \overline{\Phi_{-q}(w)} \qquad q = -k, -k-1, \ldots$$

where $\Phi_q(w)$ is as in (5—98), constitute an orthonormal basis in $\mathscr{H}(k, -)$. The operators $T_{u(\psi)}^{(k,-)}$ are diagonal on this basis with eigenvalues $e^{q\psi}$. The matrix elements of the operators $T_a^{(k,-)}$ can be deduced from the matrix elements of $T_a^{(k,+)}$ by

$$C_{q_1q_2}^{(k,-)}(a) = \overline{C_{-q_1,-q_2}^{(k,+)}(a)}$$

$$c_{q_1q_2}^{(k,-)}(\eta) = \overline{c_{-q_1,-q_2}^{(k,+)}(\eta)} = c_{-q_1,-q_2}^{(k,+)}(\eta) \qquad (5\text{-}126)$$

## 5—11   REMARKS

In Chapter 5 we have followed along the lines of Gelfand *et al.* [12 (Volume 5, Chapter VII)] when we deal with spaces of homogeneous functions and bilinear invariant functionals. This book contains the proof that the conditions on the bilinear invariant functionals are not only necessary but also sufficient. For further information on distributions, especially homogeneous distributions on the real line, we refer to Gelfand and Shilov [12 (Volume 1, Chapter I)]. Titchmarsh's textbook [32] contains all the facts on classical Fourier transformations on Banach spaces $\mathscr{L}^q(-\infty, \infty)$ needed in Section 5—4, see in particular his Theorem 74. An example of Fourier transformations on a group in a space $\mathscr{L}^q$, $q \neq 2$, has been investigated by Ehrenpreis and Mautner [9]. They study SL(2, R) and the space $\mathscr{L}^1$.

# Chapter 6

# Harmonic Analysis of Polynomially Bounded Functions on the Group SL(2, R)

Harmonic analysis on the group SL(2, R) is more complicated than on the group SL(2, C) due to the presence of the discrete series and its participation in the Plancherel formula. In order to formulate a theory of analytic functionals we proceed similarly as in Chapter 4 but with two essential changes. First we restrict our discussion to bicovariant functionals, they possess at most a finite number of contributions from the discrete series. Instead of Fourier transforms of the first kind (that means, built with functions of the first kind), we prefer moreover to use transforms of the second kind, which permit us to incorporate these terms of the discrete series into the principal series integral by a finite deformation of the integration contour. Such Fourier transforms of the second kind have a particular asymptotic behavior in the group invariant $\sigma$ or $J$, which is caused by the singularity of the representation functions of the second kind at the group unit. Since this property may strongly influence the convergence of the inverse Fourier transformation, we give some heuristic arguments on how to determine it.

In Sections 6–1 and 6–2 we define Fourier transforms as certain integral kernels for both the principal and the discrete series. In order to prepare the proof of the Plancherel theorem which we give in Section 6–3 we compute the characters of the group representations as in Section 4–1. Representation functions of the second kind are introduced in Section 6–4. In Section 6–4 we switch from the representation functions defined in Chapter 5 by means of the canonical basis to another type of functions which in the case of the principal and the discrete series differ from the former functions only by a sign factor. They are commonly used in physical literature and are closely related to the representation functions of the rotation group SU(2). We make this change mainly for the convenience of the physicists. In Section 6–5 we study formal aspects of bicovariant functions and their Fourier transforms. In Section 6–6 we introduce bicovariant distributions and analytic functionals. In order to gain a deeper insight into the asymptotic properties of Fourier transforms of the second kind mentioned earlier we study one example in detail.

## 6–1    REPRESENTATIONS OF THE GROUP ALGEBRA OF SL(2, R) AND THE FOURIER TRANSFORM IN THE PRINCIPAL SERIES

We construct representations of the group algebra $\mathscr{R}(A)$ of SL(2, R) [A is a shorthand for SL(2, R)] repeating essentially all the definitions and arguments given in connection with the analogous problem for the group SL(2, C) in Section 4–1. We consider a space $\mathscr{C}^\infty$ of functions $x(a)$ on SL(2, R) which have compact support and possess derivatives of all orders. $\mathscr{C}^\infty$ is dense in the Banach space of integrable functions on the group SL(2, R) which we denote $\mathscr{L}^1(A)$. Both $\mathscr{C}^\infty$ and $\mathscr{L}^1(A)$ can be made algebras, the latter a Banach algebra, if we define a multiplication operation as the convolution integral

$$x_1 \cdot x_2(a) = \int x_1(a_1) x_2(a_1^{-1} a) \, d\mu(a_1) \qquad (6\text{-}1)$$

The invariant measure $d\mu(a)$ on SL(2, R) was given in (1–23).

For any representation $\chi = \{\sigma, \varepsilon\}$ and for $x$ in $\mathscr{C}^\infty$ we define an operator $T_x^\chi$ in the Hilbert space $\mathscr{L}_\varepsilon{}^2(U)$ or $\mathscr{L}_{\sigma_1}{}^2(Z)$ (Section 5–3) by the integral

$$T_x^\chi = \int x(a) T_a^\chi \, d\mu(a) \qquad (6\text{-}2)$$

This operator is bounded, its norm can be estimated by means of (5–19)

$$\|T_x^\chi\| \leq N^{|\operatorname{Re}\sigma|} \int |x(a)| \, d\mu(a)$$

where $N$ is the maximum of $|a|$ on the support of $x$. These operators represent the algebra $\mathscr{C}^\infty$. We have in fact

$$T_{x_1}^\chi T_{x_2}^\chi = T_{x_1 \cdot x_2}^\chi$$

and continuity in the sense

$$\|T_x^\chi\| \to 0 \qquad \text{if} \quad x \to 0 \quad \text{in} \quad \mathscr{C}^\infty$$

In the case that $\chi$ belongs to the principal series, the representation of $\mathscr{C}^\infty$ can be extended to a representation of $\mathscr{L}^1(A)$, which is still continuous and moreover symmetric in the sense

$$T^\chi_{x\dagger} = (T^\chi_x)^\dagger$$

We use the notation

$$x^\dagger(a) = \overline{x(a^{-1})}$$

Next we pose the problem of finding an integral operator form for the abstract operator $T^\chi_x$ , $x \in \mathscr{C}^\infty$, such that

$$T^\chi_x\varphi(\psi) = \frac{1}{4\pi} \int_0^{4\pi} K_x(\psi, \psi' \mid \chi)\varphi(\psi') \, d\psi' \tag{6-3}$$

To find the kernel we proceed exactly as for the group SL(2, C). With

$$a = ku(\psi) \qquad d\mu(a) = \tfrac{1}{4} \, d\mu_l(k) \, d\psi$$

(see Appendix A-2a for the computation of the Jacobian), where $d\mu_l(k)$ is the left-invariant measure on the group K

$$d\mu_l(k) = (2\pi)^{-2} \, d\lambda \, d\mu \qquad -\infty < \lambda, \mu < \infty$$

we have

$$K_x(\psi_1, \psi_2 \mid \chi) = \pi \int x(u(\psi_1)^{-1} ku(\psi_2))$$

$$\times |\lambda|^{\sigma-1}[\text{sign } \lambda]^\varepsilon \, d\mu_l(k) \tag{6-4}$$

This kernel $K_x(\psi_1, \psi_2 \mid \chi)$ for $x \in \mathscr{C}^\infty$ has similar properties as the corresponding kernel for SL(2, C).

(1)  For fixed $\psi_1, \psi_2$, and $\varepsilon$ it is entire in $\sigma$.
(2)  It possesses derivatives of all orders with respect to $\psi_1, \psi_2$.
(3)  It is covariant as

$$K_x(\psi_1 + 2\pi, \psi_2 \mid \chi) = K_x(\psi_1, \psi_2 + 2\pi \mid \chi)$$

$$= (-1)^\varepsilon K_x(\psi_1, \psi_2 \mid \chi)$$

(4)  It represents the algebra $\mathscr{C}^\infty$ for all $\chi$,

$$K_{x_1 \cdot x_2}(\psi_1, \psi_3 \mid \chi) = \frac{1}{4\pi} \int_0^{4\pi} K_{x_1}(\psi_1, \psi_2 \mid \chi)K_{x_2}(\psi_2, \psi_3 \mid \chi) \, d\psi_2 \tag{6-5}$$

and is symmetric in the sense

$$K_{x\dagger}(\psi_1, \psi_2 \,|\, \chi) = \overline{K_x(\psi_2, \psi_1 \,|\, \chi)} \qquad (6\text{-}6)$$

if $\chi$ is in the principal series.

(5)  It is of Hilbert-Schmidt type

$$\|K_x(\chi)\|^2 = (4\pi)^{-2} \int\limits_0^{4\pi} d\psi_1 \int\limits_0^{4\pi} d\psi_2 \,|\, K_x(\psi_1, \psi_2 \,|\, \chi)|^2$$

$$< \infty \qquad (6\text{-}7)$$

Assertion (4) remains valid if we extend the definition of $K_x(\chi)$ to functions $x \in \mathcal{L}^1(A)$ and restrict $\chi$ to the principal series. We may regard the operator $T_x{}^\chi$ or this kernel $K_x(\chi)$ as the Fourier transform of the function $x(a)$. Contrary to the complex group SL(2, C), the inverse Fourier transformation involves the discrete series in addition to the principal series. The Fourier transform in the discrete series is introduced in Section 6–2 as an integral kernel of a different type.

The proof of the Plancherel theorem which we present in Section 6–3 makes use of the notion of characters just as in Section 4–2. For arbitrary $\chi$ and $x \in \mathscr{C}^\infty$ the quantity

$$\mathrm{Tr}(T_x{}^\chi) = \frac{1}{4\pi} \int\limits_0^{4\pi} K_x(\psi, \psi \,|\, \chi)\, d\psi \qquad (6\text{-}8)$$

exists and is denoted the trace of the operator $T_x{}^\chi$ . With the notation

$$x_1(a) = x \cdot x^\dagger(a)$$

we can, for example, write the Hilbert-Schmidt norm (6–7) as

$$\|K_x(\chi)\|^2 = \mathrm{Tr}(T_{x_1}^\chi)$$

It is easy to verify that the trace (6–8) is identical to the usual trace in the canonical basis (5–14)

$$\sum_{q=-\infty}^{\infty} \int C_{qq}^\chi(a) x(a)\, d\mu(a) = \sum_{q=-\infty}^{\infty} \frac{1}{4\pi} \int\limits_0^{4\pi} \overline{\varphi_q(\psi)}\, T_x{}^\chi \varphi_q(\psi)\, d\psi$$

$$= \sum_{q=-\infty}^{\infty} (4\pi)^{-2} \int_0^{4\pi} d\psi_1 \int_0^{4\pi} d\psi_2 \, \overline{\varphi_q(\psi_1)}$$

$$\times K_x(\psi_1, \psi_2 \,|\, \chi)\varphi_q(\psi_2)$$

$$= \frac{1}{4\pi} \int_0^{4\pi} d\psi K_x(\psi, \psi \,|\, \chi)$$

However, due to the properties of the kernel $K\ (\chi)$ listed earlier, this trace exists not only for all operators $T_x^\chi$, $x \in \mathscr{C}^\infty$, but is moreover linear and continuous in $x$. In other words, the trace of $T_x^\chi$ is a linear continuous functional on $\mathscr{C}^\infty$ and is therefore generated by a distribution $\xi^\chi(a)$,

$$\mathrm{Tr}(T_x^\chi) = \int x(a)\xi^\chi(a) \, d\mu(a) \tag{6-9}$$

We can identify the distribution $\xi^\chi(a)$ with the trace of the operator $T_a^\chi$

$$\xi^\chi(a) = \mathrm{Tr}(T_a^\chi) = \sum_{q=-\infty}^{\infty} C_{qq}^\chi(a) \tag{6-10}$$

where the sum over $q$ converges in a distribution sense. We call $\xi^\chi(a)$ the character of the representation $\chi$.

To compute this character we return to the definition (6–4)

$$\mathrm{Tr}(T_x^\chi) = \frac{1}{4\pi} \int_0^{4\pi} K_x(\psi, \psi \,|\, \chi) \, d\psi$$

$$= \tfrac{1}{4} \int x(u(\psi)^{-1}ku(\psi)) \,|\lambda|^{\sigma-1}$$

$$\times [\mathrm{sign}\ \lambda]^\varepsilon \, d\mu_l(k) \, d\psi$$

We put

$$u(\psi)^{-1}ku(\psi) = a \tag{6-11}$$

such that

$$|\mathrm{Tr}\ a| \geqq 2$$

In turn, if for an element $a \in \mathrm{SL}(2,\ R)\,|\,\mathrm{Tr}\ (a)| > 2$, we can decompose it as in (6–11) with

$$\lambda = \tfrac{1}{2}\{\mathrm{Tr}\ a \pm [(\mathrm{Tr}\ a)^2 - 4]^{1/2}\} \tag{6-12}$$

The Jacobian for (6–11) (Appendix A–2b) is

$$\tfrac{1}{2}|\lambda^{-1}|\,d\mu_l(k)\,d\psi = |\lambda - \lambda^{-1}|^{-1}\,\theta(|\mathrm{Tr}\,a| - 2)\,d\mu(a)$$

where $\psi$ is restricted to the interval $0 \le \psi \le 2\pi$, and $\lambda$ either to $1 \le |\lambda| < \infty$ or to $0 < |\lambda| \le 1$. This implies

$$\mathrm{Tr}(T_x^{\,x}) = \int x(a)\,\frac{|\lambda|^{\sigma} + |\lambda|^{-\sigma}}{|\lambda - \lambda^{-1}|}\,[\mathrm{sign}\,\lambda]^{\varepsilon}$$

$$\times\,\theta(|\mathrm{Tr}\,a| - 2)\,d\mu(a)$$

or finally with (6–9)

$$\xi^x(a) = \frac{|\lambda|^{\sigma} + |\lambda|^{-\sigma}}{|\lambda - \lambda^{-1}|}\,[\mathrm{sign}\,\lambda]^{\varepsilon}\,\theta(|\mathrm{Tr}\,a| - 2) \qquad (6\text{-}13)$$

where $\lambda$ is defined in terms of $a$ by (6–12).

As in the case of the group SL(2, C) we can also define a kernel $K_x(\chi)$ for the realization $\mathscr{L}_{\sigma_1}^2(Z)$. We make use of the decomposition

$$a = k\zeta \qquad d\mu(a) = d\mu_l(k)\,dz$$

(Appendix A–2c). This gives

$$K_x(z_1, z_2\,|\,\chi) = \int x(\zeta_1^{-1}k\zeta_2)\,|\lambda|^{\sigma - 1}[\mathrm{sign}\,\lambda]^{\varepsilon}\,d\mu_l(\kappa) \qquad (6\text{-}14)$$

Finally we mention that the kernel $K_x(z_1, z_2\,|\,\chi)$ satisfies a symmetry relation

$$\int_{-\infty}^{+\infty} K_x(z_1, z_2\,|\,-\chi)A(z_2, z_3\,|\,\chi)\,dz_2 = \int_{-\infty}^{+\infty} A(z_1, z_2\,|\,\chi)K_x(z_2, z_3\,|\,\chi)\,dz_2 \qquad (6\text{-}15)$$

where $x$ is in $\mathscr{C}^{\infty}$, $\chi$ is a nonanalytic representation, and the intertwining kernel can be read off (5–108).

## 6–2 FOURIER TRANSFORMS IN THE DISCRETE SERIES

In the course of deriving the discrete series by induction from the representations of the one-dimensional subgroup of unitary matrices $u$ in Section 5–8, we had to project the space of functions square integrable in the unit circle onto the Hilbert spaces $\mathscr{H}(k, \pm)$ of holomorphic or antiholomorphic functions. A similar projection occurs when we define Fourier transforms as integral operators in the spaces $\mathscr{H}(k, \pm)$.

We construct Fourier transforms in the discrete series similarly as for the principal series (Section 6–1). From the outset we restrict ourselves to the cases $k \geq 1$. In the Hilbert space $\mathscr{H}(k, \pm)$ of holomorphic functions $\Phi(w)$ we define the operator $T^{(k,+)}, x \in \mathscr{C}^\infty$, by

$$T_x^{(k, +)}\Phi(w) = \int x(a)T_a^{(k, +)}\Phi(w) \, d\mu(a) \tag{6-16}$$

Its norm in $\mathscr{H}(k, \pm)$ is finite

$$\|T_x^{(k, +)}\| \leq \int |x(a)| \, d\mu(a)$$

As in the case of the principal series we look for an integral operator form of $T_x^{(k, +)}$

$$T_x^{(k, +)}\Phi(w_1) = \int_{|w_2| < 1} K_x(w_1, w_2 \,|\, k, +)\Phi(w_2)$$
$$\times (1 - |w_2|^2)^{2k-2}Dw_2 \tag{6-17}$$

Since this integral operator need only be applied to holomorphic functions, it is not uniquely determined by the requirement (6–17). There exists, however, a unique kernel $K_x'(w_1, w_2 \,|\, k, +)$ for the corresponding operators $T_x'^{(k,+)}$ on the Hilbert space $\mathscr{H}'$ of square integrable, not necessarily holomorphic functions $\Phi'(w)$. If $P$ is the orthogonal projection we have

$$\mathscr{H}(k, +) = P\mathscr{H}'$$

we have of course

$$T_x^{(k, +)} = PT_x'^{(k, +)} = T_x'^{(k, +)}P$$

The ambiguity of the kernel $K_x(w_1, w_2 \,|\, k, +)$ in the space $\mathscr{H}(k, +)$ can therefore be removed if we project $K_x'(w_1, w_2 \,|\, k, +)$ on $\mathscr{H}(k, +)$ by means of the Cauchy kernel (5–100),

$$K_x(w_1, w_2 \,|\, k, +)$$
$$= \frac{2k - 1}{\pi} \int_{|w| < 1} K_x'(w_1, w \,|\, k, +)G(w, w_2 \,|\, k, +)$$
$$\times (1 - |w|^2)^{2k-2}Dw$$

$$= \frac{2k-1}{\pi} \int\limits_{|w| < 1} G(w_1, w \,|\, k, +) K'_x(w, w_2 \,|\, k, +)$$

$$\times (1 - |w|^2)^{2k-2} Dw \qquad\qquad (6\text{-}18)$$

The equality of the two terms on the right-hand side of (6–18) can be verified easily by direct computation. We regard either the abstract operator $T_x^{(k,+)}$ or the kernel $K_x(w_1, w_2 \,|\, k, +)$ defined by (6–18) as the Fourier transform of $x \in \mathscr{C}^\infty$.

In order to derive the kernel $K'_x(k, +)$ we consider symmetric matrices $s(w)$ (5–86) and put

$$a = u(\psi) s(w)$$

The Jacobian (Appendix A–2d) is

$$d\mu(a) = (2\pi)^{-2} (1 - |w|^2)^{-2} \, Dw \, d\psi \qquad\qquad (6\text{-}19)$$

Using (5–88), (5–91), (5–96), and (6–19) we obtain

$$T_x'^{(k,\,+)} \Phi'(w_1) = \int x(a) T_a^{(k,\,+)} \Phi'(w_1) \, d\mu(a)$$

$$= \int x(s(w_1)^{-1} a) T_{s(w_1)^{-1} a}^{(k,\,+)} \Phi'(w_1) \, d\mu(a)$$

$$= \int x(s(w_1)^{-1} a)(1 - |w_1|^2)^{-k} T_a^{(k,\,+)} \Phi'(0) \, d\mu(a)$$

$$= (2\pi)^{-2} \int x(s(w_1)^{-1} a s(w_2))(1 - |w_1|^2)^{-k}$$

$$\times (1 - |w_2|^2)^{k-2} e^{ik\psi} \Phi'(w_2) \, d\psi \, dw_2$$

and

$$K'_x(w_1, w_2 \,|\, k, +) = (2\pi)^{-2} (1 - |w_1|^2)^{-k} (1 - |w_2|^2)^{-k}$$

$$\times \int\limits_0^{4\pi} x(s(w_1)^{-1} u(\psi) s(w_2)) e^{ik\psi} \, d\psi \qquad\qquad (6\text{-}20)$$

The kernels $K_x'(k, +)$ establish a symmetric continuous representation of the algebra $\mathscr{C}^\infty$ in the Hilbert space $\mathscr{H}'$. We turn our attention now to the kernels $K_x(k, +)$.

From (6–18) we see that the kernel $K_x(k, +)$ can also be obtained directly from

$$K_x(w_1, w_2 \mid k, +) = \frac{2k-1}{\pi} \int x(a) T_a^{(k, +)}$$

$$\times G(w_1, w_2 \mid k, +) \, d\mu(a) \qquad (6\text{-}21)$$

where $T_a^{(k,+)}$ applies to the variable $w_1$ of the Cauchy kernel. The transformed Cauchy kernel is

$$T_a^{(k, +)} G(w_1, w_2 \mid k, +) = (-\bar\beta w_1 + \alpha)^{-2k} \left(1 - \frac{\bar\alpha w_1 - \beta}{-\bar\beta w_1 + \alpha} \bar w_2\right)^{-2k}$$

$$= (\alpha + \beta \bar w_2 - \bar\beta w_1 - \bar\alpha w_1 \bar w_2)^{-2k} \qquad (6\text{-}22)$$

We list the most important properties of the kernel $K_x(w_1, w_2 \mid k, +)$, $x \in \mathscr{C}^\infty$.

(1)  It is holomorphic in the variable $w_1$ in a circle $|w_1| < 1 + \varepsilon(|w_2|, N)$ provided $|w_2| < 1$, and antiholomorphic in the variable $w_2$ in a circle $|w_2| < 1 + \varepsilon'(|w_1|, N)$ provided $|w_1| < 1$. $\varepsilon$ and $\varepsilon'$ are positive numbers which shrink to zero if $w_2$, respectively $w_1$ approaches the boundary of the unit circle or if $N$ tends to infinity. $N$ is as usual the maximum of $|a|$ on the support of $x(a)$.

(2)  $K_x(w, w \mid k, +)$ is infinitely differentiable in the unit circle and on its boundary.

(3)  The kernels represent the algebra $\mathscr{C}^\infty$ symmetrically and continuously.

(4)  They are of Hilbert-Schmidt type

$$\| K_x(k, +) \|^2 = \int |K_x(w_1, w_2 \mid k, +)|^2 (1 - |w_1|^2)^{2k-2}$$

$$\times (1 - |w_2|^2)^{2k-2} \, Dw_1 \, Dw_2 < \infty \qquad (6\text{-}23)$$

In order to verify (6–23) we denote

$$x_1(a) = x \cdot x^\dagger(a)$$

and use property (3) to rewrite it as

$$\| K_x(k, +) \|^2 = \int K_{x_1}(w, w \mid k, +)(1 - |w^2|)^{2k-2} \, Dw \qquad (6\text{-}24)$$

We introduce the trace of the operator $T_x^{(k,+)}$ as usual by

$$\mathrm{Tr}(T_x^{(k,+)}) = \int K_x(w, w \mid k, +)(1 - |w|^2)^{2k-2} \, Dw \qquad (6\text{-}25)$$

This trace exists because of item (2) of our list. It remains therefore to prove only the assertion (2). With (6–21) and (6–22) it can directly be reduced to the following lemma, which in turn is easily established.

Let $x(\zeta, \varphi)$ be a function which is infinitely differentiable in $\zeta$ and $\varphi$ in the closed domain $-1 \leq \zeta \leq 1$, $-\infty < \varphi < \infty$, with the period $2\pi$ in $\varphi$. Then the function

$$K(t) = \int\limits_{-1}^{1} d\zeta \int\limits_{0}^{2\pi} d\varphi x(\zeta, \varphi)(1 - 2\zeta t e^{i\varphi} + t^2 e^{2i\varphi})^{-2k}$$

is infinitely differentiable in the interval $-1 \leq t \leq 1$ for any integer or half-integer $k$. To prove this lemma we expand $K(t)$ in a series

$$K(t) = \sum\limits_{n=0}^{\infty} a_n t^n \qquad |t| < 1$$

with

$$a_n = \int\limits_{-1}^{1} d\zeta C_n^{2k}(\zeta) \int\limits_{0}^{2\pi} d\varphi e^{in\varphi} x(\zeta, \varphi)$$

where $C_n^{2k}(\zeta)$ is a Gegenbauer polynomial (GR 8.93). Since $(k > 0)$

$$\max_{-1 \leq \zeta \leq 1} |C_n^{2k}(\zeta)| = C_n^{2k}(1) = \binom{4k + n - 1}{n}$$

and

$$\left| \int\limits_{0}^{2\pi} d\varphi e^{in\varphi} x(\zeta, \varphi) \right| \leq 2\pi n^{-q} \max_{\varphi} \left| \left( \frac{\partial}{\partial \varphi} \right)^q x(\zeta, \varphi) \right|, \qquad n > 0$$

for all $q = 0, 1, 2, \ldots$, it follows $(k > 0, n > 0)$

$$|a_n| \leq 2\pi C_n^{2k}(1) n^{-q} \max_{\zeta, \varphi} \left| \left( \frac{\partial}{\partial \varphi} \right)^q x(\zeta, \varphi) \right|$$

This completes the proof of the lemma.

The trace of $T_x^{(k,+)}$ is easily seen to be identical with the sum over the diagonal elements of $T_x^{(k,+)}$ in the basis (5–98)

$$\text{Tr}(T_x^{(k,+)}) = \int K_x(w, w \mid k, +)(1 - |w|^2)^{2k-2} Dw$$

$$= \frac{2k - 1}{\pi} \int (1 - |w|^2)^{2k-2} Dw \int d\mu(a) x(a)$$

$$\times [T_a^{(k,+)} G(w, w' \mid k, +)]_{w'=w}$$

$$= \sum_{q=k}^{\infty} \frac{2k - 1}{\pi} \int \overline{\Phi_q(w)} T_x^{(k,+)} \Phi_q(w)$$

$$\times (1 - |w|^2)^{2k-2} Dw$$

$$= \sum_{q=k}^{\infty} \langle q | T_x^{(k,+)} | q \rangle \qquad (6\text{-}26)$$

It is moreover linear in $x$ and continuous on $\mathscr{C}^\infty$. There exists therefore a distribution $\xi^{(k,+)}(a)$ with the property

$$\text{Tr}(T_x^{(k,+)}) = \int x(a) \xi^{(k,+)}(a)\, d\mu(a) \qquad (6\text{-}27)$$

We call $\xi^{(k,+)}(a)$ the character of the representation $(k, +)$. The character can be obtained, in principle, from the sum

$$\xi^{(k,+)}(a) = \sum_{q=k}^{\infty} C_{qq}^{(k,+)}(a)$$

However, since this sum converges only in the sense of the topology of the distribution space, this equation is not yet of great practical value. Nevertheless, explicit computation will show that this sum can be regularized by

$$\xi^{(k,+)}(a) = \lim_{\zeta \to 1+} \sum_{q=k}^{\infty} \zeta^{-q} C_{qq}^{(k,+)}(a)$$

$$= \lim_{\zeta \to 1+} \frac{2k - 1}{\pi} \int_{|w| < 1} [\zeta(\alpha - \bar\beta w) + (\beta - \bar\alpha w)\bar w]^{-2k}$$

$$\times (1 - |w|^2)^{2k-2} Dw \qquad (6\text{-}28)$$

such that the sum, the integral, and the limit exist. It can be verified easily that the character satisfies

$$\xi^{(k,\,+)}(a_1^{-1}aa_1) = \xi^{(k,\,+)}(a)$$

for arbitrary $a_1$. We need therefore compute it only for one representative of each equivalence class of SL(2, R) (Section 5–1).

For $|\text{Tr } a| < 2$ we take $a = u(\psi)$. We obtain from (6–28)

$$\xi^{(k,\,+)}(u(\psi)) = \lim_{\zeta \to 1+} \sum_{q=k}^{\infty} \zeta^{-q} e^{iq\psi}$$

$$= \frac{\exp\{i\psi(k - 1/2)\}}{-2i \sin (1/2)\psi} \tag{6-29}$$

For $\text{Tr } a > 2$ we choose the element $a = d$ (5–2). It yields the integral (6–28)

$$\xi^{(k,\,+)}(d) = \lim_{\zeta \to 1+} \frac{2k - 1}{\pi} \int_{|w| < 1} Dw(1 - |w|^2)^{2k-2}$$

$$\times [(\zeta - |w|^2)\cosh \tfrac{1}{2}\eta + (\bar{w} - w\zeta)\sinh \tfrac{1}{2}\eta]^{-2k}$$

whose evaluation is a bit more complicated. The imaginary part of the square bracket is nonzero, and the integral is therefore finite for all $\zeta > 1$. Changing the integration variable into

$$w = \frac{ire^{i\varphi} + 1}{ire^{i\varphi} - 1}$$

and introducing the notations

$$\xi = \frac{\zeta + 1}{\zeta - 1}$$

$$\alpha(\varphi) = \sin \varphi \cosh \tfrac{1}{2}\eta + i \cos \varphi \sinh \tfrac{1}{2}\eta$$

gives

$$\xi^{(k,\,+)}(d) = \lim_{\xi \to \infty} (2\xi)^{2k} \frac{2k - 1}{2\pi} \int_0^{\pi} d\varphi (\sin \varphi)^{2k-2}$$

$$\times \int_0^1 dr \, r^{2k-1} [r^2 + 2r\xi\alpha(\varphi) + 1]^{-2k}$$

In the square bracket we can neglect $r^2$. Integrating by parts yields finally

$$\xi^{(k,\,+)}(d) = \frac{\exp\{-\eta(k-1/2)\}}{2\sinh(1/2)\eta} \tag{6-30}$$

Together we have

$$\xi^{(k,\,+)}(a) = \frac{\exp\{-\eta(k-1/2)\}}{2\sinh(1/2)\eta}\left[\theta(\mathrm{Tr}\,a - 2) + (-1)^{2k}\right.$$

$$\left. \times\ \theta(-\mathrm{Tr}\,a - 2)\right]$$

$$+ \frac{\exp\{i\psi(k-1/2)\}}{-2i\sin(1/2)\psi}\,\theta(2 - |\mathrm{Tr}\,a|) \tag{6-31}$$

where $a$ has to be brought into the form

$$a = a_1 e_{\pm}\, da_1^{-1}$$

if $|\mathrm{Tr}\,a| > 2$, and into the form

$$a = a_1 u(\psi)a_1^{-1}$$

if $|\mathrm{Tr}\,a| < 2$. In fact, one can prove that the behavior of the character at $|\mathrm{Tr}\,a| = 2$ is correctly described by the expression (6–31). Since

$$\xi^{(k,\,-)}(a) = \overline{\xi^{(k,\,+)}(a)} \tag{6-32}$$

we get also

$$\xi^{(k,\,+)}(a) + \xi^{(k,\,-)}(a)$$

$$= \frac{\exp\{-\eta(k-1/2)\}}{\sinh(1/2)\eta}\left[\theta(\mathrm{Tr}\,a - 2) + (-1)^{2k}\,\theta(-\mathrm{Tr}\,a - 2)\right]$$

$$- \frac{\sin\psi(k-1/2)}{\sin(1/2)\psi}\,\theta\,(2 - |\mathrm{Tr}\,a|) \tag{6-33}$$

## 6–3   THE PLANCHEREL THEOREM

Let $x$ be a function of $\mathscr{C}^\infty$, $Kx(\psi_1, \psi_2 | \chi)$ be the Fourier transform (6–4) of $x$ in the principal series, $\chi = \{\sigma, \varepsilon\}$, $\sigma = i\rho$, $\rho$ real, and let $K_x(w_1, w_2 | k, \pm)$ be the Fourier transform (6–21) of $x$ in the discrete series with $k \geq 1$. An inverse

Fourier transformation allows us to express the function $x(a)$ by the kernels $K_x$ in the following fashion

$$x(a) = \sum_{\varepsilon = 0, 1} \tfrac{1}{2} \int\limits_{-\infty}^{+\infty} d\rho \zeta_\varepsilon(\rho)$$

$$\times \frac{1}{4\pi} \int\limits_0^{4\pi} K_x(\psi, \psi_a | \chi) a^{-\chi}(\psi, a)\, d\psi$$

$$+ \sum_{n = 1, 3/2, \ldots}^\infty (2n - 1) \int\limits_{|w| < 1} Dw(1 - |w|^2)^{2n - 2}$$

$$\times \{ K_x(w, w_a | n, +)(-\beta \bar{w} + \bar{\alpha})^{-2n}$$

$$+ K_x(w, w_a | n, -)(-\bar{\beta} w + \alpha)^{-2n} \} \qquad (6\text{-}34)$$

where the weight functions $\zeta_\varepsilon(\rho)$ are

$$\zeta_0(\rho) = \tfrac{1}{2}\rho \tanh \tfrac{1}{2}\pi\rho \qquad \zeta_1(\rho) = \tfrac{1}{2}\rho \coth \tfrac{1}{2}\pi\rho \qquad (6\text{-}35)$$

and the multiplier $\alpha^{-\chi}(\psi, a)$ for the representation $-\chi$ was defined in (5–12). We emphasize that in (6–34) the contributions of the principal series are counted twice, since $\rho$ and $-\rho$ describe equivalent representations (Section 5–9).

Before we set out to prove formula (6–34) we quote two other equivalent formulations of it. A form of basic theoretic interest can be obtained from (6–34) if we take into account the behavior of the kernels $K_x$ under right translations. With

$$x_1(a_1) = T_a^r x(a_1) = x(a_1 a)$$

we find

$$K_{x_1}(\psi_1, \psi_2 | \chi) = K_x(\psi_1, (\psi_2)_a | \chi) \alpha^{-\chi}(\psi_2, a)$$

$$K_{x_1}(w_1, w_2 | n, +) = K_x(w_1, (w_2)_a | n, +)(-\beta \bar{w}_2 + \bar{\alpha})^{-2n} \qquad (6\text{-}36)$$

$$K_{x_1}(w_1, w_2 | n, -) = K_x(w_1, (w_2)_a | n, -)(-\bar{\beta} w_2 + \alpha)^{-2n}$$

Since

$$x(a) = T_a^r x(e)$$

we arrive at the formula

$$x(e) = \sum_{\varepsilon=0,1} \tfrac{1}{2} \int_{-\infty}^{+\infty} d\rho \zeta_\varepsilon(\rho) \mathrm{Tr}(T_x^\chi)$$

$$+ \sum_{n=1,3/2,\ldots}^{\infty} (2n-1)[\mathrm{Tr}(T_x^{(n,\,+)}) + \mathrm{Tr}(T_x^{(n,\,-)})] \qquad (6\text{-}37)$$

which will actually be proved later on. The original formula (6–34) can be regained from (6–37) by applying a right translation to $x$. If we set

$$x_1(a) = x \cdot x^\dagger(a)$$

Parseval's formula is obtained [recall (6–23)–(6–25)]

$$\int |x(a)|^2 \, d\mu(a) = \sum_{\varepsilon=0,1} \tfrac{1}{2} \int_{-\infty}^{+\infty} d\rho \zeta_\varepsilon(\rho) \, ||| K_x(\chi) |||^2$$

$$+ \sum_{n=1,3/2,\ldots}^{\infty} (2n-1)[||| K_x(k,\,+) |||^2 + ||| K_x(k,\,-) |||^2] \qquad (6\text{-}38)$$

Another form of the relation (6–34) of some practical interest involves matrix elements of the operators $T_x^\chi$ and $T_a^\chi$, respectively $T_x^{(k,\,\pm)}$ and $T_a^{(k,\,\pm)}$. We start from (6–37), replace the traces by the sums over diagonal matrix elements, and exert a translation on $x$. This yields

$$x(a) = \sum_{\varepsilon=0,1} \tfrac{1}{2} \int_{-\infty}^{+\infty} d\rho \zeta_\varepsilon(\rho) \sum_{q_1,q_2} \overline{C_{q_1 q_2}^\chi(a)}$$

$$\times \int x(a_1) C_{q_1 q_2}^\chi(a_1) \, d\mu(a_1)$$

$$+ \sum_{n=1,3/2,\ldots}^{\infty} (2n-1) \left\{ \sum_{q_1,q_2 \geq n}^{\infty} \overline{C_{q_1 q_2}^{(n,\,+)}(a)} \right.$$

$$\times \int x(a_1) C_{q_1 q_2}^{(n,\,+)}(a_1) \, d\mu(a_1)$$

$$+ \left. \sum_{q_1 q_2 \leq -n}^{-\infty} \overline{C_{q_1 q_2}^{(n,\,-)}(a)} \int x(a_n) C_{q_1 q_2}^{(n,\,-)}(a_1) \, d\mu(a_1) \right\} \qquad (6\text{-}39)$$

As a preliminary to the proof of formula (6–37) we note that the function

$x(a) \epsilon \; \mathscr{C}^{\infty}$ can be specialized in several respects without restricting the generality of the proof. First we can expand any function $x \epsilon \mathscr{C}^{\infty}$ into a series

$$x(a) = \sum_{q=-\infty}^{\infty} x_q(a)$$

with integral $q$, which converges in the sense of $\mathscr{C}^{\infty}$, such that each term $x_q(a)$ is in $\mathscr{C}^{\infty}$ and obeys

$$x_q(u(\psi)au(\psi)^{-1}) = e^{iq\psi}x_q(a)$$

This series can be obtained by expanding $x(u(\psi)au(\psi)^{-1})$ into a Fourier series in $\psi$ and setting $\psi = 0$. Since

$$x(e) = x_0(e)$$
$$\mathrm{Tr}(T_x^{\chi}) = \mathrm{Tr}(T_{x_0}^{\chi})$$
$$\mathrm{Tr}(T_x^{(n,\,\pm)}) = \mathrm{Tr}(T_{x_0}^{(n,\,\pm)})$$

we can a priori assume that $x(a)$ is invariant in the sense

$$x(u(\psi)au(\psi)^{-1}) = x(a) \tag{6-40}$$

Similarly we can decompose $x(a)$ into two terms

$$x(a) = x_s(a) + x_{as}(a)$$

where

$$x_s(a) = \tfrac{1}{2}[x(a) + x(a^{-1})]$$

We have

$$x(e) = x_s(e)$$
$$\mathrm{Tr}(T_x^{\chi}) = \mathrm{Tr}(T_{x_s}^{\chi})$$
$$\mathrm{Tr}(T_x^{(n,\,+)}) + \mathrm{Tr}(T_x^{(n,\,-)}) = \mathrm{Tr}(T_{x_s}^{(n,\,+)}) + \mathrm{Tr}(T_{x_s}^{(n,\,-)})$$

We shall therefore assume from the outset that $x(a)$ is symmetric in the sense

$$x(a) = x(a^{-1})$$

Finally we can decompose $x(a)$ into two terms

$$x(a) = x_+(a) + x_-(a)$$

such that

$$x_\pm(e_- a) = \pm x_\pm(a)$$

Since the proof for the part $x_-$ is analogous to that for $x_+$, we shall prove the inversion formula (6–37) only for functions $x(a)$ which satisfy

$$x(e_- a) = x(a)$$

Under the expression (6–8) for the trace of $T_x^\chi$, and taking into account (6–4), (6–40), and the other restrictions on $x(a)$, we can bring the contribution of the principal series to (6–37) into the form

$$\frac{1}{2} \int\limits_{-\infty}^{+\infty} d\rho \tfrac{1}{2}\rho \tanh \tfrac{1}{2}\pi\rho \ \pi \int x(k) |\lambda|^{i\rho - 1} \, d\mu_i(k)$$

Since the function

$$\int\limits_{-\infty}^{+\infty} x(k) \, d\mu \qquad k = \begin{pmatrix} \lambda^{-1} & \mu \\ 0 & \lambda \end{pmatrix}$$

is symmetric under both the replacements

$$\lambda \to -\lambda \quad \text{and} \quad \lambda \to \lambda^{-1}$$

we may set

$$\lambda = e^t \qquad t \geq 0$$

This yields for the contribution of the principal series

$$\frac{1}{4\pi} \int\limits_{-\infty}^{+\infty} d\rho\rho \tanh \tfrac{1}{2}\pi\rho \int\limits_{0}^{\infty} dt \cos \rho t \int\limits_{-\infty}^{+\infty} x(k) \, d\mu$$

The integration over $\rho$ can be performed (*GR* 4.114.2) with the result

$$-\frac{1}{2\pi} \int\limits_{0}^{\infty} dt \, \frac{1}{\sinh t} \frac{d}{dt} \int\limits_{-\infty}^{+\infty} x(k) \, d\mu \qquad \lambda = e^t \tag{6-41}$$

The existence of the double integral (6–41) is not questionable because the function

$$\int\limits_{-\infty}^{+\infty} x(k) \, d\mu$$

is even, infinitely differentiable, and of compact support in $t$.

Next we consider the terms

$$(2n - 1)\frac{1}{\pi}\int_0^\infty dt\, e^{-t(2n-1)} \int_{-\infty}^{+\infty} x(k)\, d\mu \quad \lambda = e^t$$

These result from the contribution of the discrete series to (6–37) if we restrict the character (6–33) to $|\mathrm{Tr}\, a| > 2$ and take account of (6–11), (6–40), and the other restrictions on $x(a)$, and of the Jacobian of Appendix A–2b. Integrating by parts gives

$$-\frac{1}{\pi}\int_0^\infty dt\left[\frac{d}{dt}\, e^{-t(2n-1)}\right] \int_{-\infty}^{+\infty} x(k)\, d\mu$$

$$= \frac{1}{\pi}\int_{-\infty}^{+\infty} x(k)_{\lambda=1}\, d\mu + \frac{1}{\pi}\int_0^\infty dt\, e^{-t(2n-1)}$$

$$\times \frac{d}{dt}\int_{-\infty}^{+\infty} x(k)\, d\mu$$

We sum the second terms over $n$ and get

$$+\frac{1}{2\pi}\int_0^\infty dt\, \frac{1}{\sinh t}\frac{d}{dt}\int_{-\infty}^{+\infty} x(k)\, d\mu$$

This expression cancels the contribution (6–41) of the principal series. We are therefore left with the problem of proving the formula

$$x(e) = \sum_{n=1, 2, \ldots}^\infty \left[\frac{1}{\pi}\int_{-\infty}^{+\infty} x(k)_{\lambda=1}\, d\mu\right.$$

$$\left. -(2n-1)\int x(a)\, \frac{\sin\psi(n-1/2)}{\sin(1/2)\psi}\, \theta(2 - |\mathrm{Tr}\, a|)\, d\mu(a)\right] \quad (6\text{-}42)$$

where

$$a = ku(\psi)k^{-1}$$

The corresponding decomposition of the measure $d\mu(a)$ (Appendix A–2e) is

$$d\mu(a) = (2\pi)^{-2} \sin^2 \tfrac{1}{2}\psi \, d(\lambda^2) \, d\kappa \, d\psi$$
$$\mu = \kappa\lambda \qquad \lambda > 0$$
$$-\infty < \kappa < \infty \qquad 0 \leq \psi \leq 4\pi$$

With these variables and the notation

$$G(\psi) = \sin \tfrac{1}{2}\psi \int_0^\infty d(\lambda^2) \int_{-\infty}^{+\infty} x(ku(\psi)k^{-1}) \, d\kappa \tag{6-43}$$

we can write the second term of (6–42) as

$$-\frac{2n-1}{2\pi^2} \int_{-\pi}^{\pi} d\psi \, \sin(n - \tfrac{1}{2})\psi G(\psi)$$

We study the function $G(\psi)$ in detail.

   Because of the symmetry of $x(a)$ we see from (6–43) that $G(\psi)$ is antisymmetric

$$G(\psi) = -G(-\psi) \tag{6-44}$$

and infinitely differentiable in $\psi$ except possibly at the point $\psi = 0$. We denote

$$\tau = \lambda^2 |\sin \tfrac{1}{2}\psi|$$

The argument of $x$ for $\psi > 0$ is the matrix

$$ku(\psi)k^{-1} = a(\tau, \kappa, \psi)$$
$$= \begin{pmatrix} \cos \tfrac{1}{2}\psi + \kappa\tau & -\kappa^2\tau - \tau^{-1} \sin^2 \tfrac{1}{2}\psi \\ \tau & \cos \tfrac{1}{2}\psi - \kappa\tau \end{pmatrix}$$

It possesses the limit

$$a(\tau, \kappa, 0+) = \begin{pmatrix} 1 + \kappa\tau & -\kappa^2\tau \\ \tau & 1 - \kappa\tau \end{pmatrix}$$
$$= u(\varphi)\begin{pmatrix} 1 & -\mu \\ 0 & 1 \end{pmatrix} u(\varphi)^{-1}$$

where

$$\kappa = \cot \varphi \qquad \tau = \mu \sin^2 \varphi \qquad d\kappa \, d\tau = d\mu \, d\varphi$$

$$0 \leq \varphi \leq \pi \qquad 0 \leq \mu \leq \infty$$

The invariance relation (6–40) implies therefore

$$G(0+) = \int\limits_0^\infty d\tau \int\limits_{-\infty}^{+\infty} d\kappa x(a(\tau, \kappa, 0+))$$

$$= \tfrac{1}{2}\pi \int\limits_{-\infty}^{+\infty} x(k)_{\lambda=1} \, d\mu$$

Due to (6–44) $G(\psi)$ is discontinuous at $\psi = 0$ with a jump

$$G(0+) - G(0-) = \pi \int\limits_{-\infty}^{\infty} x(k)_{\lambda=1} \, d\mu \qquad (6\text{-}45)$$

If the limit

$$\lim_{\psi \to 0+} \frac{d}{d\psi} G(\psi) = G'(0+)$$

exists, the derivative $G'(\psi)$ can be made a continuous function because the anti-symmetry (6–44) implies

$$G'(0+) = G'(0-)$$

In this case we obtain the following formula from (6–42), (6–43), and (6–45) by partial integrations and performing the summation

$$x(e) = -\frac{1}{\pi} \lim_{n \to \infty} \frac{1}{2\pi} \int\limits_{-\pi}^{\pi} \frac{\sin n\psi}{\sin(1/2)\psi} G'(\psi) \, d\psi \qquad (6\text{-}46)$$

We are therefore interested to know the value $G'(0+)$ and whether $G'(\psi)$ is of bounded variation in a neighborhood of $\psi = 0$.

To gain this information we apply the same trick as in the proof of the Plancherel theorem for the group SL(2, C) (Section 4–2). We continue $x(a)$ off the shell $\det(a) = 1$ such that the continued function $x^*(a)$ is infinitely differentiable and of compact support in the four real variables $a_{ij}$, $i, j = 1, 2$. Let $f(b)$ be the Fourier transform of $x^*(a)$

$$x^*(a) = \int \prod_{i,\,j=1.\,2} db_{ij} \, f(b) \exp\left\{ i \sum_{ij} a_{ij} b_{ji} \right\}$$

Replacing $\kappa$ by $\kappa/\tau$ we obtain from (6–43)

$$G(\psi) = \int \prod_{i,j} db_{ij}\, f(b) \int_0^\infty \frac{d\tau}{\tau} \int_{-\infty}^\infty d\kappa$$

$$\times \exp\{ib_{11}(\cos \tfrac{1}{2}\psi + \kappa) + ib_{22}(\cos \tfrac{1}{2}\psi - \kappa)$$
$$+ ib_{12}\tau - ib_{21}\tau^{-1}(\kappa^2 + \sin^2 \tfrac{1}{2}\psi)\}$$

$$= \int \prod_{i,j} db_{ij}\, f(b) K(b, \psi)$$

The integrals defining the kernel $K(b,\ \psi)$ are elementary. From *GR* 8.421.8, 6.677.3, 6.677.4, 6.677.5 we find for $\psi > 0$

$$K(b, \psi) = 2\pi|\Omega|^{-1/2} \exp\{i(b_{11} + b_{22}) \cos \tfrac{1}{2}\psi\}$$
$$\times [\theta(\Omega) \exp\{-|\Omega|^{1/2}|\sin \tfrac{1}{2}\psi|\}$$
$$+ \theta(-\Omega)\theta(b_{12})i \exp\{+i|\Omega|^{1/2}|\sin \tfrac{1}{2}\psi|\}$$
$$- \theta(-\Omega)\theta(-b_{12})i \exp\{-i|\Omega|^{1/2}|\sin \tfrac{1}{2}\psi|\}]$$

where

$$\Omega = (b_{11} - b_{22})^2 + b_{12} b_{21}$$

It follows immediately

$$G'(0+) = -\pi \int \prod_{i,j} db_{ij}\, f(b) \exp\{i(b_{11} + b_{22})\}$$

$$= -\pi x(e) \tag{6-47}$$

The boundedness of the variation of $G'(\psi)$ around $\psi = 0$ can also be inspected from the kernel $K(b, \psi)$. We may therefore apply Dirichlet's lemma to (6–46) and get

$$x(e) = -\frac{1}{\pi} G'(0+)$$

Recalling (6–47) completes the proof.

The Fourier transformation and its inversion have been restricted so far to functions of $\mathscr{C}^\infty$, however, the extension to $\mathscr{L}^2(A)$ is easily achieved by means of Parseval's formula (6–38). The main result of this extension is the statement that the Fourier transformation constitutes an isometric mapping of the Hilbert space $\mathscr{L}^2(A)$ on the Hilbert space $\mathscr{L}^2(K)$ of kernels of Hilbert-Schmidt type. This space $\mathscr{L}^2(K)$ is the direct orthogonal sum of two subspaces

$$\mathscr{L}^2(K) = \mathscr{L}_p{}^2(K) \oplus \mathscr{L}_d{}^2(K)$$

The subspace $\mathscr{L}_p{}^2(K)$ consists of kernels $K(z_1, z_2 | \chi)$ which are measurable with respect to $dz_1\, dz_2\, d\rho$ ($\chi$ is restricted to the principal series), satisfy the symmetry constraint (6–15) almost everywhere, and have finite norm $\|K\|_p = (K, K)_p^{1/2}$, where $(K_1, K_2)_p$ denotes the scalar product

$$(K_1, K_2)_p = \sum_{\varepsilon = 0,\, 1} \tfrac{1}{2} \int\limits_{-\infty}^{+\infty} d\rho \zeta_\varepsilon(\rho)$$

$$\times \int \overline{K_1(z_1, z_2 | \chi)} K_2(z_1, z_2 | \chi)\, dz_1\, dz_2$$

The subspace $\mathscr{L}_d{}^2(K)$ decomposes into the direct orthogonal sum

$$\mathscr{L}_d{}^2(K) = \sum_{k = 1,\, 3/2,\, 2,\, \ldots}^{\oplus} (2k - 1)[\mathscr{L}_{k,\,+}^2(K) \oplus \mathscr{L}_{k,\,-}^2(K)]$$

A space $\mathscr{L}_{k,+}^2(K)$ consists, for example, of kernels $K(w_1, w_2 | k, +)$, which are defined for $|w_{1,2}| < 1$, are holomorphic in $w_1$, antiholomorphic in $w_2$, and have finite norm (6–23). The integral (6–4) converges in the $\mathscr{L}_p{}^2(K)$ mean, the integral (6–21) for all $w_1$, $w_2$, and $k$, and the inverse transformation (6–34) in the $\mathscr{L}^2(A)$ mean. These assertions constitute the Plancherel theorem for the group SL(2, R).

## 6–4    REPRESENTATION FUNCTIONS OF THE FIRST AND SECOND KIND

In Section 5–3 we defined the functions $c_{q_1 q_2}^\chi(\eta)$ by the integral representation (5–24), which after an expansion into a binomial series led us to hypergeometric functions (5–26), (5–27) of the argument $\tanh^2 \tfrac{1}{2}\eta$. Instead of these functions of argument $\tanh^2 \tfrac{1}{2}\eta$, one prefers often to deal with hypergeometric functions of the arguments $-\sinh^2 \tfrac{1}{2}\eta$ or $-\sinh^{-2} \tfrac{1}{2}\eta$. The latter functions are particularly suited for an asymptotic expansion at $\eta = \infty$ and are therefore denoted functions of the second kind (Section 4–5). Relations between hypergeometric functions of such different types are tabulated under the heading of "the 20 relations between Kummer's 24 solutions of the hypergeometric differential equation." For convenience we use the table in Erdélyi et al. [10] and refer to it as *EMOT*. In all our formulas we place the cut of the hypergeometric function $_2F_1(\dots;z)$ in $z$ on the real axis between 1 and $+\infty$. From now on we shall moreover replace the parameter $\sigma$ by $J$ (5–31).

But apart from the issue of defining functions of the second kind, in this section we want to deal with new functions $d_{q_1 q_2}^J(\cosh \eta)$ of the first kind. Whereas the functions $c_{q_1 q_2}^J(\eta)$ (note that the label $\varepsilon$ can be dropped, because $2q_1 - \varepsilon$ is even) are entire in $J$, these functions $d_{q_1 q_2}^J(\cosh \eta)$ will exhibit singularaties in $J$ which we shall have to study later. We start our investigations anticipating a list of the properties of $d_{q_1 q_2}^J(\cosh \eta)$ compared with the functions $c_{q_1 q_2}^J(\eta)$.

(1)   The functions $d_{q_1 q_2}^J(\cosh \eta)$ differ from $c_{q_1 q_2}^J(\eta)$ by a factor that depends on $J$ but not on $\eta$. This factor reduces to a phase on the principal series. If we set $J = -k$ and restrict $q_{1,2}$ to $q_{1,2} \geq k$ ($q_{1,2} \leq -k$), the function $d_{q_1 q_2}^J(\cosh \eta)$ is equal to $c_{q_1 q_2}^{(k,+)}(\eta)$ $[c_{q_1 q_2}^{(k,-)}(\eta)]$ up to a sign factor. The functions $d_{q_1 q_2}^J(\cosh \eta)$ allow therefore a simultaneous description of the principal and the discrete series. This has to be compared with the more complicated relation between (5–24) and (5–124).

(2)   The functions $d_{q_1 q_2}^J(\cosh \eta)$ represent the group SL(2, R) in the same fashion as the functions $c_{q_1 q_2}^J(\eta)$. With (5–1), (5–2) we define

$$D_{q_1 q_2}^J(a) = D_{q_1 q_2}^J(u(\psi_1)\, du(\psi_2))$$
$$= \exp\{i(q_1 \psi_1 + q_2 \psi_2)\}\, d_{q_1 q_2}^J(\cosh \eta) \qquad (6\text{-}48)$$

and get

$$\sum_{q_2 = -\infty}^{\infty} D_{q_1 q_2}^J(a_1) D_{q_2 q_3}^J(a_2) = D_{q_1 q_3}^J(a_1 a_2)$$

For the proof of the last identity it suffices to show

$$\frac{c_{q_1 q_2}^J(\eta)}{d_{q_1 q_2}^J(\cosh \eta)} = \frac{c_{q_1 q_3}^J(\eta)}{d_{q_1 q_3}^J(\cosh \eta)} \, \frac{c_{q_3 q_2}^J(\eta)}{d_{q_3 q_2}^J(\cosh \eta)}$$

for arbitrary $q_1$, $q_2$, $q_3$.

(3)   Under the replacement $J \to -J - 1$ (i.e., $\chi \to -\chi$) the functions $d_{q_1 q_2}^J(\cosh \eta)$ are symmetric as

$$d_{q_1 q_2}^J(\cosh \eta) = (-1)^{q_1 - q_2}\, d_{q_1 q_2}^{-J-1}(\cosh \eta) \qquad (6\text{-}49)$$

as compared with

$$c_{q_1 q_2}^J(\eta) = (\beta^{q_1})^{-1} \beta^{q_2} c_{q_1 q_2}^{-J-1}(\eta)$$

where $\beta^q$ was given in (5–112), (5–113).

    (4) If $J = j = 0, \frac{1}{2}, 1, \ldots, |q_1| \leqq j, |q_2| \leqq j$, and $j - q_1$ integral, the functions $d^J_{q_1 q_2}(\cosh \eta)$ reduce to polynomials. These polynomials are identical to (2–30) if we continue analytically in $\eta$, $\eta \to i\varphi$. Under the same continuation the matrices of $SU(1, 1)$ obtained from $SL(2, R)$ by the standard isomorphism (5–3) go into the matrices of $SU(2)$.

Combinations of the properties (1) to (4) can in turn be used to define the functions $d^J_{q_1 q_2}(\cosh \eta)$.

    From (5–26) and (5–27) and *EMOT* Eqs. (3) and (4) we derive

$$c^J_{q_1 q_2}(\eta) = (-1)^{q_1 - q_2}(\cosh \tfrac{1}{2}\eta)^{q_1 + q_2}(\sinh \tfrac{1}{2}\eta)^{q_1 - q_2}\binom{J - q_2}{q_1 - q_2}$$

$$\times \, _2F_1(-J + q_1, J + q_1 + 1, q_1 - q_2 + 1, -\sinh^2 \tfrac{1}{2}\eta)$$

$$= (-1)^{q_1 - q_2}(\cosh \tfrac{1}{2}\eta)^{-q_1 - q_2}(\sinh \tfrac{1}{2}\eta)^{q_1 - q_2}\binom{J - q_2}{q_1 - q_2}$$

$$\times \, _2F_1(-J - q_2, J - q_2 + 1, q_1 - q_2 + 1, -\sinh^2 \tfrac{1}{2}\eta)$$

if $q_1 \geqq q_2$, and the same expression with $q_1, q_2$ replaced by $-q_1, -q_2$ if $q_2 \geqq q_1$. On the principal series we have $\bar{J} = -J - 1$ and therefore

$$\left|\binom{J - q_2}{q_1 - q_2}\right|^2 = \frac{(-1)^{q_1 - q_2}}{[(q_1 - q_2)!]^2}\frac{\Gamma(J - q_2 + 1)\Gamma(J + q_1 + 1)}{\Gamma(J + q_2 + 1)\Gamma(J - q_1 + 1)}$$

Taking this into account we define the function $d^J_{q_1 q_2}(z)$ by

$$d^J_{q_1 q_2}(z) = \frac{1}{(q_1 - q_2)!}\left\{\frac{\Gamma(J - q_2 + 1)\Gamma(J + q_1 + 1)}{\Gamma(J + q_2 + 1)\Gamma(J - q_1 + 1)}\right\}^{1/2}$$

$$\times [\tfrac{1}{2}(1 + z)]^{(q_1 + q_2)/2}[\tfrac{1}{2}(1 - z)]^{(q_1 - q_2)/2}$$

$$\times \, _2F_1(-J + q_1, J + q_1 + 1, q_1 - q_2 + 1; \tfrac{1}{2}(1 - z))$$

$$(6\text{-}50)$$

if $q_1 \geqq q_2$, and by the requirement

$$d^J_{q_1 q_2}(z) = (-1)^{q_1 - q_2} d^J_{q_2 \bar{q}_1}(z) \qquad (6\text{-}51)$$

if $q_1 \leqq q_2$. Equation (6–50) implies the identity

$$d^J_{q_1 q_2}(z) = d^J_{-q_2, -q_1}(z) \qquad (6\text{-}52)$$

In order to make the definition (6–50) unique we must still fix the phases of the square root factors in front of the hypergeometric functions. The ratio of $\Gamma$-functions is equal to the product of monomials

$$[(J + q_1)(J + q_1 - 1)\cdots(J + q_2 + 1)][(J - q_2)(J - q_2 - 1)\cdots(J - q_1 + 1)]$$

if $q_1 \geqq q_2$, and to the same expression with $q_1$ replaced by $q_2$ and vice versa if $q_1 \leqq q_2$. If $q_1 q_2 \geqq 0$, this product has simple zeros in $J$ at

$$- M, \ - M + 1, \ldots, \ - m - 1 \quad \text{and} \quad m, m + 1, \ldots, M - 1$$

where

$$m = \min(|q_1| \, |q_2|) \quad M = \max(|q_1|, |q_2|)$$

In the case $q_1 q_2 < 0$ there are, in addition, double zeros at

$$-m, \ - m + 1, \ldots, m - 1$$

We recall that analytic representations correspond to values $J$ for which $J - m$ is integral. For convenience we call these values the integral points in the $J$-plane. The square root of the ratio of $\Gamma$-functions then has square root branch points at the integral points in the intervals $-M \leq J \leq -m - 1$ and $m \leq J \leq M - 1$. We place the cuts on the real axis between $M - 1$ and $M - 2$, $M - 3$ and $M - 4$, and so forth. A cut between $m$ and $-m - 1$ results only if $q_1 - q_2$ is odd. For $J \to \infty$ we adjust the phase such that the square root is asymptotically equal to $J^{|q_1 - q_2|}$. In the remaining square roots

$$[\tfrac{1}{2}(1 + z)]^{1/2} \qquad [\tfrac{1}{2}(1 - z)]^{1/2}$$

we take

$$\arg(1 + z) = \arg(1 - z) = 0$$

in the interval $-1 < z < 1$ and cut the $z$-plane at

$$- \infty < z \leqq - 1 \quad \text{and} \quad 1 \leqq z < \infty$$

If the functions $d^J_{q_1, q_2}(z)$ are to describe representations of SL(2, R) we set

$$z = \cosh \eta + i0 + \quad \text{and} \quad J + i0 + \quad \text{if } J \text{ is real}$$

It is then easy to verify that these functions indeed possess the properties (1) to (4) listed earlier.

Hypergeometric functions with argument $- \sinh^{-2} \frac{1}{2}\eta$ can be introduced by means of *EMOT* Eqs. (34), (9), (13). For $q_1 \geqq q_2$ we get

$$
\begin{aligned}
d^J_{q_1 q_2}(z) = \Big\{ &\frac{\Gamma(J+q_1+1)\Gamma(J-q_2+1)}{\Gamma(J-q_1+1)\Gamma(J+q_2+1)} \Big\}^{1/2} \\
&\times [\tfrac{1}{2}(1+z)]^{-(q_1+q_2)/2}[\tfrac{1}{2}(1-z)]^{(q_1-q_2)/2} \\
&\times \Big\{ [\tfrac{1}{2}(z-1)]^{J+q_2} \frac{\Gamma(2J+1)}{\Gamma(J-q_2+1)\Gamma(J+q_1+1)} \\
&\times {}_2F_1(-J-q_2,\,-J-q_1,\,-2J;\,2(1-z)^{-1}) \\
&+ [\tfrac{1}{2}(z-1)]^{-J+q_2-1} \frac{\Gamma(-2J-1)}{\Gamma(-J-q_2)\Gamma(-J+q_1)} \\
&\times {}_2F_1(J-q_2+1,\,J-q_1+1,\,2J+2;\,2(1-z)^{-1}) \Big\}
\end{aligned}
$$

In the big bracket the argument of $z-1$ is defined to be zero for $z > 1$ and the $z$-plane is cut at $-\infty < z \leq 1$. The ratio of the $\Gamma$-functions in front of the second term of the big bracket can be transformed as

$$
\frac{\Gamma(-2J-1)}{\Gamma(-J-q_2)\Gamma(-J+q_1)} = \frac{(-1)^{q_1-q_2}\Gamma(J+q_2+1)\Gamma(J-q_1+1)}{2\pi \cot \pi(J-q_1)\Gamma(2J+2)}
$$

Defining the function $e^J_{q_1 q_2}(z)$ of the second kind by

$$
\begin{aligned}
e^J_{q_1 q_2}(z) = \tfrac{1}{2}\{ &\Gamma(J+q_1+1)\Gamma(J-q_1+1)\Gamma(J+q_2+1)\Gamma(J-q_2+1)\}^{1/2} \\
&\times [\tfrac{1}{2}(1+z)]^{-(q_1+q_2)/2}[\tfrac{1}{2}(1-z)]^{-(q_1-q_2)/2} \\
&\times [\tfrac{1}{2}(z-1)]^{-J+q_1-1}[\Gamma(2J+2)]^{-1} \\
&\times {}_2F_1(J-q_2+1,\,J-q_1+1,\,2J+2;\,2(1-z)^{-1}) \qquad (6\text{-}53)
\end{aligned}
$$

we obtain

$$
\pi \cot \pi(J-q_1)\, d^J_{q_1 q_2}(z) = e^J_{q_1 q_2}(z) - (-1)^{q_1-q_2} e^{-J-1}_{q_1 q_2}(z) \qquad (6\text{-}54)
$$

The inverse formula which allows us to express $e^J_{q_1 q_2}(z)$ as a linear combination of functions $d^J_{q_1 q_2}(z)$ can be obtained from *EMOT* Eq. (25)

$$
2 \sin \pi(J-q_1) e^J_{q_1 q_2}(z) = \pi[d^J_{q_1 q_2}(z)\exp\{\mp i\pi(J-q_1)\} - d^J_{q_1,\,-q_2}(-z)] \qquad (6\text{-}55)
$$

where the minus (plus) sign holds in the upper (lower) half $z$-plane. Recalling (6–51), (6–52) we deduce from (6–55) the identities

$$e^J_{q_1 q_2}(z) = (-1)^{q_1 - q_2} e^J_{q_2 q_1}(z) = e^J_{-q_2, -q_1}(z) \tag{6-56}$$

Due to their relation with the functions $c^J_{q_1 q_2}(\eta)$ the singularities in $J$ of both $d^J_{q_1 q_2}(z)$ and $e^J_{q_1 q_2}(z)$ are situated at integral points, at all other points these functions are holomorphic. The singularities of $d^J_{q_1 q_2}(z)$ are due to the square root of the ratio of $\Gamma$-functions and have been discussed already in connection with the definition (6–50). The nature of the singularities of $e^J_{q_1 q_2}(z)$ follows from the properties of $d^J_{q_1 q_2}(z)$ and the relations (6–54), (6–55).

In the half-plane

$$\mathrm{Re}(J + \tfrac{1}{2}) > M - \tfrac{1}{2}$$

$e^J_{q_1 q_2}(z)$ is holomorphic. This result and (6–54) imply that $e^J_{q_1 q_2}(z)$ is meromorphic in the half-plane

$$\mathrm{Re}(J + \tfrac{1}{2}) < -(M - \tfrac{1}{2})$$

with simple poles at the integral points. The residues of these poles are the functions $d^J_{q_1 q_2}(z)$. In the strip

$$|\mathrm{Re}(J + \tfrac{1}{2})| < m + \tfrac{1}{2}$$

$e^J_{q_1 q_2}(z)$ is meromorphic with poles at the integral points, which have residue $\tfrac{1}{2} d^J_{q_1 q_2}(z)$ if $q_1 q_2 > 0$ and $\tfrac{1}{2}(-1)^{J-m+1} d_{q_1, -q_2}(-z)$ if $q_1 q_2 < 0$. This follows from (6–55). In the strips

$$m + \tfrac{1}{2} \leqq |\mathrm{Re}(J + \tfrac{1}{2})| \leqq M - \tfrac{1}{2}$$

$e^J_{q_1 q_2}(z)$ exhibits square root branch points at the integral points such that the product $e^J_{q_1 q_2}(z) d^J_{q_1 q_2}(z)$ is holomorphic. In this context we mention also the asymptotic expansion of $e^J_{q_1 q_2}(z)$ at $J = \infty$ in the angle $-\pi < \arg J < \pi$, which can be derived from (6–53) with some labor. Its first term is

$$e^J_{q_1 q_2}(z) = \left(\frac{\pi}{2J}\right)^{1/2} (z^2 - 1)^{-1/4} \exp\{i\tfrac{1}{2}\pi(q_1 - q_2)\,\mathrm{sign}(\mathrm{Im}\,z)\}$$

$$\times [z + (z^2 - 1)^{1/2}]^{-J-1/2}(1 + O(J^{-1})) \tag{6-57}$$

where the square bracket and $z^2 - 1$ are defined to have argument zero for $z > 1$,

We know that the asymptotic behavior of the functions $d^J_{q_1 q_2}(z)$ and $e^J_{q_1 q_2}(z)$ at $z = 1$, that is, at the group unit of SL(2, R), is crucial for the existence of Fourier transforms of the second kind on the whole space $\mathscr{C}^\infty$. The definition (6–50) implies that

$$[\tfrac{1}{2}(1 - z)]^{-|q_1 - q_2|/2} \, d^J_{q_1 q_2}(z)$$

is holomorphic for $\operatorname{Re} z > -1$ and nonnull at $z = 1$ in general. Further one can show that $e^J_{q_1 q_2}(z)$ possesses an asymptotic expansion in $|\arg(z - 1)| < \pi$ of the kind

$$
\begin{aligned}
e^J_{q_1 q_2}(z) \cong \exp\{&i\tfrac{1}{2}\pi(q_1 - q_2)\operatorname{sign}(\operatorname{Im} z)\} \\
&\times \left[ (\tfrac{1}{2}(z - 1))^{-|q_1 - q_2|/2} \sum_{n=0}^{\infty} a_n (\tfrac{1}{2}(z - 1))^n \right. \\
&\quad + (\tfrac{1}{2}(z - 1))^{+|q_1 - q_2|/2} \log(\tfrac{1}{2}(z - 1)) \\
&\quad \left. \times \sum_{n=0}^{\infty} b_n (\tfrac{1}{2}(z - 1))^n \right]
\end{aligned}
\tag{6-58}
$$

The zero of $d^J_{q_1 q_2}(z)$ suffices just to cancel the worst singularity of $e^J_{q_1 q_2}(z)$ at $z = 1$, and Fourier transforms of the second kind can indeed be formed for the whole space $\mathscr{C}^\infty$ (Section 6–5).

For an element $a \in SL(2, R)$ decomposed as in (5–1), (5–2) we denote

$$
\begin{aligned}
E^J_{q_1 q_2}(a) &= E^J_{q_1 q_2}(u(\psi_1) \, du(\psi_2)) \\
&= \exp\{i(q_1\psi_1 + q_2\psi_2)\} e^J_{q_1 q_2}(\cosh \eta + i0+)
\end{aligned}
\tag{6-59}
$$

These functions $E^J_{q_1 q_2}(a)$ are called the representation functions of the second kind.

## 6–5    FOURIER TRANSFORMATIONS OF BICOVARIANT FUNCTIONS

A function $x_{q_1 q_2}(a)$ with integral or half-integral subscripts $q_1$ and $q_2$ is called bicovariant if

$$x_{q_1 q_2}(u(\psi_1)au(\psi_2)) = \exp\{i(q_1\psi_1 + q_2\psi_2)\}x_{q_1 q_2}(a) \tag{6-60}$$

for all $\psi_1$, $\psi_2$. The formal theory of such bicovariant functions on SL(2, R) parallels the theory of bicovariant functions on SL(2, C) so greatly that we can be rather brief in our arguments and refer to Section 4–7 for more details. In the sequel we shall assume that the bicovariant functions considered are in $\mathscr{C}^\infty$ if the opposite is not explicitly stated.

If we set $\psi_1 = \psi_2 = 2\pi$, (6–60) implies that $q_1 - q_2$ is integral. If we decompose the group element $a$ as in (5–1), (5–2) we get

$$x_{q_1 q_2}(a) = \exp i(q_1 \psi_1 + q_2 \psi_2) x_{q_1 q_2}(\eta) \tag{6-61}$$

where we used the notation

$$x_{q_1 q_2}(d) \equiv x_{q_1 q_2}(\eta)$$

For convenience we study the Fourier transforms of the complex conjugate of a bicovariant function $x_{q_1 q_2}(a)$. We denote the Fourier transforms (6–4) and (6–21) of $\overline{x_{q_1 q_2}(a)}$ by $K^*_{q_1 q_2}(\psi_1, \psi_2 | \chi)$ and $K^*_{q_1 q_2}(w_1, w_2 | k, \pm)$, respectively. In addition we introduce reduced kernels (or "Fourier coefficients") $\mathscr{F}^*_{q_1 q_2}(\chi)$ (6–64), $\mathscr{F}^*_{q_1 q_2}(k, \pm)$ (6–66) for $\overline{x_{q_1 q_2}(a)}$ and, similarly, reduced kernels $\mathscr{F}_{q_1 q_2}(\chi)$ and $\mathscr{F}_{q_1 q_2}(k, \pm)$ for $x_{q_1 q_2}(a)$ itself. On the principal and discrete series both types of reduced kernels are connected by complex conjugation. Nevertheless it is convenient to introduce independent notations for them, since we want to regard both types as analytic functions of $\rho$ if possible. The symmetry under complex conjugation depends on the type of function used [$c$-functions $c^\chi_{q_1 q_2}(\eta)$ or $d$-functions $d^J_{q_1 q_2}(\cosh \eta)$] and is therefore given later.

The integral kernels (6–4) representing the Fourier transforms in the principal series are

$$K^*_{q_1 q_2}(\psi_1, \psi_2 | \chi) = \exp\{i(q_1\psi_1 - q_2\psi_2)\}$$

$$\times \pi \int \overline{x_{q_1 q_2}(k)} |\lambda|^{i\rho - 1} [\text{sign } \lambda]^\varepsilon \, d\mu_l(k) \tag{6-62}$$

To simplify the integral we set

$$k = u(\psi) \, du(\psi_d)^{-1} \qquad 0 \leqq \eta < \infty \qquad 0 \leqq \psi < 4\pi$$

and (Appendix A-2f)

$$d\mu_l(k) = \frac{1}{8\pi^2} \sinh \eta \, d\eta \, d\psi$$

This yields

$$K^*_{q_1 q_2}(\psi_1, \psi_2 | \chi) = \exp\{i(q_1\psi_1 - q_2\psi_2)\}\mathscr{F}^*_{q_1 q_2}(\chi) \tag{6-63}$$

where the Fourier coefficient $\mathscr{F}^*_{q_1 q_2}(\chi)$ is defined as

$$\mathscr{F}^*_{q_1 q_2}(\chi) = \frac{1}{2} \int_0^\infty \overline{x_{q_1 q_2}(\eta)} c^\chi_{q_1 q_2}(\eta) \sinh \eta \, d\eta \tag{6-64}$$

Using the form (5–26), (5–27) of the functions $c^{\chi}_{q_1 q_2}(\eta)$ we see that the following symmetry relation holds

$$\overline{\mathscr{F}^{*}_{q_1 q_2}(\chi)} = \mathscr{F}_{q_1 q_2}(-\chi)$$

The contribution of the principal series to the inverse Fourier transformation (6–34) is correspondingly

$$\tfrac{1}{2} \int\limits_{-\infty}^{+\infty} d\rho \zeta_{\varepsilon}(\rho) C^{\chi}_{q_1 q_2}(a) \mathscr{F}_{q_1 q_2}(-\chi)$$

where

$$\varepsilon \cong 2m \bmod 2 \qquad m = \min(|q_1|, |q_2|)$$

Now we derive the Fourier coefficients in the discrete series. We start from (6–21), (6–22)

$$K^{*}_{q_1 q_2}(w_1, w_2 \,|\, k, +) = \frac{2k-1}{\pi} \int \overline{x_{q_1 q_2}(a)} [\alpha - \bar{\beta} w_1 + \beta \bar{w}_2 - \bar{\alpha} w_1 \bar{w}_2]^{-2k} \, d\mu(a)$$

and insert (5–69)

$$a = u(\psi_1) \, du(\psi_2)$$
$$\alpha = \cosh \tfrac{1}{2}\eta \, \exp\{-\tfrac{1}{2}(\psi_1 + \psi_2)\}$$
$$\beta = \sinh \tfrac{1}{2}\eta \, \exp\{-\tfrac{1}{2}(\psi_1 - \psi_2)\}$$

and (Appendix A–2f)

$$d\mu(a) = \tfrac{1}{2} \sinh \eta \, d\eta (4\pi)^{-2} \, d\psi_1 \, d\psi_2$$

This yields the integral

$$K^{*}_{q_1 q_2}(w_1, w_2 \,|\, k, +) = \frac{2k-1}{\pi} \frac{1}{2} \int\limits_{0}^{\infty} d\eta \, \sinh \eta \, \overline{x_{q_1 q_2}(\eta)}$$

$$\times (4\pi)^{-2} \int\limits_{0}^{4\pi} d\psi_1 \int\limits_{0}^{4\pi} d\psi_2 \, \exp\{-i(q_1 - k)\psi_1$$

$$- i(q_2 - k)\psi_2\}$$

$$\times \left[\cosh \tfrac{1}{2}\eta - w_1 \sinh \tfrac{1}{2}\eta e^{i\psi_1} + \bar{w}_2 \sinh \tfrac{1}{2}\eta e^{i\psi_2}\right.$$
$$\left. - w_1 \bar{w}_2 \cosh \tfrac{1}{2}\eta \exp\{i(\psi_1 + \psi_2)\}\right]^{-2k}$$

The elementary integrations over $\psi_1$ and $\psi_2$ can be performed and we obtain

$$K^*_{q_1 q_2}(w_1, w_2 \,|\, k, +) = \begin{cases} \overline{\Phi_{q_1}(w_1)}\,\overline{\Phi_{q_2}(w_2)} \dfrac{2k-1}{\pi} \mathscr{F}^*_{q_1 q_2}(k, +) \\[2mm] \text{if } \ 1 \le k \le \min(q_1, q_2) \ \text{ and } \ k - q_1 \text{ integral} \\ \quad 0 \ \ \text{in all other cases} \end{cases}$$

$$(6\text{-}65)$$

The Fourier coefficient $\mathscr{F}^*_{q_1 q_2}(k, +)$ is defined by

$$\mathscr{F}^*_{q_1 q_2}(k, +) = \tfrac{1}{2} \int\limits_0^\infty \overline{x_{q_1 q_2}(\eta)}\, c^{(k,\,+)}_{q_1 q_2}(\eta) \sinh \eta \; d\eta \qquad (6\text{-}66)$$

and $\Phi_q(w)$ are the orthonormal basis elements (5–98). Due to the reality of the functions $c^{(k,+)}_{q_1 q_2}(\eta)$ (5–126) we have the symmetry relation

$$\overline{\mathscr{F}^*_{q_1 q_2}(k, +)} = \mathscr{F}_{q_1 q_2}(k, +)$$

The contribution of the discrete series to the inverse Fourier transformation (6–34) of $x_{q_1 q_2}(a)$ is

$$\sum_{k \ge 1}^{\min(q_1, q_2)} (2k-1) C^{(k,\,+)}_{q_1 q_2}(a) \mathscr{F}_{q_1 q_2}(k, +)$$

$$+ \sum_{k \ge 1}^{-\max(q_1, q_2)} (2k-1) C^{(k,\,-)}_{q_1 q_2}(a) \mathscr{F}_{q_1 q_2}(k, -)$$

Void sums are zero of course. The inverse Fourier transformation (6–34) looks altogether

$$x_{q_1 q_2}(a) = \tfrac{1}{2} \int\limits_{-\infty}^{+\infty} d\rho\, \zeta_\varepsilon(\rho) C^\chi_{q_1 q_2}(a) \mathscr{F}_{q_1 q_2}(-\chi)$$

$$+ \delta_{\tau_1, \tau_2} \sum_{k \ge 1}^{m} (2k-1) C^{(k,\,\tau_1)}_{q_1 q_2}(a) \mathscr{F}_{q_1 q_2}(k, \tau_1) \qquad (6\text{-}67)$$

where

$$\tau_i = \text{sign } q_i \qquad i = 1, 2$$

$$m = \min(|q_1|, |q_2|)$$

$$\varepsilon \cong 2k \cong 2m \bmod 2$$

and the sum is void if $m = 0$ or $m = \frac{1}{2}$.

We reformulate Eq. (6–67) replacing the $c$-functions by the $d$-functions and $\rho, k$ by $J$,

$$J = \tfrac{1}{2}(i\rho - 1) \qquad J = k - 1$$

Since the $c$-functions and the $d$-functions differ only by a phase factor on the representations concerned, we obtain

$$x_{q_1 q_2}(\eta) = \frac{1}{2\pi i} \int_{-(1/2)-i\infty}^{-(1/2)+i\infty} dJ(2J+1)\pi \cot \pi(J-m)$$

$$\times\ d_{q_1 q_2}^J(\cosh \eta)\overline{\mathscr{F}_{q_1 q_2}^*(J)}$$

$$+ \delta_{\tau_1, \tau_2} \sum_{J \geq 0}^{m-1} (2J+1)\, d_{q_1 q_2}^J(\cosh \eta)\overline{\mathscr{F}_{q_1 q_2}^*(J)} \qquad J - q_1 \text{ integral} \tag{6-68}$$

where instead of (6–64), (6–66) we use the Fourier coefficient

$$\mathscr{F}_{q_1 q_2}^*(J) = \tfrac{1}{2} \int_0^\infty \overline{x_{q_1 q_2}(\eta)}\, d_{q_1 q_2}^J(\cosh \eta)\sinh \eta\, d\eta \tag{6-69}$$

This coefficient is symmetric as

$$\overline{\mathscr{F}_{q_1 q_2}^*(J)} = \mathscr{F}_{q_1 q_2}(J) \tag{6-70}$$

To prove (6–70) we recall that $d_{q_1 q_2}(\cosh \eta)$ is defined as the value along the upper edge of the cut caused by the factor

$$[\tfrac{1}{2}(1 - z)]^{(q_1 - q_2)/2}$$

Due to this factor the $d$-function would change its phase by a factor $(-1)^{q_1-q_2}$ under complex conjugation. If $J$ is in the principal series, complex conjugation turns $J$ into $-J-1$, and (6–49) proves that $d_{q_1 q_2}^J(\cosh \eta)$ is indeed real. In the case of the discrete series the branch cut between $m$ and $-m-1$ in the $J$-plane causes a second phase change $(-1)^{q_1-q_2}$, which makes $d_{q_1 q_2}^J(\cosh \eta)$ real.

Next we use (6–54) to replace the $d$-functions by the functions of the second kind. The integral appearing in (6–68) then takes the form

$$\frac{1}{2\pi i} \int_{-(1/2)-i\infty}^{-(1/2)+i\infty} dJ(2J+1) d^J_{q_1 q_2}(\cosh \eta) \mathscr{F}^{(+)}_{q_1 q_2}(J)$$

where the "Fourier coefficient of the second kind" is defined by

$$\mathscr{F}^{(+)}_{q_1 q_2}(J) = \int_0^\infty x_{q_1 q_2}(\eta) e^J_{q_1 q_2}(\cosh \eta)\sinh \eta \, d\eta \qquad (6\text{-}71)$$

This Fourier coefficient of the second kind $\mathscr{F}^{(+)}_{q_1 q_2}(J)$ exists in fact for all $x \in \mathscr{C}^\infty$ and is analytic in $J$. We read off (6–68) that $x_{q_1 q_2}(\eta)$ has a zero at $\eta = 0$ at least of the same order as $d^J_{q_1 q_2}(\cosh \eta)$. The integral (6–71) converges therefore at $\eta = 0$ because of (6–58). The Fourier coefficient $\mathscr{F}^{(+)}_{q_1 q_2}(J)$ is singular in $J$ if and only if $e^J_{q_1 q_2}(\cosh \eta)$ is singular. If $q_1 q_2 > 0$ it exhibits poles at (Section 6–4)

$$-m, -m+1, -m+2, \ldots, m-1$$

with residues $\mathscr{F}_{q_1 q_2}(J)$. In addition it possesses first order poles at

$$-M-1, -M-2, -M-3, \ldots$$

with residues $2\mathscr{F}_{q_1 q_2}(J)$. All other singularities of $\mathscr{F}^{(+)}_{q_1 q_2}(J)$ are canceled if we multiply with $d^J_{q_1 q_2}(\cosh \eta)$.

If $q_1 q_2 \geq 0$, we may replace the integral over the line $\mathrm{Re}\, J = -\frac{1}{2}$ by an integral over a contour $C_+(C_-)$, which is defined such that the contributions of the discrete series to (6–68) appear incorporated in the principal series integral. The contour $C_+(C_-)$ consists of two infinite intervals on the line $\mathrm{Re}\, J = -\frac{1}{2}$. In addition it encircles those poles in the strip $|\mathrm{Re}(J + \frac{1}{2})| < m + \frac{1}{2}$ which lie to the right (left) of $\mathrm{Re}\, J = -\frac{1}{2}$ once in the positive (negative) sense (Fig. 6–1). All the poles in the half-plane $\mathrm{Re}\, J < -M$ stay to the left of $C_+$ and $C_-$. In the case $q_1 q_2 \leq 0$ there are no contributions of the discrete series. Nevertheless we may extend the integral over the contour $C_+$ or $C_-$, since each of these contours can be deformed into the line $\mathrm{Re}\, J = -\frac{1}{2}$ without crossing a singularity of the integrand. The inverse Fourier transformation (6–68) can therefore always be brought into the form

$$x_{q_1 q_2}(\eta) = \frac{1}{2\pi i} \int_{C_\pm} dJ(2J+1) d^J_{q_1 q_2}(\cosh \eta) \mathscr{F}^{(+)}_{q_1 q_2}(J) \qquad (6\text{-}72)$$

In general the contour $C_+$ or $C_-$ can be deformed further.

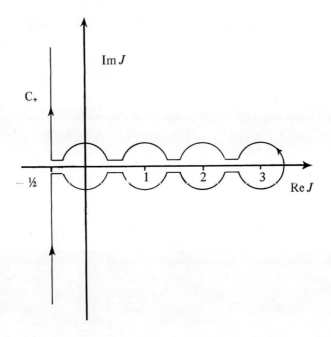

**Figure 6–1.**   The contour $C_+$ in a case with $m = 4$

In a similar fashion we may formulate Parseval's formula (6–38) for two functions $x_{q_1 q_2}(a)$ and $y(a)$ of $\mathscr{C}^\infty$, where the function $y(a)$ is expanded into a series of bicovariant functions

$$y(a) = \sum_{q_1 q_2} y_{q_1 q_2}(a)$$

We have

$$(x_{q_1 q_2} \cdot y^\dagger)(e) = \int x_{q_1 q_2}(a)\overline{y(a)}\, d\mu(a)$$

$$= \tfrac{1}{2} \int_0^\infty x_{q_1 q_2}(\eta)\overline{y_{q_1 q_2}(\eta)}\, \sinh \eta \, d\eta$$

(Appendix A–2f) and, correspondingly, Parseval's formula with the notations (6–64), (6–66), (6–67)

$$\tfrac{1}{2} \int_0^\infty x_{q_1 q_2}(\eta)\overline{y_{q_1 q_2}(\eta)}\sinh \eta \, d\eta$$

$$= \tfrac{1}{2} \int\limits_{-\infty}^{+\infty} d\rho \zeta_\varepsilon(\rho) \mathscr{F}_{q_1 q_2}(-\chi)_x \mathscr{F}^*_{q_1 q_2}(\chi)_y$$

$$+ \delta_{\tau_1, \tau_2} \sum_{k \geq 1}^{m} (2k - 1) \mathscr{F}_{q_1 q_2}(k, \tau_1)_x \mathscr{F}^*_{q_1 q_2}(k, \tau_1)_y$$

By means of the same reasoning as earlier and with the notation (6–69) we can bring this equation into the form

$$\tfrac{1}{2} \int\limits_0^\infty x_{q_1 q_2}(\eta) \overline{y_{q_1 q_2}(\eta)} \sinh \eta \, d\eta$$

$$= \frac{1}{2\pi i} \int\limits_{C_\pm} dJ (2J + 1) \mathscr{F}_{q_1 q_2}(J)_x \mathscr{F}^{*(+)}_{q_1 q_2}(J)_y$$

$$= \frac{1}{2\pi i} \int\limits_{C_\pm} dJ (2J + 1) \mathscr{F}^{(+)}_{q_1 q_2}(J)_x \mathscr{F}^*_{q_1 q_2}(J)_y \tag{6-73}$$

Finally we consider the problem of how to extend the results of this section from $\mathscr{C}^\infty$ to $\mathscr{L}^2(A)$. The simple arguments presented which guarantee the existence of the Fourier transforms of the second kind on $\mathscr{C}^\infty$ cannot be carried over to $\mathscr{L}^2(A)$. To cut the discussion short let us therefore introduce the following technical term. We call a function $x_{q_1 q_2}(a)$ which is bicovariant and in $\mathscr{L}^2(A)$ *regular* if it possesses a Fourier transform of the second kind that is measurable on the line $\operatorname{Re} J = -\tfrac{1}{2}$ and can be continued analytically off this line in such a fashion that the discrete series terms can be incorporated in the principal series integral as in (6–72). The convergence of the integral (6–72) does not cause any troubles for a regular function $x_{q_1 q_2}(a)$ because we can go back through our arguments to formula (6–68) and apply the Plancherel theorem. The situation changes, however, if we consider integration contours whose infinite parts do not lie on the line $\operatorname{Re} J = -\tfrac{1}{2}$. In this case the Plancherel theorem cannot be made responsible for the convergence of the inverse Fourier transformation, but we need independent information on the asymptotic behavior of the Fourier transforms of the second kind. Such problems arise in connection with analytic functionals, which are defined in the subsequent section.

## 6–6 GENERALIZED FUNCTIONS OF THE GROUP SL(2, R)

We consider continuous linear functionals on the space $\mathscr{C}^\infty$ of infinitely differentiable functions $x(a)$ with compact support on SL(2, R). These functionals form a closed space $\mathscr{C}^{\infty\prime}$. We denote such a functional $p$ by

$$(p, x) = \int p(a)x(a)\, d\mu(a)$$

and say that $p$ is generated by the distribution or generalized function $p(a)$. The characters (6–13) and (6–31) of the representations of SL(2, R) were distributions in this sense. We call a distribution bicovariant and write $p(a) = p_{q_1 q_2}(a)$ if, for any element $x(a) \in \mathscr{C}^\infty$ with the expansion

$$x(a) = \sum_{q_1, q_2} x_{q_1 q_2}(a)$$

where $x_{q_1 q_2}(a)$ is bicovariant, we have

$$(p_{q_1 q_2}, \bar{x}) = (p_{q_1 q_2}, \overline{x_{q_1 q_2}}) \tag{6-74}$$

Any distribution $p(a)$ can be expanded into a series of bicovariant distributions

$$p(a) = \sum_{q_1 q_2} p_{q_1 q_2}(a)$$

which converges in the sense of the topology of $\mathscr{C}^{\infty\prime}$. Moreover, we find

$$(p, \bar{x}) = \sum_{q_1 q_2} (p_{q_1 q_2}, \overline{x_{q_1 q_2}})$$

(see the arguments of Section 4–7).

We know that Fourier transformations form one-to-one mappings of the Hilbert space $\mathscr{L}^2(A)$ of square integrable functions on SL(2, R) onto the Hilbert space $\mathscr{L}^2(K)$ of Hilbert-Schmidt kernels $K = \{K(\chi), K(k, \pm)\}$ (Section 6–3). The subspace $\mathscr{C}^\infty$ of $\mathscr{L}^2(A)$ is mapped this way on a dense subspace $\mathscr{X}$ of $\mathcal{L}^2(K)$, in which we define a convergence by carrying it over from $\mathscr{C}^\infty$. The linear continuous functionals on $\mathscr{X}$ form a closed linear space $\mathscr{X}'$. Let $K_x = \{K_x(\chi), K_x(k, \pm)\}$ be the element of $\mathscr{X}$ corresponding to $x \in \mathscr{C}^\infty$. By the prescription

$$(p, x) = (p^*, K_x) \tag{6-75}$$

for all $x \in \mathscr{C}^\infty$, we establish a one-to-one correspondence of any element $p^*$ of $\mathscr{X}'$ with a distribution $p \in \mathscr{C}^{\infty\prime}$.

We are mainly interested in one class of functionals $p^*$ of $\mathscr{X}'$, which we call bicovariant analytic functionals $p^*_{q_1 q_2}$. Their images in $\mathscr{C}^{\infty\prime}$ are bicovariant distributions $p_{q_1 q_2}(a)$ in the sense (6–74). We define them in terms of a function $\mathscr{F}^{(+)}_{q_1 q_2}(J)$, which is measurable on a contour C in the $J$-plane such that [recall (6–73)]

$$(p^*_{q_1q_2}, K_{\bar{x}}) = \frac{1}{2\pi i} \int_C dJ(2J + 1)\mathscr{F}^{(+)}_{q_1q_2}(J)\mathscr{F}^*_{q_1q_2}(J)_x \qquad (6\text{-}76)$$

for all $x \in \mathscr{C}^\infty$, whose expansion into bicovariant components is

$$x(a) = \sum_{q_1q_2} x_{q_1q_2}(a)$$

In addition we require that the functional (6–76) is continuous, and that the contour C has no point in common with the interval $-m \leq J \leq m - 1$ if $q_1q_2 > 0$, but it may be arbitrary otherwise. If a functional $p$ is generated by a regular function $p_{q_1q_2}(a)$, we call it itself a *regular functional* (Section 6–5). A regular functional $p_{q_1q_2}(a)$ corresponds to a bicovariant analytic functional $p^*_{q_1q_2}$ with $C = C_\pm$ and a function $\mathscr{F}^{(+)}_{q_1q_2}(J)$ which can be identified with the Fourier coefficient of the second kind of $p_{q_1q_2}(a)$. In this case the definition (6–76) reduces to Parseval's formula (6–73) with $x_{q_1q_2}(a) = p_{q_1q_2}(a)$.

We may now raise the question as to under which circumstances the analytic bicovariant functional $p^*_{q_1q_2} \in \mathscr{Z}'$ corresponds to a distribution $p_{q_1q_2}(a)$, which is a function in the usual sense though not necessarily regular. Remembering (6–72) we attempt to define such a function $p_{q_1q_2}(\eta)$ by

$$p_{q_1q_2}(\eta) = \frac{1}{2\pi i} \int_C dJ(2J + 1) d^J_{q_1q_2}(\cosh \eta)\mathscr{F}^{(+)}_{q_1q_2}(J) \qquad (6\text{-}77)$$

We assume that (6–77) defines a function $p_{q_1q_2}(\eta)$ at least for almost all $\eta$, and apply it tentatively to an element $\bar{x} \in \mathscr{C}^\infty$. We get

$$(p_{q_1q_2}, \bar{x}) = \tfrac{1}{2} \int_0^\infty d\eta \, \sinh \eta \, \overline{x_{q_1q_2}(\eta)} p_{q_1q_2}(\eta)$$

$$= \tfrac{1}{2} \int_0^\infty d\eta \, \sinh \eta \, \overline{x_{q_1q_2}(\eta)}$$

$$\times \frac{1}{2\pi i} \int_C dJ(2J + 1) d^J_{q_1q_2}(\cosh \eta)\mathscr{F}^{(+)}_{q_1q_2}(J)$$

A change in the order of integrations leads us then to

$$(p_{q_1q_2}, \bar{x}) = \frac{1}{2\pi i} \int_C dJ(2J + 1)\mathscr{F}^{(+)}_{q_1q_2}(J)\mathscr{F}^*_{q_1q_2}(J)_x$$

$$= (p^*_{q_1q_2}, K_{\bar{x}})$$

If all these operations, that is, the integration over $\eta$ and the change of order of integrations, are allowed, the function $p_{q_1 q_2}(a)$ corresponds in fact to the functional $p^*_{q_1 q_2} \in \mathscr{L}'$. In this case we may regard the function $\mathscr{F}^{(+)}_{q_1 q_2}(J)$ as the Fourier coefficient of the second kind of $p_{q_1 q_2}(a)$ defined on a contour C, and consider the integral representation (6–77) as the inverse Fourier transformation of $p_{q_1 q_2}(a)$. We could of course justify the existence and the integrability of the function $p_{q_1 q_2}(a)$ and the possibility of changing the order of integrations by imposing some sufficient condition on the function $\mathscr{F}^{(+)}_{q_1 q_2}(J)$. But in most applications it is easy to verify the correctness of these premises anyway, and therefore we skip over this problem here.

Distributions $p_{q_1 q_2}(a)$ may depend analytically on a parameter $\sigma$ such that for $\sigma$ in a certain "regular" subdomain of the domain of analyticity, $p_{q_1 q_2}(a)$ is a regular distribution. If $\rho_1$ is a value which lies outside this regular domain, the functional $p^*_{q_1 q_2}$ can often by obtained as an analytic functional by analytic continuation in $\sigma$ from a regular value to $\sigma_1$ in such a fashion that the contour C is obtained from $C_+$ by a deformation which evades the singularities moving from the half-plane Re $J < -\frac{1}{2}$ towards $C_+$ in the course of the continuation. This way of handling Fourier transforms of polynomially bounded functions on SL(2, R) is of great practical interest.

To illustrate these definitions and conventions we investigate the distribution

$$p(a) = |a|^{2\sigma} = p(\eta) = (2 \cosh \eta)^\sigma \quad \sigma \text{ complex}$$

in detail. This distribution is bi-invariant and entire analytic in $\sigma$. It is regular in the domain Re $\sigma < -\frac{1}{2}$. The Fourier coefficient $\mathscr{F}^{(+)}(J)$ is

$$\mathscr{F}^{(+)}(J) = \int_1^\infty (2z)^\sigma e^J_{00}(z)\, dz \qquad (6\text{-}78)$$

In order to evaluate the integral (6–78) we recall the definition (6–53) of the function $e^J_{00}(z)$

$$e^J_{00}(z) = \frac{1}{2} \frac{\Gamma(J+1)^2}{\Gamma(2J+2)} [\tfrac{1}{2}(z-1)]^{-J-1}$$
$$\times {}_2F_1(J+1, J+1, 2J+2; 2(1-z)^{-1})$$

By means of an identity for certain hypergeometric functions (GR 9.134.1 and 8.703) we can rewrite this function as

$$e^J_{00}(z) = \frac{1}{2} \frac{\Gamma(J+1)^2}{\Gamma(2J+2)} \left(\frac{1}{2} z\right)^{-J-1}$$
$$\times {}_2F_1(\tfrac{1}{2}J + \tfrac{1}{2}, \tfrac{1}{2}J + 1, J + \tfrac{3}{2}; z^{-2})$$

This hypergeometric series allows a termwise integration of (6–78) with the result

$$\mathcal{F}^{(+)}(J) = \pi^{1/2} 2^{\sigma - J - 1} \sum_{n=0}^{\infty} \frac{\Gamma(J + 2n + 1)}{2^{2n} n! (J + 2n - \sigma) \Gamma(J + 3/2 + n)} \qquad (6\text{-}79)$$

The function $\mathcal{F}^{(+)}(J)$ (6–79) possesses two types of poles in $J$. A first series of poles at the positions

$$J = -n - 1 \quad n = 0, 1, 2, \ldots$$

which are independent of $\sigma$ (fixed poles), are due to the poles of the function $e_{00}^J(z)$ itself. A second series of poles

$$J = \sigma - 2n \quad n = 0, 1, 2, \ldots$$

have positions which depend on $\sigma$ (moving poles). They lie to the left of the line $\text{Re } J = -\frac{1}{2}$ if $\sigma$ is in the regular domain, touch this line if $\sigma$ is on the boundary of the regular domain, and move beyond it if $\sigma$ leaves the regular domain.

The contour $C_+$ can be identified with the line $\text{Re } J = -\frac{1}{2}$ (oriented upwards). If we continue in $\sigma$ off the regular domain, we can always deform the contour $C_+$ so that it evades the poles and so that two infinite intervals on the line $\text{Re } J = -\frac{1}{2}$ remain unchanged. This completes the construction of an analytic functional which presents the Fourier transform of the distribution $p(a)$ for any complex $\sigma$.

Let us now attempt to define an analytic functional by the function $\mathcal{F}^{(+)}(J)$ (6–79), and a contour C, which is equal to any upward oriented line $\text{Re } J = \text{Re } \sigma + \varepsilon$, $\text{Re } \sigma > -1$, $\varepsilon > 0$. We must first verify that this defines a continuous functional indeed. Second we would like to know if there is a function $p(a)$ to which this analytic functional belongs as the Fourier transform. Though the latter question will be answered in the affirmative in a rather trivial fashion, that is, $p(a)$ is the function $|a|^{2\sigma}$ just considered, it is possible to gain some general insights by the following discussion.

The differential operator $D(z)$

$$D(z) = (z^2 - 1) \frac{d^2}{dz^2} + 2z \frac{d}{dz} \qquad (6\text{-}80)$$

which is related to the Casimir operator (5–30), satisfies

$$D(z) e_{00}^J(z) = J(J + 1) e_{00}^J(z)$$

$$D(z) \, d_{00}^J(z) = J(J + 1) \, d_{00}^J(z)$$

as can be verified easily with (6–53), (6–54). If $J$ varies on a given contour C, we define the bound $M(N, \text{C})$ by

$$M(N, \text{C}) = \sup_{\substack{J \in \text{C} \\ 1 \le z \le (1/2)N}} |d_{00}^J(z)|$$

For any bi-invariant function $x(a) \in \mathscr{C}^\infty$, which vanishes for $|a| > N$, we get in this fashion the Paley-Wiener estimate (Section 3–11)

$$|\mathscr{F}(J)_x| \le \tfrac{1}{2} M(N, \text{C})|J(J + 1)|^{-q}$$
$$\times \sup_{0 \le \eta < \infty} |D(\cosh \eta)^q x(\eta)| \qquad (6\text{-}81)$$

for each $q = 0, 1, 2, \ldots$ . In particular it can be seen from (5–19), (6–50) that the bound $M(N, \text{C})$ exists on the lines Re $J$ = constant. If $\mathscr{F}^{(+)}(J)$ is bounded by any power of $|J|$ on the contour C, the continuity of the functional follows from (6–81).

We show next that the asymptotic behavior of the Fourier coefficient $\mathscr{F}^{(+)}(J)$ (6–79) on any line Re $J$ = Re $\sigma + \varepsilon$ is determined by the properties of $p(\eta)$ in any interval $1 \le \cosh \eta \le z_0$, $1 < z_0$. In fact, we define two infinitely differentiable functions $g_{1, 2}(z)$ such that

$$g_1(z) = 0 \qquad \text{for} \quad z \ge z_1 > z_0 > 1$$
$$g_2(z) = 0 \qquad \text{for} \quad 1 \le z \le z_0$$

and

$$g_1(z) + g_2(z) = (2z)^\sigma$$

Then

$$|\mathscr{F}^{(+)}(J)_{g_2}| = \left| \int_{z_0}^{\infty} g_2(z) e_{00}^J(z) \, dz \right| \le c_q |J(J + 1)|^{-q}$$

on any line Re $J$ = Re $\sigma + \varepsilon$ and for any $q = 0, 1, 2, \ldots$ , where $c_q$ is independent of $J$. This estimate follows from the uniform validity of the asymptotic expansion (6–57) on the support of $g_2(z)$, and from similar arguments as used for the derivation of the Paley-Wiener estimate (6–81). We have thus obtained the result that $\mathscr{F}^{(+)}(J)$ differs from $\mathscr{F}^{(+)}(J)_{g_1}$ only by a term that vanishes at infinity faster than any power of $|J|$.

Finally we make the following proposition. Let $g(z)$ be an infinitely differentiable function on $1 < z < \infty$ with bounded support. At $z = 1$ it has the one-sided asymptotic expansion

with
$$g(z) \cong \sum_{n=0}^{\infty} a_n(z-1)^{v_n}$$

$v_n$ complex          $\mathrm{Re}\, v_{n+1} > \mathrm{Re}\, v_n,$          $\mathrm{Re}\, v_0 > -1$

The Fourier coefficient of the second kind of $g(z)$ has then the asymptotic expansion at $J = \infty$ (GR 7.133.1 and 8.703)

$$\mathscr{F}^{(+)}(J) \cong \sum_{n=0}^{N} a_n 2^{v_n}[\Gamma(v_n+1)]^2 \frac{\Gamma(J-v_n)}{\Gamma(J+v_n+2)} + R_N(J) \qquad (6\text{-}82)$$

on any line $\mathrm{Re}\, J = \mathrm{Re}\, v_N + \varepsilon$. This proposition can be denoted Watson's lemma, in analogy with a similar lemma occurring in the theory of one-sided Laplace transforms.

We shall not prove Watson's lemma but rather just apply it. With

$$(2z)^\sigma = 2^\sigma \sum_{n=0}^{\infty} \binom{\sigma}{n}(z-1)^n \qquad 1 \le z < 2$$

we get asymptotic expansions of $\mathscr{F}^{(+)}(J)_{g_1}$ on lines $\mathrm{Re}\, J = n + \varepsilon$. Taking into account the result on $\mathscr{F}^{(+)}(J)_{g_2}$ we have

$$\mathscr{F}^{(+)}(J) \cong \sum_{n=0}^{N} 2^{\sigma+n}(n!)^2 \binom{\sigma}{n} \frac{\Gamma(J-n)}{\Gamma(J+n+2)} + R_N(J) \qquad (6\text{-}83)$$

valid on any line $\mathrm{Re}\, J = \max(N, \mathrm{Re}\, \sigma) + \varepsilon$. One can show, however, that the validity of (6–83) extends also to the lines $\mathrm{Re}\, J \ge -\frac{1}{2}$. If $\sigma$ is a nonnegative integer, the series (6–83) breaks off at $N = \sigma$, that is, $R_\sigma(J) = 0$. The series can be summed and represents the function $\mathscr{F}^{(+)}(J)$.

The estimate (6–83) allows us to identify the analytic functional $\mathscr{F}^{(+)}(J)$ (6–79), whose contour is any upward oriented line $\mathrm{Re}\, J = \mathrm{Re}\,\sigma + \varepsilon$, with the Fourier transform of the distribution $p(a) = |a|^{2\sigma}$. We mention that Watson's lemma can of course also be formulated for functions which are bicovariant of general type.

## 6–7   REMARKS

In Sections 6–1 and 6–2 we presented the Fourier transforms as integral kernels following an idea of Gelfand. The proof of the Plancherel theorem presented in Section 6–3 goes back to Harish-Chandra [16]. The first proof of this theorem was given by Bargmann [5]. In Section 6–4 we made use of the notations of Andrews and Gunson [3]. Their article contains many further formulas on the functions $d^J_{q_1 q_2}(z)$ and $e^J_{q_1 q_2}(z)$ and their relations with Jacobi and Legendre functions. In particular we refer to this article for the derivation of the asymptotic formula (6–57).

# Chapter 7

# Harmonic Analysis of Representations of the Group SL(2, C) on the Subgroup SU(1, 1)

The decomposition of a representation of a group G into irreducible representations of a subgroup $G_1$ is a standard problem if G is finite or a compact simple Lie group. In practice one may, for example, apply the theory of characters to actually find the solution. If $G_1$ is a compact subgroup of a noncompact simple Lie group, the situation is not much different. In the two cases $G = SL(2, C)$, $G_1 = SU(2)$ and $G = SU(1, 1)$, $G_1 = U(1)$ the issue has been settled completely by the construction of the canonical basis (in Sections 3–3 and 5–3).

If the subgroup $G_1$ is, however, noncompact and the representation of G considered is a unitary representation of $G_1$, we are in general left with Theorem 3 of Section 1–3c, which guarantees the uniqueness of the decomposition once it is obtained, but does not tell us how to gain it actually. Let us consider the case in which the representation of G is unitary and induced from a unitary representation of a subgroup H of G (Section 1–4b). Let the cosets G/H be representable as

$$G/H = \bigcup_i a_i G_1$$

where $a$ runs over a certain manifold A of G such that the quasi-invariant measure on G/H factorizes as

$$d\mu(\xi) = d\mu(a)\, d\mu(g)$$

where $d\mu(a)$ is any measure on A and $d\mu(g)$ is the invariant measure on $G_1$. The Plancherel theorem of the subgroup $G_1$ provides us then with the method looked for. Of course this technique may work also if $G_1$ is compact. In fact it was this method that we used to find the canonical basis. But for representations of the principal series of SL(2, C) with $H = K$ and $G_1 = SU(1, 1)$, these premises are also fulfilled. The corresponding decomposition of the principal series of SL(2, C) is presented in Sections 7–1 and 7–2. Since this method is based on the Plancherel theorem of the subgroup,

221

it can be extended in a natural way to nonunitary representations of SL(2, C) by applying the concepts of harmonic analysis of distributions on SU(1, 1) displayed in Chapter 6. This extension is dealt with in the remaining sections of this chapter.

## 7–1    A NEW REALIZATION OF THE SPACES OF HOMOGENEOUS FUNCTIONS

The subgroup SU(1, 1) of SL(2, C) consists by definition of those complex $2 \times 2$ matrices which satisfy (1–18)

$$v^\dagger \sigma_3 = \sigma_3 v^{-1} \qquad \det v = 1 \qquad (7\text{-}1)$$

The two subgroups SU(1, 1) and SL(2, R) are isomorphic to each other. The transition from one of these subgroups to the other

$$v \to a = a(v) \quad \text{and} \quad a \to v = v(a)$$

is accomplished by the standard isomorphism (1–19), (5–3). We shall always refer to this specific isomorphism. We call a function $x(v)$ on SU(1, 1) bicovariant of type $q_1, q_2$ and write

$$x(v) = x_{q_1 q_2}(v)$$

if $x(v(a))$ is bicovariant on SL(2, R) of type $q_1, q_2$. This means that

$$v = \exp\left\{\frac{i}{2}\psi_1\sigma_3\right\}\exp\{\tfrac{1}{2}\eta\sigma_2\}\exp\left\{\frac{i}{2}\psi_2\sigma_3\right\} \qquad \eta \geq 0 \qquad (7\text{-}2)$$

implies

$$x_{q_1 q_2}(v) = \exp\{i(q_1\psi_1 + q_2\psi_2)\}x_{q_1 q_2}(\eta) \qquad (7\text{-}3)$$

For the particular problem studied in this chapter it is more convenient to use the group SU(1, 1) instead of SL(2, R) because bicovariant functions on SL(2, C) with respect to SU(2) (Section 4–7) remain bicovariant if restricted to the subgroup SU(1, 1). In fact, with the decomposition (7–2) of $v$ we have for a bicovariant function $x_{j_1 q_1 j_2 q_2}(a)$ on SL(2, C)

$$x_{j_1 q_1 j_2 q_2}(v) = \exp\{i(q_1\psi_1 + q_2\psi_2)\}x_{j_1 q_1 j_2 q_2}(\exp\{\tfrac{1}{2}\eta\sigma_2\})$$

The fact that these bicovariant functions are obtained from bicovariant functions of SL(2, C) allows us to express them linearly by $2 \min(j_1, j_2) + 1$ "diagonal" functions. Because of

$$\exp\left\{\frac{1}{2}\eta\sigma_2\right\} = \exp\left\{\frac{i}{4}\pi\sigma_1\right\}\exp\left\{\frac{1}{2}\eta\sigma_3\right\}\exp\left\{-\frac{i}{4}\pi\sigma_1\right\}$$

$$\exp\left\{\pm\frac{i}{4}\pi\sigma_1\right\} = \exp\left\{\pm\frac{i}{4}\pi\sigma_3\right\}\exp\left\{\frac{i}{4}\pi\sigma_2\right\}\exp\left\{\mp\frac{i}{4}\pi\sigma_3\right\}$$

and

$$D_{q_1 q_2}^{j_1}\left(\exp\left\{\pm\frac{i}{4}\pi\sigma_1\right\}\right) = \exp\left\{\pm\frac{i}{2}\pi(q_1 - q_2)\right\} d_{q_1 q_2}^{j}(0) \qquad (7\text{-}4)$$

we have indeed

$$x_{j_1 q_1 j_2 q_2}\left(\exp\left\{\frac{1}{2}\eta\sigma_2\right\}\right) = \exp\left\{\frac{i}{2}\pi(q_1 + q_2)\right\} \sum_q e^{-i\pi q} d_{q_1 q}^{j_1}(0)$$

$$\times d_{q q_2}^{j_2}(0) x_{j_1 j_2 q}(\eta) \qquad (7\text{-}5)$$

Remember that $x_{j_1 j_2 q}(\eta)$ was defined by (4–81) as

$$x_{j_1 j_2 q}(\eta) = x_{j_1 q j_2 q}(\exp\{\tfrac{1}{2}\eta\sigma_3\})$$

The quantities $d_{q_1 q_2}^{j}(0)$ have been tabulated by Edmonds [8] up to $j = 2$.

We return now to the definition of representations of SL(2, C) on closed topological spaces $\mathscr{D}_\chi$ (Sections 3–1, 3–2). These spaces consist of homogeneous functions $F(z_1, z_2)$, in two complex variables, whose homogeneity is described by the symbol $\chi = \{n_1, n_2\} = (m, \rho)$. A transformation in $\mathscr{D}_\chi$ which represents the group element $a \ \varepsilon$ SL(2, C) was defined by (3–3)

$$T_a^\chi F(z_1, z_2) = F(z_1', z_2')$$

$$(z_1', z_2') = (z_1, z_2)\begin{pmatrix} a_{11} & a_{12} \\ a_{21} & a_{22} \end{pmatrix} \qquad (7\text{-}6)$$

Because of the homogeneity and continuity of the functions $F$ it suffices to know their values on the hyperboloids

$$|z_2|^2 - |z_1|^2 = \tau \quad \tau = \pm 1$$

For $\pi = +1$ we denote

$$z_1 = v_{21}(|z_2|^2 - |z_1|^2)^{1/2}$$

$$z_2 = v_{22}(|z_2|^2 - |z_1|^2)^{1/2} \qquad (7\text{-}7)$$

and for $\tau = -1$

$$z_1 = iv_{11}(|z_1|^2 - |z_2|^2)^{1/2}$$
$$z_2 = iv_{12}(|z_1|^2 - |z_2|^2)^{1/2} \tag{7-8}$$

We fit these parameters $v_{ij}$ into a matrix

$$v = \begin{pmatrix} v_{11} & v_{12} \\ v_{21} & v_{22} \end{pmatrix} \qquad v_{11} = \overline{v_{22}} \qquad v_{12} = \overline{v_{21}}$$

which belongs to SU(1, 1). In this fashion we can realize $\mathscr{D}_\chi$ by means of pairs of functions $\varphi_\tau(v), \tau = \pm 1$, on SU(1, 1). Explicitly we obtain for $\tau = +1$

$$F(z_1, z_2) = (|z_2|^2 - |z_1|^2)^{(i/2)\rho - 1} \varphi_+(v) \tag{7-9}$$

Introducing the variables $z = z_1/z_2 = v_{21}/v_{22}$ and recalling (3–4) we can write

$$\varphi_+(v) = (1 - |z|^2)^{-(i/2)\rho + 1} e^{im\psi} \theta(1 - |z|)f(z) \tag{7-10}$$

Here

$$\psi = -\arg z_2 \tag{7-11}$$

is an arbitrary phase and we have made use of $n_2 - n_1 = m$. Similarly we get for $\pi = -1$ and $z = z_1/z_2 = v_{11}/v_{12}$

$$F(z_1, z_2) = (|z_1|^2 - |z_2|^2)^{(i/2)\rho - 1} \varphi_-(v) \tag{7-12}$$

and

$$\varphi_-(v) = (|z^2| - 1)^{-(i/2)\rho + 1} e^{im\psi} \theta(|z| - 1)f(z) \tag{7-13}$$

With the parameters

$$v_{11} = s^{1/2} \exp\{i\,\theta_1\} \qquad v_{12} = (s - 1)^{1/2} \exp\{i\,\theta_2\}$$
$$1 \leq s < \infty \qquad 0 \leq \theta_{1,2} \leq 2\pi \tag{7-14}$$
$$d\mu(v) = (2\pi)^{-2} ds\, d\theta_1\, d\theta_2$$

we get for $\pi = +1$

$$s = (1 - |z|^2)^{-1} \qquad \theta_1 = \psi \qquad \theta_2 = \psi - \arg z$$

$$d\mu(v) = 2(2\pi)^{-2} \theta(1 - |z|)(1 - |z|^2)^{-2} \, Dz \, d\psi \qquad (7\text{-}15)$$

and for $\tau = -1$

$$s = |z|^2(|z|^2 - 1)^{-1}$$

$$\theta_1 = -\frac{\pi}{2} - \psi + \arg z \qquad \theta_2 = -\frac{\pi}{2} - \psi \qquad (7\text{-}16)$$

$$d\mu(v) = 2(2\pi)^{-2} \theta(|z| - 1)(|z|^2 - 1)^{-2} \, Dz \, d\psi$$

Both functions $\varphi_\tau(v)$ satisfy covariance constraints

$$\varphi_\tau(\gamma v) = e^{i\tau m \omega} \varphi_\tau(v) \qquad (7\text{-}17)$$

where we used the notation (3–11).

Apart from a null set of SL(2, C) any element $a \in$ SL(2, C) can be decomposed in one of the two fashions

$$a = k\varepsilon^\zeta v$$

$$k = \begin{pmatrix} \lambda^{-1} & \mu \\ 0 & \lambda \end{pmatrix} \qquad \varepsilon = \begin{pmatrix} 0 & i \\ i & 0 \end{pmatrix} \qquad \zeta = 0, 1 \qquad (7\text{-}18)$$

The null set is characterized by

$$|a_{21}| = |a_{22}|$$

and corresponds also to a null set of the right cosets SL(2, C)/K. From

$$\begin{pmatrix} 1 & 0 \\ z & 1 \end{pmatrix} = k\varepsilon^\zeta v$$

we derive

$$|z| < 1 \qquad \text{for} \quad \zeta = 0 \quad \text{and} \quad v_{21} = zv_{22} = z(1 - |z|^2)^{-1/2}e^{-i\psi}$$

and

$$|z| > 1 \qquad \text{for} \quad \zeta = 1 \quad \text{and} \quad v_{11} = zv_{12} = -iz(|z|^2 - 1)^{-1/2}e^{-i\psi}$$

The phase $\psi = \arg \lambda$ is arbitrary. These relations become identical with (7–7), (7–8) if we set

$$\psi = -\arg z_2 \qquad 2\zeta = 1 - \tau$$

This proves that the two sets $\tau = \pm 1$ of matrices $v$ can be regarded as parameters for the cosets $SL(2, C)/K$. In the same sense the transformations (7–6) of $\mathscr{D}_\chi$ can be given the form of multiplier representations (Section 1–4a)

$$T_a{}^\chi \varphi_\tau(v) = \alpha(v, \tau, a)\varphi_{\tau_a}(v_a) \tag{7-19}$$

$$\varepsilon^\zeta va = k\varepsilon^{\zeta_a}v_a \qquad 2\zeta_a = 1 - \tau_a \tag{7-20}$$

$$k = \begin{pmatrix} \lambda^{-1}(v, \tau, a) & \mu \\ 0 & \lambda(v, \tau, a) \end{pmatrix} \tag{7-21}$$

$$\alpha(v, \tau, a) = \lambda(v, \tau, a)^{n_1 - 1}\overline{\lambda(v, \tau, a)}^{n_2 - 1} \tag{7-22}$$

## 7–2    THE DECOMPOSITION OF THE PRINCIPAL SERIES OF SL(2, C)

In the case of the principal series the space $\mathscr{D}_\chi$ is dense in the Hilbert space $\mathscr{L}^2(Z)$ of measurable functions with finite invariant norm

$$\|f\|^2 = \int |f(z)|^2 \, Dz$$

We use (7–10), (7–13) to replace $f(z)$ by the pair of functions $\varphi_\tau(v)$ and (7–15), (7–16) to replace the measure $Dz$ by $d\mu_\tau(v)$. This gives

$$\|f\|^2 = \pi \sum_\tau \int |\varphi_\tau(v)|^2 \, d\mu_\tau(v) \tag{7-23}$$

We denote the Hilbert space of measurable functions $\varphi_\tau(v)$ ($\tau$ fixed) on $SU(1, 1)$ which are covariant in the same way as (7–17) and have finite norm squared (and a corresponding scalar product)

$$\|\varphi\|_\tau{}^2 = \int |\varphi_\tau(v)|^2 \, d\mu(v)$$

by $\mathscr{L}^2_{\tau m}(V)$. As (7–23) shows, the direct orthogonal sum

$$\mathscr{L}^2_{|m|}(V) = \sum_\tau{}^\oplus \mathscr{L}^2_{\tau m}(V)$$

possesses an invariant scalar product under the operations (7–19), which can therefore be extended from $\mathscr{D}_\chi$ onto $\mathscr{L}^2_{|m|}(V)$. The space $\mathscr{L}^2_{|m|}(V)$ can in this way be made to carry the representation $\chi$ of the principal series. The space $\mathscr{L}^2_{|m|}(V)$ can in this way be made to carry the representation $\chi$ of the principal series.

Either Hilbert space $\mathscr{L}^2_{\tau m}(V)$ can be regarded as a subspace of the Hilbert space $\mathscr{L}^2(V)$ of square integrable functions on $SU(1, 1)$. If we restrict the group $SL(2, C)$ to the subgroup $SU(1, 1)$, the transformation (7–19) is

$$T_v{}^\chi \varphi_\tau(v_1) = \varphi_\tau(v_1 v) \tag{7-24}$$

This means that the representation $\chi$ of SL(2, C) restricted to SU(1, 1) leaves the subspaces $\mathscr{L}^2_{\tau m}(V)$ of $\mathscr{L}^2(V)$ invariant, and that this representation of SU(1, 1) is identical to the right-regular representation of SU(1, 1) restricted to the invariant subspace $\mathscr{L}^2_{\tau m}(V)$.

If we decompose the space $\mathscr{L}^2(V)$ with the help of the Plancherel theorem into a direct integral and direct orthogonal sum of Hilbert spaces such that the right-regular representation induces in each of these Hilbert spaces an irreducible representation of SU(1, 1), then the spaces $\mathscr{L}^2(V)$ embedded into $\mathscr{L}^2(V)$ are submitted to the same decomposition. For an arbitrary vector $\varphi(v)$ of $\mathscr{L}^2(V)$ we obtain from (6—68)

$$\varphi(v) = \frac{1}{2i} \int_{-1/2-i\infty}^{-1/2+i\infty} dJ(2J + 1) \sum_{q_1 q_2} \cot \pi(J - q_1)$$

$$\times D^J_{q_1 q_2}(v) \int \varphi(v') \overline{D^J_{q_1 q_2}(v')} \, d\mu(v')$$

$$+ \sum_{J=0, \, 1/2, \, 1, \, \ldots}^{\infty} (2J + 1) \sum_{q_1 q_2} D^J_{q_1 q_2}(v)$$

$$\times \int \varphi(v') \overline{D^J_{q_1 q_2}(v')} \, d\mu(v') \tag{7-25}$$

where in the discrete $J$ sum both $q_1$ and $q_2$ are restricted either to values not smaller than $J + 1$ or to values not bigger than $-J-1$, and in both cases to values for which $J-q_{1,2}$ are integers. The Fourier coefficients

$$\int \varphi(v) \overline{D^J_{q_1 q_2}(v)} \, d\mu(v)$$

exist for almost all $J$, the integrals and sums over $J$ and $q_{1,2}$ converge in the sense of the norm of $\mathscr{L}^2(V)$. If we replace the integration variable $v$ in the Fourier coefficient of $\varphi_\tau(v)$ by $\gamma v$, $\gamma$ as in (3—11), the covariance constraint (7—17) of the function $\gamma_\tau(v)$ implies that the Fourier coefficient gets multiplied with the factor

$$\exp\{i\omega(\tau m - 2q_1)\}$$

and we find

$$\int \varphi_\tau(v) \overline{D^J_{q_1 q_2}(v)} \, d\mu(v) = 0 \qquad \text{if} \quad q_1 \neq \tfrac{1}{2}\tau m$$

With the notation

$$v = \text{sign}(\tau m)$$

the decomposition of the vector $\varphi_\tau(v) \varepsilon \mathscr{L}^2_{\tau m}(V)$ is therefore

$$\varphi_\tau(v) = \frac{1}{2i} \int\limits_{-1/2-i\infty}^{-1/2+i\infty} dJ(2J+1)\cot \pi(J - \tfrac{1}{2}m)$$

$$\times \sum_q D^J_{(1/2)\tau m, q}(v) \int \varphi_\tau(v') \overline{D^J_{(1/2)\tau m, q}(v')} \, d\mu(v')$$

$$+ \sum_{J \geqq 0}^{(1/2)|m|-1} (2J+1) \sum_{vq=J+1}^{\infty} D^J_{(1/2)\tau m, q}(v)$$

$$\times \int \varphi_\tau(v') \overline{D^J_{(1/2)\tau m, q}(v')} \, d\mu(v') \tag{7-26}$$

where the range of $q$ and $J$ in the discrete $J$ sum is further restricted to integral values $\tfrac{1}{2}m-q$, $\tfrac{1}{2}m-J$. Formula (7–26) contains the complete solution of the problem of decomposing any representation $\chi$ of the principal series of SL(2, C) into unitary representations of SU(1, 1). Only those representations of SU(1, 1) appear for which $\varepsilon \cong m$ mod 2. The representations of the principal series of SU(1, 1) appear twice, the degeneracy happens to be removed by $\tau$. The discrete series contributes at most a finite number of representations. If $v$ is positive (negative) these belong to the positive (negative) discrete series.

It is sometimes convenient to regard the functions

$$\varphi_q^{J,\,\tau}(v) = D^J_{(1/2)\tau m, q}(v)$$

where $J$ ranges over the interval $0 < -i(J + \tfrac{1}{2}) < \infty$ and $q$ over those values for which $\tfrac{1}{2}m - q$ is integral, or $J$ is one number of the finite set

$$\tfrac{1}{2}|m| - 1, \tfrac{1}{2}|m| - 2, \ldots \geqq 0$$

and $q$ is such that $vq-J$ is positive integral, as the elements of a pseudobasis in $\mathscr{L}^2_{\tau m}(V)$. The elements of the $J$-continuous part of this basis are nonnormalizable; we call them pseudovectors and the whole set pseudobasis for this reason. The basis elements belonging to the discrete series have norm $(2J + 1)^{-1/2}$. This pseudobasis resembles the canonical basis in several respects. The transition between these two bases is studied in the subsequent section.

## 7–3  THE TRANSITION FROM THE CANONICAL BASIS TO THE PSEUDOBASIS

In the representation space $\mathscr{L}^2_{|m|}(V)$ of the principal series of SL(2, C) the transition from the canonical basis to the pseudobasis can be accomplished by means of coefficients $M(\chi|j, J, \tau, q)$ that are defined as scalar products of the elements $\varphi_q^{j,\tau}$ of the pseudobasis with elements $\varphi_q^j$ of the canonical basis in the improper sense. The first step towards the construction of these coefficients consists in finding the pair of functions $\varphi_q^{j,\tau}(v)$ of $\mathscr{L}_{\tau m}^2(V)$ which corresponds to the element $\varphi_q^j(u)$ of the canonical basis in $\mathscr{L}_m^2(U)$.

Let

$$u = \begin{pmatrix} u_{11} & u_{12} \\ u_{21} & u_{22} \end{pmatrix} \qquad v = \begin{pmatrix} v_{11} & v_{12} \\ v_{21} & v_{22} \end{pmatrix}$$

be elements of SU(2) and SU(1, 1), respectively. We denote

$$t = |u_{22}|^2 = \cos^2 \tfrac{1}{2}\vartheta \qquad s = |v_{22}|^2 = \cosh^2 \tfrac{1}{2}\eta \qquad (7\text{-}27)$$

Replacing the homogeneous coordinates $z_1$, $z_2$ first by $u$ and then by $v$ in the fashion described by (3–13), respectively (7–9), (7–12), yields for $\tau = +1$

$$\frac{u_{21}}{u_{22}} = \frac{v_{21}}{v_{22}} \qquad \arg u_{21} = \arg v_{21} \qquad \arg u_{22} = \arg v_{22}$$

and for $\tau = -1$

$$\frac{u_{21}}{u_{22}} = \frac{v_{11}}{v_{12}} \qquad \arg u_{21} = \arg(iv_{11}) \qquad \arg u_{22} = \arg(iv_{12})$$

These relations can also be brought into the form

$$\tau = +1: \qquad u_+(v) = \exp\left\{\frac{i}{2}\psi_1\sigma_3\right\}\exp\left\{\frac{i}{2}\vartheta\sigma_1\right\}\exp\left\{\frac{i}{2}\psi_2\sigma_3\right\}$$

$$\tanh \tfrac{1}{2}\eta = \tan \tfrac{1}{2}\vartheta \qquad t = s(2s-1)^{-1} \qquad 0 \leq \eta < \infty$$

$$0 \leq \vartheta \leq \tfrac{1}{2}\pi \qquad 1 \leq s < \infty \qquad \tfrac{1}{2} \leq t \leq 1$$

$$\tau = -1: \qquad u_-(v) = \exp\left\{-\frac{i}{2}\psi_1\sigma_3\right\}\exp\left\{\frac{i}{2}\vartheta\sigma_1\right\}\exp\left\{\frac{i}{2}\psi_2\sigma_3\right\}$$

$$\coth \tfrac{1}{2}\eta = \tan \tfrac{1}{2}\vartheta \qquad t = (s-1)(2s-1)^{-1} \qquad 0 \leq \eta < \infty$$

$$\tfrac{1}{2}\pi \leq \vartheta \leq \pi \qquad 1 \leq s < \infty \qquad 0 \leq t \leq \tfrac{1}{2}$$

where $v$ has been parametrized as in (7–2).

Inserting these expressions $u_\tau(v)$ into (7–9), (7–12) we get

$$\varphi_q^{j,\tau}(v) = (2j+1)^{1/2}(2s-1)^{(i/2)\rho-1}D_{(1/2)m, q}^j(u_\tau(v)) \qquad (7\text{-}28)$$

where we have used

$$\frac{|z_1|^2 + |z_2|^2}{\tau(|z_2|^2 - |z_1|^2)} = 2s - 1$$

If $\chi$ is in the principal series of SL(2, C) and $J$ in the principal or discrete series of SU(1, 1), we define the transition coefficients by the absolutely convergent integrals

$$\delta_{qq'} M(\chi \,|\, j, J, \tau, q) = \int \overline{\varphi_q^{j, \tau}(v)} \varphi_{q'}^{J, \tau}(v) \, d\mu(v) \qquad (7\text{-}29)$$

which yields

$$M(\chi \,|\, j, J, \tau, q) = (2j + 1)^{1/2} \exp\{i\tfrac{1}{2}\pi(q - \tfrac{1}{2}m)\}$$

$$\times \int_1^\infty ds(2s - 1)^{-(i/2)\rho - 1}$$

$$\times d_{(1/2)m, q}^j(\cos \vartheta_\tau(s)) \, d_{(1/2)\tau m, q}^J(2s - 1) \qquad (7\text{-}30)$$

The phase $\exp\{i\tfrac{1}{2}\pi(q - \tfrac{1}{2}m)\}$ stems from the decomposition

$$\exp\!\left(\frac{i}{2}\vartheta\sigma_1\right) = \exp\!\left(\frac{i}{4}\pi\sigma_3\right)\exp\!\left(\frac{i}{2}\vartheta\sigma_2\right)\exp\!\left(-\frac{i}{4}\pi\sigma_3\right)$$

The two angles $\vartheta_\tau(s)$ are related by

$$\vartheta_+(s) + \vartheta_-(s) = \pi$$

so that the third identity (2–31) allows us to express $M$ as

$$M(\chi \,|\, j, J, \tau, q) = (2j + 1)^{1/2} \exp\!\left\{\frac{i}{2}\pi(q - \tfrac{1}{2}m + (1 - \tau)(j + q))\right\}$$

$$\times \int_1^\infty ds(2s - 1)^{-(i/2)\rho - 1}$$

$$\times d_{(1/2)\tau m, q}^j((2s - 1)^{-1}) \, d_{(1/2)\tau m, q}^J(2s - 1) \qquad (7\text{-}31)$$

The evaluation of the integral (7–31) is rather laborious. We skip over the long computations and give the result, which takes a different form for each of the following four cases

$$\text{(A)} \qquad \tfrac{1}{2}\tau m \geq |q|$$

$$\text{(B)} \qquad q \geq \tfrac{1}{2}|\tau m|$$

$$\text{(C)} \quad -\tfrac{1}{2}\tau m \geq |q|$$

$$\text{(D)} \quad -q \geq \tfrac{1}{2}|\tau m|$$

In case (A) we get

$$M(\chi\,|\,j, J, \tau, q)$$

$$= 2^{(i/2)\rho}\,\exp\!\left\{\frac{i}{2}\,\pi\delta_\tau^{(A)}\right\}\gamma_{(1/2)\tau m,\, q}^J$$

$$\times \sum_{s=0}^{\tau m}\sum_{v=0}^{j+q}(-1)^s\binom{\tau m}{s}\,\Delta_v(j, \tfrac{1}{2}\tau m, q)$$

$$\times (-2J - 1 + \tau m - 2s)\,\frac{\Gamma(-2J - 1 - s)}{\Gamma(-2J + \tau m - s)}$$

$$\times \frac{\Gamma((1/2)((i/2)\rho - J + \tau m + v - s))\Gamma((1/2)((i/2)\rho + J + 1 + v + s))}{\Gamma(1/2)\Gamma((i/2)\rho - q + v + 1)}$$

$$(7\text{-}32)$$

$\Delta_v(j, q_1, q_2)$ is the square root of a rational number and is independent of $J$ and $\rho$. We have for $q_1 \geq |q_2|$

$$\Delta_v(j, q_1, q_2) = [(2j + 1)(j - q_1)!(j + q_1)!(j - q_2)!(j + q_2)!]^{1/2}$$

$$\times \sum_{\alpha,\,\beta}(-1)^{\alpha+\beta+v}2^{\beta+v-j+2q_1-2}(j + q_2 - \beta)!$$

$$\times [\alpha!\,\beta!(q_1 - q_2 + \alpha)!(j - q_1 - \alpha)!(j + q_2 - \alpha - \beta)!$$

$$\times (v - \beta)!(j + q_2 - v)!]^{-1} \qquad (7\text{-}33)$$

where the double sum extends over all those values $\alpha, \beta$ for which none of the factorials in the denominator is infinite. If we make use of the quantities

$$\Delta_{q_1 q_2}^j = d_{q_1 q_2}^j(0)$$

which are tabulated in Edmonds [8] up to $j = 2$, we are able to perform one summation in (7–33) and get

$$\Delta_v(j, q_1, q_2) = (2j + 1)^{1/2}\sum_{k=0}^{v}(-1)^{k+v}2^{v+2q_1+(1/2)k-2}$$

$$\times \binom{j + q_2 - k}{v - k}\left[\binom{j + q_1}{k}\binom{j + q_2}{k}\right]^{1/2}\Delta_{q_1 - (1/2)k,\, q_2 - (1/2)k}^{j - (1/2)k}$$

$$(7\text{-}34)$$

The function $\gamma^J_{q_1 q_2}$ appearing in (7–32) is defined as

$$\gamma^J_{q_1 q_2} = \left\{ \frac{\Gamma(J + q_1 + 1)\Gamma(J - q_2 + 1)}{\Gamma(J - q_1 + 1)\Gamma(J + q_2 + 1)} \right\}^{1/2} \tag{7-35}$$

with the phase convention as in (6–50). The phase $\delta^{(A)}_\tau$ is

$$\delta^{(A)}_\tau = (1 + \tau)(q - \tfrac{1}{2}m) + (1 - \tau)(j + 2q) \tag{7-36}$$

In case (B) the result can be obtained from (7–32) by replacing

$$q \to \tfrac{1}{2}\tau m \qquad \tfrac{1}{2}\tau m \to q$$

and the phase $\delta^{(A)}_\tau$ by $\delta^{(B)}_\tau$

$$\delta^{(B)}_\tau = (1 - \tau)(j + q - \tfrac{1}{2}m) \tag{7-37}$$

In case (C) we replace in (7–32)

$$q \to -q \qquad \tfrac{1}{2}\tau m \to -\tfrac{1}{2}\tau m$$

and take the phase $\delta^{(C)}_\tau = \delta^{(B)}_\tau$. Finally in case (D) we leave the phase unchanged, $\delta^{(D)}_\tau = \delta^{(A)}_\tau$, and replace

$$q \to -\tfrac{1}{2}\tau m \qquad \tfrac{1}{2}\tau m \to -q$$

In all cases (A) to (D) the function $M$ exhibits the symmetry

$$M(m, \rho \,|\, j, J, \tau, q) = \exp\{i\pi(-\tfrac{1}{2}m + j + q)\}$$
$$\times M(-m, \rho \,|\, j, J, -\tau, q) \tag{7-38}$$

which can be read off its integral representation (7–31). The analytic properties of $M$ are studied in detail in the subsequent section.

   We close this section with a heuristic consideration. The matrix $M$ is usually applied to problems of the following kind. An operator $A$ in the Hilbert space $\mathcal{L}^2_{|m|}(V)$ carrying the representation $\chi$ of SL(2, C) may be given as a matrix with respect to the canonical basis

$$\langle j_1 q_1 \,|\, A \,|\, j_2 q_2 \rangle = \sum_{\tau, \tau_2} \langle j_1 \tau_1 q_1 \,|\, A \,|\, j_2 \tau_2 q_2 \rangle$$

$$\langle j_1 \tau_1 q_1 \,|\, A \,|\, j_2 \tau_2 q_2 \rangle = \int \overline{\varphi^{j_1 \tau_1}_{q_1}(v)} A \varphi^{j_2 \tau_2}_{q_2}(v) \, d\mu(v)$$

and is to be cast into a matrix form with respect to the pseudobasis. Such change of basis is of particular interest if $A$ is partially diagonal on the pseudobasis as

$$A\varphi_q^{J,\,\tau}(v) = \sum_{q'} A_{q'q}^{J;\,\tau}\,\varphi_{q'}^{J;\,\tau}(v) \tag{7-39}$$

The Plancherel theorem (provided it applies of course) solves this change of basis problem

$$\langle j_1\tau_1 q_1 | A | j_2\tau_2 q_2\rangle$$

$$= -\tfrac{1}{4} \int_{-1/2-i\infty}^{-1/2+i\infty} dJ_1(2J_1 + 1)\cot \pi(J_1 - q_1) \int_{-1/2-i\infty}^{-1/2+i\infty} dJ_2(2J_2 + 1)$$

$$\times \cot \pi(J_2 - q_2)\langle J_1\tau_1 q_1 | A | J_2\tau_2 q_2\rangle$$

$$\times M(\chi | j_1, J_1, \tau_1, q_1)\overline{M(\chi | j_2, J_2, \tau_2, q_2)} \tag{7-40}$$

$$+ \text{ sums over the discrete series}$$

If $A$ is partially diagonal as defined by (7–39), (7–40) reduces to

$$\langle j_1, \tau, q_1 | A | j_2, \tau, q_2\rangle = \frac{1}{2i} \int_{-1/2-i\infty}^{-1/2+i\infty} dJ(2J + 1)\cot \pi(J - q_1)A_{q_1q_2}^{J,\,\tau}$$

$$\times M(\chi | j_1, J, \tau, q_1)\overline{M(\chi | j_2, J, \tau, q_2)}$$

$$+ \sum_{J\geqq 0}^{J'} (2J + 1)A_{q_1q_2}^{J;\,\tau} M(\chi | j_1, J, \tau, q_1)$$

$$\times \overline{M(\chi | j_2, J, \tau, q_2)} \tag{7-41}$$

where $J' = \min (\tfrac{1}{2}|m|, |q_1|, |q_2|) - 1$. If $A$ is the unit operator, Eq. (7–41) reduces further to the orthogonality relation

$$\frac{1}{2i} \int_{-1/2-i\infty}^{-1/2+i\infty} dJ(2J + 1)\cot \pi(J - q)$$

$$\times \sum_{\tau} M(\chi | j_1, J, \tau, q)\overline{M(\chi | j_2, J, \tau, q)}$$

$$+ \sum_{J\geqq 0}^{J'} (2J + 1) \sum_{\tau} M(\chi | j_1, J, \tau, q)\overline{M(\chi | j_2, J, \tau, q)}$$

$$= \delta_{j_1j_2} \tag{7-42}$$

where $|q| \leqq \min(j_1, j_2)$. In order to prove (7–42) directly we apply Parseval's formula (6–38) to the square integrable functions (7–28)

$$x_\tau(v) = \varphi_{q_1}^{j_1, \tau}(v) \qquad y_\tau(v) = \varphi_{q_2}^{j_2, \tau}(v)$$

We find

$$\int \overline{x_\tau(v)} y_\tau(v) \, d\mu(v) = \delta_{q_1 q_2}(2j_1 + 1)^{1/2}(2j_2 + 1)^{1/2}$$

$$\times \tfrac{1}{2} \int\limits_0^1 dz \, d_{(1/2)m, q_1}^{j_1}(\tau z) \, d_{(1/2)m, q_1}^{j_2}(\tau z)$$

If we sum over $\tau$, the right-hand side gives $\delta_{q_1 q_2} \, \delta_{j_1 j_2}$, which leads us to the asserted relation (7–42).

## 7–4    THE DECOMPOSITION OF A GENERAL REPRESENTATION OF SL(2, C) ON THE SUBGROUP SU(1, 1) BY MEANS OF CONTOUR INTEGRALS

After the preliminaries of Section 7–3 we return to our problem proper, the harmonic analysis of a representation $\chi$ of $SL(2, C)$ on the subgroup $SU(1, 1)$. The results of Sections 7–1 and 7–2 suggest an investigation of the function

$$p_{j_1 q_1 j_2 q_2}^{\chi, \tau}(v) = \langle j_1 \tau q_1 | T_v^\chi | j_2 \tau q_2 \rangle$$

This bicovariant function on $SU(1, 1)$ can be reduced as in (7–3) to the function

$$p_{j_1 q_1 j_2 q_2}^{\chi, \tau}(\eta) = \langle j_1 \tau q_1 | T_{\exp\{(1/2)\eta \sigma_2\}}^\chi | j_2 \tau q_2 \rangle$$

The sum over $\tau$ of $p_{j_1 q_1 j_2 q_2}^{\chi, \tau}(\exp\{\tfrac{1}{2}\eta \sigma_2\})$ is related to the function $d_{j_1 j_2 q}^\chi(\eta)$ (3–32) by (7–5)

$$\sum_\tau p_{j_1 q_1 j_2 q_2}^{\chi, \tau}(\eta) = \exp\left\{\frac{i}{2}\pi(q_1 + q_2)\right\} \sum_q e^{-i\pi q} d_{q_1 q}^{j_1}(0)$$

$$\times d_{q q_2}^{j_2}(0) \, d_{j_1 j_2 q}^\chi(\eta) \qquad\qquad (7\text{–}43)$$

The functions $p_{j_1 q_1 j_2 q_2}^{\chi, \tau}(v)$ are square integrable on $SU(1, 1)$ if $\chi$ is in the principal series of $SL(2, C)$ and polynomially bounded for other $\chi$. To prove this we compute the matrix elements in $\mathcal{L}_m^2(U)$. In this space the states $|j, \tau, q \rangle$ are realized as the functions $\varphi_q^j(u)$ of the canonical basis (3–20) multiplied with the factor $\theta(\tau(t - \tfrac{1}{2}))$. These functions are bounded by $(2j + 1)^{1/2}$. Let $\varphi_1(u)$, $\varphi_2(u)$ be any two essentially bounded functions of $\mathcal{L}_m^2(U)$. For any such pair we get

$$\left| \int \overline{\varphi_1(u)} T_a^\chi \varphi_2(u) \, d\mu(u) \right| \leqq \text{const} \int |\alpha(u, a)| \, d\mu(u)$$

$$= \text{const} \frac{\sinh((1/2)\eta \, \text{Im} \, \rho)}{(\sinh \eta)((1/2)\text{Im} \, \rho)}$$

where $\alpha(u, a)$ is the multiplier (3–19) and

$$|a|^2 = 2 \cosh \eta \qquad \eta \geqq 0$$

On the principal series this estimate gives

$$\leqq \text{const} \frac{\eta}{\sinh \eta}$$

We may therefore Fourier analyze $p_{j_1 q_1 j_2 q_2}^{\chi, \tau}(v)$ for $\chi$ in the principal series of SL(2, C) in the $\mathcal{L}^2$ sense. With the notations (6–69), (6–70) we have (6–68)

$$p_{j_1 q_1 j_2 q_2}^{\chi, \tau}(\eta) = \frac{1}{2i} \int_{-1/2-i\infty}^{-1/2+i\infty} dJ(2J + 1) \cot \pi(J - q_1)$$

$$\times d_{q_1 q_2}^J(\cosh \eta) \mathscr{F}_{q_1 q_2}(J)$$

$$+ \sum_{J \geqq 0}^{J'} (2J + 1) \, d_{q_1 q_2}^J(\cosh \eta) \mathscr{F}_{q_1 q_2}(J) \qquad (7\text{-}44)$$

where $J' = \min(|q_1|, |q_2|) - 1$. In the Fourier coefficients we dropped the labels $\chi, \tau, j_1, j_2$. In (7–44) the sum over the discrete series is void if $q_1 q_2 \leqq 0$. Comparing (7–44) with (7–41), where $A$ is identical with $T_{\exp\{(1/2)\eta\sigma_2\}}^\chi$, that is,

$$A_{q_1 q_2}^{J, \tau} = d_{q_1 q_2}^J(\cosh \eta)$$

we get

$$\mathscr{F}_{q_1 q_2}(J) = M(\chi \mid j_1, J, \tau, q_1) \overline{M(\chi \mid j_2, J, \tau, q_2)} \qquad (7\text{-}45)$$

In fact from (7–45) and (7–31) we obtain

$$\mathscr{F}_{q_1 q_2}(-J - 1) = (-1)^{q_1 - q_2} \mathscr{F}_{q_1 q_2}(J)$$

as we should have expected for a Fourier coefficient of the first kind. The integral representation (7–31) and the reality of the functions $d^J_{q_1 q_2}(z + i0 +)$ on the principal and discrete series of SU(1, 1) and $1 \leqq z < \infty$ [see the arguments given after (6–70)] imply

$$\overline{M(m, \rho \mid j, J, \tau, q)} = (-1)^{(1/2)m - q} M(m, -\rho \mid j, J, \tau, q)$$

and therefore finally

$$\mathscr{F}_{q_1 q_2}(J) = (-1)^{(1/2)m - q_2} M(m, \rho \mid j_1, J, \tau, q_1) M(m, -\rho \mid j_2, J, \tau, q_2) \qquad (7\text{-}46)$$

The Fourier coefficient (7–46) can now be continued analytically off the real $\sigma$-axis. Its analytic properties can be read off (7–31) or (7–32).

     As can be inspected from (7–32) the function $M(J)$ is meromorphic in $J$ apart from the square root branch points introduced by the factor $\gamma^J_{1/2\tau m, q}$ (7–35). This factor has been completely discussed in Section 6–4, and we may omit it from our further investigations. $M(J)$ exhibits in addition two sequences of first order poles at

$$J = \frac{i}{2}\rho + n_1 \qquad n_1 = 0, 1, 2, \ldots \qquad (7\text{-}47)$$

and

$$J = -\frac{i}{2}\rho - 1 - n_2 \qquad n_2 = 0, 1, 2, \ldots \qquad (7\text{-}48)$$

Coincidences between these sequences can be avoided if we require that

$$i\rho + 1 \neq -n \qquad n = 0, 1, 2, \ldots$$

There is a further finite sequence of first order poles whose positions are independent of $\rho$ ("fixed poles"), namely at integral and half-integral points in the interval

$$-\max(\tfrac{1}{2}|m|, |q|) \leqq J \leqq \max(\tfrac{1}{2}|m|, |q|) - 1$$

If we require that

$$\pm i\rho + 1 \neq -n \qquad n = 0, 1, 2, \ldots$$

the function $\mathscr{F}_{q_1 q_2}(J)$ (7–46) possesses first order poles at

$$J = \frac{i}{2}\rho + n_1 \qquad J = -\frac{i}{2}\rho - 1 - n_2 \qquad J = -\frac{i}{2}\rho + n_3$$

$$J = \frac{i}{2}\rho - 1 - n_4 \qquad n_{1, 2, 3, 4} = 0, 1, 2, \ldots \qquad (7\text{-}49)$$

and also a finite number of fixed poles.

The function $p_{j_1 q_1 j_2 q_2}^{\chi, \tau}(\eta)$ admits also Fourier coefficients $\mathscr{F}_{q_1 q_2}^+(J)$ of the second kind, since $p_{j_1 q_1 j_2 q_2}^{\chi, \tau}(v)$ is infinitely differentiable and bicovariant on the group SU(1, 1). If $\chi$ is in the principal series of SL(2, C), this Fourier coefficient of the second kind is, as usual, holomorphic to the right of the contour $C_+$, that is, it exhibits only the poles of $\mathscr{F}_{q_1 q_2}(J)$ labeled $n_2$ and $n_4$ in (7−49) and fixed poles on the real $J$-axis. The inverse Fourier transformation of $p_{j_1 q_1 j_2 q_2}^{\chi, \tau}(v)$ can therefore be cast into the form (6−72)

$$p_{j_1 q_1 j_2 q_2}^{\chi, \tau}(\eta) = \frac{1}{2\pi i} \int_{C_+} dJ(2J + 1) d_{q_1 q_2}^J (\cosh \eta) \mathscr{F}_{q_1 q_2}^{(+)}(J) \qquad (7\text{-}50)$$

We may then abandon the premise that $\chi$ is in the principal series of SL(2, C) and continue $\mathscr{F}_{q_1 q_2}^+(J)$ in $\rho$ off the real axis. In the course of this continuation the two sequences of poles labeled $n_2$ and $n_4$ in (7−49) may impinge on the contour $C_+$ and force us to deform it, for example, exerting on it a parallel displacement to the right. The analog of Watson's lemma (Section 6−6) guarantees a falloff of $\mathscr{F}_{q_1 q_2}^+(J)$ at Im $J \to \pm\infty$, which permits us to make this displacement. This accomplishes the decomposition of any irreducible representation of SL(2, C) (with exception of the cases $\pm i\sigma + 1 = -n$) into irreducible representations of SU(1, 1) by means of the method of analytic functionals.

### 7−5   THE DECOMPOSITION OF A GENERAL REPRESENTATION OF SL(2, C) ON THE SUBGROUP SU(1, 1) BY MEANS OF SERIES EXPANSIONS

We emphasize that the method of decomposing a representation of SL(2, C) into representations of SU(1, 1), which was displayed in Section 7−4, is not the only one possible. In fact, we shall now investigate a concept which lies outside the scope of analytic functionals though it is closely related to it. Whereas the method of analytic functionals expresses functions of the first kind of SL(2, C) as a contour integral (7−50) over functions of the first kind of the subgroup SU(1, 1), we are now going to derive a series expansion (7−60) of the functions of the second kind of SL(2, C) into functions of the second kind of SU(1, 1). The kind of functions used in either case is, to be sure, only a natural and not an obligatory choice.

We start from (7−44) and rewrite this formula as

$$p_{j_1 q_1 j_2 q_2}^{\chi, \tau}(\eta) = \frac{1}{\pi i} \int_{-1/2 - i\infty}^{-1/2 + i\infty} dJ(2J + 1) \mathscr{F}_{q_1 q_2}(J) e_{q_1 q_2}^J (\cosh \eta)$$

$$+ \text{ terms of the discrete series identical with those}$$
$$\text{in (7−44)} \qquad (7\text{-}51)$$

In (7–51) $\chi$ is still regarded as belonging to the principal series. If we consider general representations $\chi$, restricting $\varphi$, however, as in Section 7–4 to

$$\pm i\rho + 1 \neq -n \qquad n = 0, 1, 2, \ldots$$

we must deform the integration path and obtain a contour $C_0$ as depicted in Fig. 7–1.

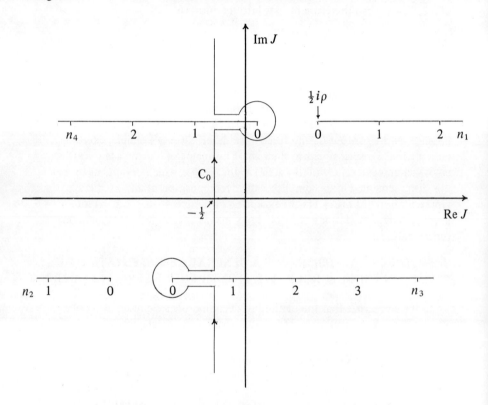

**Figure 7–1.** The integration contour $C_0$

We try to evaluate the integral (7–51) by the residue method, that is, we replace the contour $C_0$ by a sequence of contours $C_n, n = 1, 2, 3, \ldots$, which in the limit $n \to \infty$ encircle all poles of the integrand to the right of $C_0$ once in the positive sense. Among these poles are

(1) two sequences of first order poles labeled $n_1$ and $n_3$ in (7–49);
(2) a finite set of fixed poles of $\mathscr{F}_{q_1 q_2}(J)$ as discussed in Section 7–4;
(3) a finite set of fixed first order poles of $e_{q_1 q_2}^J(\cosh \eta)$ as studied in Section 6–4.

The contributions of the finite sets (1) and (2) converge in a trivial fashion. We lump them together with the finite number of terms of the discrete series and denote the result $R_f^{\chi,\tau}(\eta)$. Later in this section we shall prove that a certain sequence of contours $C_n$ exists for which

$$\lim_{n\to\infty} \frac{1}{\pi i} \int_{C_n} dJ(2J+1)\mathscr{F}_{q_1q_2}(J)e^J_{q_1q_2}(\cosh \eta) = 0$$

uniformly in any interval $0 < \eta_1 \leq \eta \leq \eta_2 < \infty$. The two sequences (1) give therefore a convergent series. This assertion is referred to as proposition I. In fact, either sequence $n_1$ or $n_3$ yields a separately convergent series as can be inspected easily from the proof of proposition I.

We denote

$$W_n^{\chi,\tau}(j', q) = \operatorname*{Res}_{J=-(i/2)\rho+n} M(m, -\rho \,|\, j, J, \tau, q) \tag{7-52}$$

and in particular

$$W_n^{\chi}(j, q) = W_n^{\chi,+}(j, q)$$

As "proposition II" we shall later prove the symmetry relation

$$W_n^{\chi,\tau}(j, q) = \exp\{-i\pi(\tfrac{1}{2}\tau m + n)\}W_n^{\chi,-\tau}(j, q) \tag{7-53}$$

Using (7–52) and (7–46) we can express the residues of $\mathscr{F}_{q_1q_2}(J)$ as

$$\operatorname*{Res}_{J=-(i/2)\rho+n} \mathscr{F}_{q_1q_2}(J) = (-1)^{-(1/2)m+q_2}W_n^{\chi,\tau}(j_2, q_2)$$

$$\times M\left(\chi \,|\, j_1, -\frac{i}{2}\rho + n, \tau, q_1\right) \tag{7-54}$$

$$\operatorname*{Res}_{J=(i/2)\rho+n} \mathscr{F}_{q_1q_2}(J) = (-1)^{(1/2)m+q_1+j_1-j_2}W_n^{-\chi,-\tau}(j_1, q_1)$$

$$\times M\left(-\chi \,|\, j_2, \frac{i}{2}\rho + n, -\tau, q_2\right) \tag{7-55}$$

where we used (7–38) to change the sign of $m$ in (7–55). We obtain this way

$$p_{j_1q_1j_2q_2}^{\chi,\tau}(\eta) = 2 \sum_{n_3=0}^{\infty} [(i\rho - 2n_3 - 1) \operatorname*{Res}_{J=-(i/2)\rho+n_3} \mathscr{F}_{q_1q_2}(J)$$

$$\times e_{q_1q_2}^{-(i/2)\rho+n_3}(\cosh \eta)$$

$$+ 2 \sum_{n_1=0}^{\infty} [(-i\rho - 2n_1 - 1) \operatorname*{Res}_{J=(i/2)\rho+n_1} \mathscr{F}_{q_1 q_2}(J)]$$

$$\times e_{q_1 q_2}^{(i/2)\rho+n_1}(\cosh \eta) + R_f^{\chi,\tau}(\eta) \tag{7-56}$$

where each sum converges uniformly in the interval $0 < \eta_1 \leq \eta \leq \eta_2 < \infty$.

The simplification which arises if we sum $(7-56)$ over $\tau$ is considerable. First we can show that

$$\sum_{\tau=\pm 1} R_f^{\chi,\tau}(\eta) = 0 \tag{7-57}$$

The proof of this "proposition III" is postponed. Next we have from $(7-43)$, $(4-56)$, $(4-66)$

$$\sum_{\tau=\pm 1} P_{j_1 q_1 j_2 q_2}^{\chi,\tau}(\eta)$$

$$= \exp\left\{\frac{i}{2}\pi(q_1 + q_2)\right\} \sum_q e^{-i\pi q} d_{q_1 q}^{j_1}(0) \, d_{q q_2}^{j_2}(0)$$

$$\times [e_{j_1 j_2 q}^{\chi}(\eta) + (-1)^{j_1 - j_2} e_{j_2 j_1 q}^{-\chi}(\eta)]$$

$$= E_{j_1 q_1 j_2 q_2}^{\chi}(\exp\{\tfrac{1}{2}\eta\sigma_2\}) + (-1)^{j_1 - j_2 + q_1 - q_2} E_{j_2 q_2 j_1 q_1}^{-\chi}(\exp\{\tfrac{1}{2}\eta\sigma_2\})$$

Taking into account $(7-54)$–$(7-57)$ we have

$$D_{j_1 q_1 j_2 q_2}^{\chi}(\exp\{\tfrac{1}{2}\eta\sigma_2\}) = 2 \sum_{n=0}^{\infty} [(i\rho - 2n + 1) \operatorname*{Res}_{J=-(i/2)\rho+n}$$

$$\times (\mathscr{F}_{q_1 q_2}(J)_{\tau=+1} + \mathscr{F}_{q_1 q_2}(J)_{\tau=-1})]$$

$$\times e_{q_1 q_2}^{-(i/2)\rho+n}(\cosh \eta)$$

$$+ (-1)^{j_1 - j_2 + q_1 - q_2} \text{ times the previous sum in which the replacements } \chi \leftrightarrow -\chi, \; j_1 \leftrightarrow j_2, \; q_1 \leftrightarrow q_2 \text{ have been made}$$

Remembering the fact that the decomposition of functions of the first kind into functions of the second kind is unique, and using the asymptotic properties $(4-63)$, $(6-57)$ to identify the right terms, we obtain

$$E_{j_1 q_1 j_2 q_2}^{\chi}(\exp\{\tfrac{1}{2}\eta\sigma_2\})$$

$$= 2 \sum_{n=0}^{\infty} [(i\rho - 2n - 1) \operatorname*{Res}_{J=-(i/2)\rho+n} (\mathscr{F}_{q_1 q_2}(J)_{\tau=+1} + \mathscr{F}_{q_1 q_2}(J)_{\tau=-1})]$$

$$\times e_{q_1 q_2}^{-(i/2)\rho+n}(\cosh \eta)$$

$$= \sum_{n=0}^{\infty} (-1)^{(1/2)m - q_2} W_n^{\chi}(j_2, q_2) V_n^{\chi}(j_1, q_1)$$

$$\times e_{q_1 q_2}^{-(i/2)\rho + n}(\cosh \eta) \tag{7-58}$$

Here we made use of the notation

$$V_n^{\chi}(j, q) = 2(i\rho - 2n - 1)\left[ M\left(\chi | j, -\frac{i}{2}\rho + n, +1, q\right)\right.$$

$$\left. + \exp\{i\pi(\tfrac{1}{2}m + n)\} M\left(\chi | j, -\frac{i}{2}\rho + n, -1, q\right)\right] \tag{7-59}$$

and of (7–53). With (6–59) the series (7–58) can finally be presented in the alternative form

$$E_{j_1 q_1 j_2 q_2}^{\chi}(v) = \sum_{n=0}^{\infty} (-1)^{(1/2)m - q_2} V_n^{\chi}(j_1, q_1) W_n^{\chi}(j_2, q_2) E_{q_1 q_2}^{-(i/2)\rho + n}(v) \tag{7-60}$$

The series (7–60) converges uniformly in any domain

$$2 < N_1^2 \leq |v|^2 \leq N_2^2 < \infty$$

The same series (7–60) serves as an asymptotic expansion at $|v| \to \infty$. To prove this we need an asymptotic estimate of its remainders, which can be obtained by replacing the contours $C_n$ in the proof of proposition I by a sequence of straight lines $\operatorname{Re} J = c_n$, with $c_n$ tending to $+\infty$. An asymptotic expansion of

$$\sum_{\tau = \pm 1} R_f^{\chi, \tau}(\eta)$$

in powers of $e^{\eta}$ can obviously contain only powers that are independent of $\sigma$. Since the left-hand side of (7–60) and the sum on the right-hand side do not involve such powers, proposition III must be correct. We need therefore prove only the less trivial propositions I and II.

We choose real numbers $\varepsilon_1$, $\varepsilon_2$, $\varepsilon_3$ such that

$$0 < \varepsilon_{1,2} < \tfrac{1}{2}\pi \qquad 0 < \varepsilon_3 < \tfrac{1}{2}$$

In the $J$-plane we consider open circles of radius $\varepsilon_3$ with centers in

$$J = \frac{i}{2}\rho + n \quad \text{and} \quad J = -\frac{i}{2}\rho - n - 1 \quad n = 0, 1, 2, \dots$$

We call the union of all these open circles $Q_1$. In the same fashion we construct a sequence of open circles with radius $\varepsilon_3$ with centers in

$$J = -\frac{i}{2}\rho + n \quad \text{and} \quad J = \frac{i}{2}\rho - n - 1 \qquad n = 0, 1, 2, \ldots$$

whose union is $Q_2$, and a finite number of open circles with the same radius around the fixed pole positions of $\mathscr{F}_{q_1 q_2}(J)$ whose union is $Q_3$. Next we define two sectors

$$S_1 = \{J \mid |\arg J| \leq \varepsilon_1\}$$
$$S_2 = \{J \mid \varepsilon_1 \leq |\arg J| \leq \pi - \varepsilon_2\}$$

and the unions

$$Q = \bigcup_{i=1}^{3} Q_i \qquad S = S_1 \cup S_2$$

$R_n$ is a sequence of positive numbers such that

$$\lim_{n \to \infty} R_n = \infty$$

A sequence of contours $C_n$ which suffice to prove proposition I can be constructed in the following fashion.

(1) On the straight line $\operatorname{Re} J = -\frac{1}{2}$ we take two infinite and not overlapping intervals extending from $|J| = R_n$ till infinity.

(2) In the intersection of $S_2$ with the half-plane $\operatorname{Re} J \geq -\frac{1}{2}$ we take the arcs of the circle $|J| = R_n$.

(3) Eventually a finite number of arcs defined in (2) may cut the set $Q$ within the sector $S_2$. We simply drop these contours and renumber $R_n$ correspondingly.

(4) In the sector $S_1$ we run along the shortest path connecting the points $R_n e^{\pm i \varepsilon_1}$ which lies completely within $S_1 - Q$.


In order to verify that proposition I holds for this sequence $C_n$, we derive an asymptotic formula for $\mathscr{F}_{q_1 q_2}(J)$ from (7–32). An asymptotic formula for $M(\chi \mid j, J, \pi, q)$ in the sector $S_2$ is obtained as

$$M(\chi \mid j, J, \tau, q)$$

$$= (2\pi)^{1/2} 2^{2M - j} (2j + 1)! \exp\left\{\frac{i}{2} \pi \zeta_\tau^{\pm}\right\} \left[\Gamma\left(\frac{i}{2}\rho + j + 1\right)\right]^{-1}$$

$$\times \ [(2j+1)(j+q)! \ (j-q)! \ (j+\tfrac{1}{2}m)! \ (j-\tfrac{1}{2}m)!]^{-1/2}$$

$$\times \ \exp\left\{-\tfrac{1}{2}\pi|J| \ |\sin\varphi| - \tfrac{i}{2}\pi\kappa\left(\tfrac{i}{2}\rho + j - 1 - |J|\cos\varphi\right)\right\}$$

$$\times \ J^{(i/2)\rho+j-1/2}\left(1 + O\!\left(\tfrac{1}{J}\right)\right) \tag{7-61}$$

where

$$M = \max(|q|, \tfrac{1}{2}|m|)$$

$$J = |J|e^{i\varphi} \qquad -\pi \leq \varphi \leq \pi$$

$$\pm = \operatorname{sign}\varphi$$

$$\zeta_\tau^{\,+} = (1-\tau)(j+q) + q - \tfrac{1}{2}m$$

$$\zeta_\tau^{\,-} = \zeta_\tau^{\,+} + 2(\tfrac{1}{2}\tau m - q)$$

Formula (7–61) is independent of whether the pair of subscripts ($\frac{1}{2}\tau m$, $q$) is in domain (A), (B), (C), or (D) (Section 7–4). For the set $S_1 - Q_1 - Q_3$ we have the estimate

$$|M(\chi \mid j, J, \tau, q)| \leq c(\varepsilon_3)|J|^{-(1/2)J \operatorname{Im}\rho + j - 1/2} \tag{7-62}$$

From (7–61) and (7–62) follow the estimates for $\mathscr{F}_{q_1 q_2}(J)$

$$|\mathscr{F}_{q_1 q_2}(J)| \leq c_1(\varepsilon_3)|J|^{j_1+j_2-1} \ \exp\{-\pi|J| \ |\sin\varphi|\} \qquad \text{in} \quad S_2 - Q \tag{7-63}$$

$$|\mathscr{F}_{q_1 q_2}(J)| \leq c_2(\varepsilon_3)|J|^{j_1+j_2-1} \qquad \text{in} \quad S_1 - Q \tag{7-64}$$

From (6–57) we have moreover the estimate

$$|e_{q_1 q_2}^J(\cosh\eta)| \leq c_3 |J|^{-1/2} \ \exp\{-\eta(|J|\cos\varphi + \tfrac{1}{2})\} \tag{7-65}$$

for $J$ in S and uniformly in any interval

$$0 < \eta_1 \leq \eta \leq \eta_2 < \infty$$

Proposition I follows immediately from (7–63)–(7–65).

Finally we prove proposition II. We introduce a function $M^{(+)}(\chi|j, J, \tau, q)$ replacing $d_{(1/2)\tau m, q}^J$ in (7–31) by $2e_{(1/2)\tau m, q}^J$,

$$M^{(+)}(\chi \mid j, J, \tau, q)$$

$$= \exp\left\{\tfrac{i}{2}\pi(q - \tfrac{1}{2}m + (1-\tau)(j+q)\right\}$$

$$\times \ (2j+1)^{1/2} \int_{C_1} z^{-(i/2)\rho-1} \ d_{(1/2)\tau m, q}^j(z^{-1}) e_{(1/2)\tau m, q}^J(z) \ dz \tag{7-66}$$

The contour $C_1$ and all other contours to occur in this proof are drawn in Fig. 7–2. We place the cut of the function $z^{-(i/2)\rho-1}$ at $-\infty < z \leqq 0$. Moreover we assume that Im $\rho$ is sufficiently big negative. We may later continue in $\rho$. We replace the contour $C_1$ first by the contour $C_2$, which yields a total change of phase $(-1)^{(1/2)\tau m-q}$. Next we replace the contour $C_2$ by the contour $C_3$ and the variable $z$ by $-z$ using the relations (Im $z < 0$)

$$d^j_{(1/2)\tau m, q}(z) = (-1)^{j+q} \, d^j_{-(1/2)\tau m, q}(-z)$$

$$e^J_{(1/2)\tau m, q}(z) = -e^{i\pi(J+q)} e^J_{-(1/2)\tau m, q}(-z)$$

$$z^{-(i/2)\rho-1} = e^{i\pi((i/2)\rho+1)}(-z)^{-(i/2)\rho-1}$$

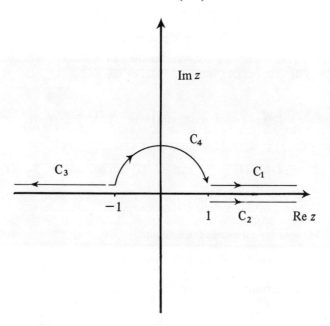

**Figure 7–2.**    The integration contours $C_1$, $C_2$, $C_3$, $C_4$

which follow from (2–31) and (6–55). The total phase factor up to this point is

$$\exp\left\{i\pi\left(\tfrac{1}{2}\tau m + J + \frac{i}{2}\rho + 1 - j - q\right)\right\}$$

Finally we replace $C_3$ by $C_4 \cup C_1$. The integral over $C_1$ can be expressed by $M^{(+)}(\chi|j, J, -\tau, q)$ yielding an additional phase factor $(-1)^{J+q}$. In this fashion we arrive at

$$M^{(+)}(\chi \mid j, J, \tau, q)$$

$$= \exp\left\{i\pi\left(\tfrac{1}{2}\tau m + J + \frac{i}{2}\rho + 1\right)\right\} M^{(+)}(\chi \mid j, J, -\tau, q)$$

$$+ (2j + 1)^{1/2} e^{i\pi\delta} \int_{C_4} z^{-(i/2)\rho - 1} \, d^j_{-(1/2)\tau m, q}(z^{-1}) e^J_{-(1/2)\tau m, q}(z) \, dz \qquad (7\text{-}67)$$

where $\delta$ is of no further interest. The integral over $C_4$ exhibits only the fixed singularities of the function $e^J_{-(1/2)\tau m, q}(z)$ itself but is otherwise regular in $J$. Since both functions $M^{(+)}(\chi \mid j, J, \pm\tau, q)$ have first order poles at

$$J = -\frac{i}{2}\rho - n - 1 \qquad n = 0, 1, 2, \ldots$$

we end up with

$$\operatorname*{Res}_{J = -(i/2)\rho - n - 1} M^{(+)}(\chi \mid j, J, \tau, q)$$

$$= \exp\{i\pi(\tfrac{1}{2}\tau m - n)\} \operatorname*{Res}_{J = -(i/2)\rho - n - 1} M^{(+)}(\chi \mid j, J, -\tau, q) \qquad (7\text{-}68)$$

Because of (6–54) we have

$$M(\chi \mid j, J, \tau, q)$$

$$= (2\pi)^{-1} \tan \pi(J + \tfrac{1}{2}m)$$

$$\times \left[M^{(+)}(\chi \mid j, J, \tau, q) - (-1)^{(1/2)\tau m - q} M^{(+)}(\chi \mid j, -J - 1, \tau, q)\right] \qquad (7\text{-}69)$$

It follows therefore

$$\operatorname*{Res}_{J = (i/2)\rho + n} M(\chi \mid j, J, \tau, q) = (2\pi)^{-1}(-1)^{(1/2)\tau m - q + 1} \tan \pi\left(\tfrac{1}{2}m + \frac{i}{2}\rho\right)$$

$$\times \operatorname*{Res}_{J = (i/2)\rho + n} M^{(+)}(\chi \mid j, -J - 1, \tau, q)$$

$$= \exp\{-i\pi(\tfrac{1}{2}\tau m + n)\} \operatorname*{Res}_{J = (i/2)\rho + n} M(\chi \mid j, J, -\tau, q)$$

which implies proposition II, formula (7–53).


## 7–6   REMARKS

The problem investigated in this chapter was first attacked by Sciarrino and Toller [30] in the context of relativistic scattering theory, and independently by Rühl [28] in dealing with the issue of diagonalizing covariant operators of SL(2, C) by algebraic methods. The relation (7–53), that is, proposition II, has been proved here for the first time. The convergence of the series (7–60), that is proposition I, has also first been established in these lectures. The expression (7–32) of the transition coefficient is due to Akyeampong et al. [1].

# Chapter 8

# Regge and Toller Poles

As an application of the results derived in the preceding chapters we want to discuss now a problem of physical interest: the group theoretical aspects of Regge and Toller poles. Regge and Toller poles are used in the phenomenologic analysis of scattering processes in elementary particle physics. It turns out that they may be identified with poles of the Fourier transforms of certain distributions describing the scattering process.

In Section 8–1 we introduce the scattering amplitude as a distribution acting on certain spaces of test functions. As test functions we can use wave functions in momentum space which are infinitely differentiable and have compact support. The most important feature of the scattering amplitude for our purposes is its invariance with respect to the inhomogeneous SL(2, C) group. The symmetry group of elementary particle physics in the realm of strong interactions, which we always have in mind, is in fact bigger. In addition there exist internal symmetries which can be described by compact Lie groups, the parity invariance (i.e., invariance against space reflection $I_{st}$, Section 1–1a), and the *CPT* symmetry. We take only the *CPT* transformation into account because it is responsible for the "signature," a characteristic property of Regge and Toller poles. Parity and internal symmetries imply refinements of the Regge and Toller pole schemes which are almost trivial. We neglect them for this reason.

Section 8–2 is devoted to *CPT* invariance and the extension of the inhomogeneous SL(2,C) group which involves the *CPT* transformation. In Sections 8–3 and 8–4 we present unitary irreducible representations of this extended group and the reduction of tensor products of such representations. These are the main tools in the decomposition of the scattering amplitude by what is usually called "phase analysis." The standard argument on which phase analysis is based is Schur's lemma, which we deal with in Section 8–5. Schur's lemma is a result of the theory of Banach algebras and we use it in this connection. An interesting feature of Schur's lemma is the possibility of extending, with its help, the domain of definition of Fourier transformations from an $\mathscr{L}^1$-space over a Lie group to certain distributions over such a group. It turns out, however, that this is not yet an extension sufficient to treat Regge and Toller poles. After a

247

preparation of suitable spaces of test functions in Section 8–6, we present Regge and Toller poles in Sections 8–7 and 8–8.

## 8–1   THE SCATTERING AMPLITUDE

We consider the scattering process of two elementary particles 1, 2 into two particles 3, 4, formally

$$1 + 2 \rightarrow 3 + 4 \qquad (8\text{-}1)$$

Such particles can be described by wave functions $\varphi_{q_i}(p_i)$, $i = 1, 2, 3, 4$ (2–16) on positive timelike orbits with masses squared $M_i^2$ and spins $S_i$. The wave functions $\varphi_{q_i}(p_i)$ span Hilbert spaces $\mathscr{H}_i$. These Hilbert spaces possess dense linear subspaces $\mathscr{C}_i^\infty$ which consist of wave functions $\varphi_{q_i}(p_i)$ that are infinitely differentiable and have compact support. The spaces $\mathscr{C}_i^\infty$ are invariant with respect to transformations of the inhomogeneous SL(2, C) group.

The quantum mechanical transition amplitude for the process (8–1) is a quadrilinear functional on the spaces $\mathscr{C}_1^\infty$, $\mathscr{C}_2^\infty$ and $\overline{\mathscr{C}_3^\infty}$, $\overline{\mathscr{C}_4^\infty}$ which is continuous in each of its arguments with respect to the convergence defined in the $\mathscr{C}^\infty$-spaces. By $\overline{\mathscr{C}^\infty}$ we mean the space of functions $\overline{\varphi_q(p)}$ if $\varphi_q(p)$ is in $\mathscr{C}^\infty$. We denote this functional

$$\Xi(\bar{\varphi}_3, \bar{\varphi}_4 ; \varphi_1, \varphi_2)$$

[for the normalization see (8–14)]. According to a classic theorem in the theory of distributions this functional can be presented in the form [with $C$ as in (2–36)]

$$\Xi(\bar{\varphi}_3, \bar{\varphi}_4 ; \varphi_1, \varphi_2)$$
$$= \sum_{q_i} \int \prod_{j=1}^{4} d\mu(p_j) \left( \overline{\sum_{q_3'} C_{q_3 q_3'}^{S_3} \varphi_{q_3'}(p_3)} \right) \left( \overline{\sum_{q_4'} C_{q_4 q_4'}^{S_4} \varphi_{q_4'}(p_4)} \right)$$
$$\times \varphi_{q_1}(p_1) \varphi_{q_2}(p_2) T_{q_3 q_4; q_1 q_2}(p_3, p_4 ; p_1, p_2) \qquad (8\text{-}2)$$

where $T$ is a distribution on a $\mathscr{C}^\infty$-space of test functions $\varphi_{q_1 q_2 q_3 q_4}(p_1, p_2, p_3, p_4)$. We call $T$ the scattering amplitude. The functional $\Xi$ can also be applied to covariant wave functions $\varphi_{q_i}(a_i)$ (2–18) and can then be presented as

$$\Xi(\bar{\varphi}_3, \bar{\varphi}_4 ; \varphi_1, \varphi_2)$$
$$= \sum_{q_i} \int \prod_{j=1}^{4} d\mu(p_j) \left( \overline{\sum_{q_3'} C_{q_3 q_3'}^{S_3} \varphi_{q_3'}(a_3)} \right) \left( \overline{\sum_{q_4'} C_{q_4 q_4'}^{S_4} \varphi_{q_4'}(a_4)} \right)$$
$$\times \varphi_{q_1}(a_1) \varphi_{q_2}(a_2) T_{q_3 q_4; q_1 q_2}(a_3, a_4 ; a_1, a_2) \qquad (8\text{-}3)$$

with

$$(p_j)_\mu = (p_j{}^R)_\nu \Lambda_\mu{}^\nu(a_j)$$

We require that the distribution $T$ (8–3) be covariant as

$$T_{q_3 q_4;\, q_1 q_2}(u_3 a_3,\, u_4 a_4\, ;\, u_1 a_1,\, u_2 a_2)$$

$$= \sum_{q_i'} T_{q_3' q_4';\, q_1' q_2'}(a_3,\, a_4\, ;\, a_1,\, a_2) \prod_{j=1}^{4} D_{q_j' q_j}^{S_j}(u_j^{-1}) \qquad (8\text{-}4)$$

to compensate for the covariance (2–19).

Finally we can use spinor wave functions (2–43), (2–44) as test functions. This gives, for example, with (2–41), (2–43)

$$\Xi(\bar\varphi_3,\, \bar\varphi_4\, ;\, \varphi_1,\, \varphi_2)$$

$$= \sum_{A_i} \int \prod_{j=1}^{4} d\mu(p_j) \overline{\psi^{\dot A_3}(p_3) \psi^{\dot A_4}(p_4)} \psi_{A_1}(p_1) \psi_{A_2}(p_2)$$

$$\times T_{\dot A_3 \dot A_4}^{A_1 A_2}(p_3,\, p_4\, ;\, p_1,\, p_2) \qquad (8\text{-}5)$$

where the spinor scattering amplitude is related to the scattering amplitude (8–2) by

$$T_{\dot A_3 \dot A_4}^{A_1 A_2}(p_3,\, p_4\, ;\, p_1,\, p_2)$$

$$= \sum_{q_i} D_{q_3 A_3}^{S_3}((a(p_3)^{-1})^\dagger)\, D_{q_4 A_4}^{S_4}((a(p_4)^{-1})^\dagger)\, D_{q_1 A_1}^{S_1}(a(p_1))$$

$$\times D_{q_2 A_2}^{S_2}(a(p_2)) T_{q_3 q_4;\, q_1 q_2}(p_3,\, p_4\, ;\, p_1,\, p_2) \qquad (8\text{-}6)$$

and $a(p)$ is a boost on a positive timelike orbit (Section 2–3). Instead of the spinor amplitude (8–5) we could as well have defined spinor amplitudes with other kinds of indices. These spinor amplitudes can also be extended to the whole group SL(2, C) by (2–52), (2–53)

$$T_{\dot A_3 \dot A_4}^{A_1 A_2}(a_3,\, a_4\, ;\, a_1,\, a_2)$$

$$= \sum_{q_i} D_{q_3 A_3}^{S_3}((a_3^{-1})^\dagger)\, D_{q_4 A_4}^{S_4}((a_4^{-1})^\dagger)\, D_{q_1 A_1}^{S_1}(a_1)\, D_{q_2 A_2}^{S_2}(a_2)$$

$$\times T_{q_3 q_4;\, q_1 q_2}(a_3,\, a_4\, ;\, a_1,\, a_2) \qquad (8\text{-}7)$$

The spinor amplitude (8–7) is invariant on right cosets of SU(2). The transition from (8–5) to (8–7), that is, the replacement of $p_i$ by $a_i$, can be accomplished by the substitution

$$\mathbf{p}_i = a_i^\dagger \mathbf{p}_i{}^R a_i \qquad (8\text{-}8)$$

which illustrates the invariance on the right cosets.

That property of the functional $\Xi$ which is basic for our investigations in this chapter is its invariance against transformations of SL(2, C) $\times$ $T_4$. We require

$$\Xi(\bar{\varphi}_3, \bar{\varphi}_4 ; \varphi_1, \varphi_2) = \Xi(\overline{U^3_{(a, y)} \varphi_3}, \overline{U^4_{(a, y)} \varphi_4} ; U^1_{(a, y)} \varphi_1, U^2_{(a, y)} \varphi_2) \qquad (8\text{-}9)$$

for four unitary representations $U^i$ in $\mathscr{H}_i$ and for any element $(a, y) \in$ SL(2, C) $\times$ $T_4$. If we put $a = e$ and $y = 0$, (8–9) implies that the total number of particles with half-integral spin ("fermions") involved in the scattering process is even. Inserting (2–17) into (8–3) and using (8–9) yields the invariance condition for the scattering amplitude

$$T_{q_3 q_4; q_1 q_2}(a_3, a_4 ; a_1, a_2)$$

$$= \exp i\{(p_1{}^R \Lambda(a_1) + p_2{}^R \Lambda(a_2) - p_3{}^R \Lambda(a_3) - p_4{}^R \Lambda(a_4))\Lambda(a^{-1})y\}$$

$$\times T_{q_3 q_4; q_1 q_2}(a_3 a^{-1}, a_3 a^{-1}; a_1 a^{-1}, a_2 a^{-1})$$

The support of the scattering amplitude must therefore be restricted (by "momentum conservation") to the manifold

$$\sum_i \pm p_i = p_1{}^R \Lambda(a_1) + p_2{}^R \Lambda(a_2) - p_3{}^R \Lambda(a_3) - p_4{}^R \Lambda(a_4)$$

$$= 0 \qquad (8\text{-}10)$$

and we may define a distribution $R_{q_3 q_4; q_1 q_2}(a_3, a_4 ; a_1, a_2)$ on the manifold (8–10) by

$$T_{q_3 q_4; q_1 q_2}(a_3, a_4 ; a_1, a_2) = \delta\left(\sum_i \pm p_i\right) R_{q_3 q_4; q_1 q_2}(a_3, a_4 ; a_1, a_2) \qquad (8\text{-}11)$$

In the same fashion we get for the spinor amplitudes

$$T^{A_1 A_2}_{A_3 A_4}(a_3, a_4 ; a_1, a_2) = \delta\left(\sum_i \pm p_i\right) M^{A_1 A_2}_{A_3 A_4}(a_3, a_4 ; a_1, a_2) \qquad (8\text{-}12)$$

where the distribution $M^{A_1 A_2}_{A_3 A_4}$ is called an "$M$-function."

In quantum field theory one can prove that the $M$-functions (8–12) with $a_i = a(p_i)$ are the boundary values of analytic functions in the complexified components of the momenta $p_i$ restricted to the mass shells $(p_i)^2 = M_i^2$ and the manifold (8–10). We introduce real parameters $\alpha_\mu, \beta_\mu$ for each matrix $a$ by

$$a = a_0 e + \sum_{k=1}^{3} a_k \sigma_k \qquad a_\mu = \alpha_\mu + i\beta_\mu \qquad \mu = 0, 1, 2, 3$$

such that the constraints

$$\alpha_0{}^2 - \beta_0{}^2 - \sum_{k=1}^{3} (\alpha_k{}^2 - \beta_k{}^2) = 1$$

$$\alpha_0 \beta_0 - \sum_{k=1}^{3} \alpha_k \beta_k = 0$$

(8-13)

are fulfilled. Using (8–8) we may regard the $M$-functions (8–12) as boundary values of analytic functions in the complexified variables $\alpha_\mu, \beta_\mu$ submitted to the constraints (8–10), (8–13). In addition we make the assumption that these boundary values are continuous functions. Quantum field theory tells us further that the $M$-functions are polynomially bounded on SL(2, C). By means of (8–6), (8–11), (8–12) we can deduce corresponding properties for the distribution $R_{q_3 q_4 ; q_1 q_2}(a_3, a_4 ; a_1, a_2)$.

For heuristic reasons it is sometimes comfortable to remember the existence of a scattering operator $S$, which generates the functional $\Xi$ in the following sense. We construct a Hilbert space $\mathscr{H}_{1\times2}$ ($\mathscr{H}_{3\times4}$) as the tensor product of the spaces $\mathscr{H}_1$, $\mathscr{H}_2$ ($\mathscr{H}_3$, $\mathscr{H}_4$). The functional $\Xi$ can be extended onto the spaces $\mathscr{H}_{1\times2}$ and $\mathscr{H}_{3\times4}$ such that (in hopefully obvious notation)

$$\Xi(\bar{\varphi}_3, \bar{\varphi}_4 ; \varphi_1, \varphi_2) = -i(2\pi)^{-4} \langle C\varphi_3 \times C\varphi_4 | S - 1 | \varphi_1 \times \varphi_2 \rangle \quad (8\text{-}14)$$

If the scattering process is "elastic," that is to say, either both pairs $(1, 3)$, $(2, 4)$ or both pairs $(1, 4)$, $(2, 3)$ consist of identical particles, the spaces $\mathscr{H}_{1\times2}$ and $\mathscr{H}_{3\times4}$ can be identified. $S$ is then an operator bounded by 1 and "1" is the unit operator in $\mathscr{H}_{1\times2}$. In the case of an inelastic (that is to say, not elastic) scattering process, $S$ is also bounded by 1 but "1" is now the null operator. $S$ is interpreted as the restriction to the spaces $\mathscr{H}_{1\times2}$ and $\mathscr{H}_{3\times4}$ of a unitary operator in a hypothetical space, which is to describe the scattering of any number of any particles into any other number of any other particles.

## 8–2   THE *CPT* TRANSFORMATION

The physical interpretation of the *CPT* transformation considers this transformation as the product of two separate operations: a charge conjugation $C$, which maps particles onto antiparticles (note that antiparticles have the same spin and mass as the particles), and a total inversion $PT$ which was denoted $I_{st}$ in Section 1–1a and inverts all vectors in Minkowski space. The total inversion changes the roles of incoming and outgoing particles in a scattering process. It must therefore be represented by an antilinear operator. This carries over to the *CPT* transformation, which we shall denote $\theta$ from now on. Consequently we introduce the antiparticles ("charge conjugates") $1c$, $2c$, $3c$, $4c$ in addition to the particles 1, 2, 3, 4 and describe them by wave functions in Hilbert spaces

$\mathcal{H}_{ic}$. For convenience we write the process (8–1) in the form

$$1 + 2 \to 3c + 4c$$

which allows a more symmetric treatment in the sequel. It may happen that particles and antiparticles are identical, nevertheless we may distinguish between them formally.

The group SL(2, C) $\times$ T$_4$ can be extended to include the operation $\Theta$ in two different fashions. We have the possibilities

$$\Theta(a, y) = (a, -y)\Theta \qquad \Theta\Theta = (e_{\pm}, 0) \tag{8-15}$$

and denote these two groups F$_{\pm}$ $\times$ T$_4$, respectively. Quantum field theory suggests the use of the group F$_-$ $\times$ T$_4$ in physics. By $\mathcal{H} \oplus \mathcal{H}_c$ we denote the direct orthogonal sum of the particle space $\mathcal{H}$ and the antiparticle space $\mathcal{H}_c$ with the elements $\varphi_q(p)$ and $\varphi_q{}^c(p)$, respectively. The group F$_-$ $\times$ T$_4$ can be represented on $\mathcal{H} \oplus \mathcal{H}_c$ by (2–16) for the subgroup SL(2, C) $\times$ T$_4$ on both subspaces $\mathcal{H}$ and $\mathcal{H}_c$, and by

$$U_\Theta \varphi_q(p) = \sum_{q'} \overline{C^S_{qq'} \varphi_{q'}(p)} \in \mathcal{H}_c$$

$$U_\Theta \varphi_q{}^c(p) = \sum_{q'} \overline{C^S_{qq'} \varphi_{q'}^c(p)} \in \mathcal{H} \tag{8-16}$$

In (8–16) we chose that phase which is most convenient for our further investigations. Because of (2–36)

$$(C^S C^S)_{qq'} = D^S_{qq'}(e_-) = (-1)^{2S} \delta_{qq'}$$

we have

$$U_\Theta U_\Theta \varphi_q(p) = (-1)^{2S} \varphi_q(p) \tag{8-17}$$

as required. The operator $U_\Theta$ is obviously antilinear.

The unitary representations of F$_{\pm}$ $\times$ T$_4$, however, also play an important role in the sequel. We construct these representations by means of the technique of induction (Section 1–4). In the present section we need representations on the timelike orbits only, which can be obtained by a slight modification of our treatment of the representations of SL(2, C) $\times$ T$_4$ in Chapter 2. Representations on other orbits are studied in Section 8–3.

Since

$$\Lambda_\mu{}^\nu(\Theta) = -\delta_\mu{}^\nu$$

the timelike orbits are unions of a positive and a negative timelike orbit of $SL(2, C) \times T_4$ both to the same mass squared. We choose the reference momentum on such an orbit as

$$p^R = (M, 0, 0, 0)$$

so that the little group is $SU(2)$. Denoting the elements of $F_\pm$ by $f$, we use functions $\Phi_q(f)$ submitted to the covariance constraint

$$\Phi_q(uf) = \sum_{q'} D^S_{qq'}(u)\Phi_{q'}(f) \tag{8-18}$$

as a basis for a unitary irreducible representation. Such representation is obtained from $(1-45)$

$$V_{(f, y)} \Phi_q(f_1) = \exp i\{p_v{}^R \Lambda_\mu{}^v(f_1)y^\mu\}\Phi_q(f_1 f) \tag{8-19}$$

In addition we may define the functions

$$\Phi_q(p) = \Phi_q(a(p)) \qquad \Phi_q(-p) = \Phi_q(a(p)\Theta) \tag{8-20}$$

where $a(p)$ is any boost on a positive timelike orbit of $SL(2, C) \times T_4$ (Section $2-3$). We require that $\Phi_q(p)$ be measurable on the orbit with the finite norm

$$\|\Phi\| = (\Phi, \Phi)^{1/2}$$

$$(\Phi, \Phi') = \sum_q \int d\mu(p)\overline{\Phi_q(p)}\Phi'_q(p) \tag{8-21}$$

The invariant measure on the orbit is defined as

$$d\mu(p) = d\mu_+(p) + d\mu_-(p)$$

$$d\mu_\pm(p) = d^4p\, \delta((p)^2 - M^2)\, \delta_{\text{sign } po,\, \pm 1} \tag{8-22}$$

We are free to identify the Hilbert space of such functions $\Phi_q(f)$ or $\Phi_q(p)$ obeying $(8-18)$, $(8-21)$ with the direct orthogonal sum $\mathscr{H} \oplus \overline{\mathscr{H}_c}$, where $\overline{\mathscr{H}_c}$ consists of the elements $\overline{\varphi_q{}^c(p)}$ such that $\varphi_q{}^c(p)$ is in $\mathscr{H}_c$. In explicit terms we set

$$\Phi_q(p) = \varphi_q(p) \qquad \Phi_q(-p) = \sum_{q'} \overline{C^S_{qq'}\, \varphi_{q'}^c(p)} \tag{8-23}$$

Instead of (8–23) we can also establish a relation between the covariant wave functions.

$$\Phi_q(f) = \begin{cases} \varphi_q(a) & f = a \\ \sum_{q'} \overline{C^S_{qq'}}\, \varphi^c_{q'}(a) & f = a\Theta \end{cases} \tag{8-24}$$

(8–24) is compatible with (8–18) because of (2–19), (2–35), and (2–36). This way we get from (8–16) and (8–19)

$$V_\Theta \Phi_q(f) = \Phi_q(f\Theta)$$

$$= \begin{cases} \sum_{q'} \overline{C^S_{qq'}}\, \varphi^c_{q'}(a) = U_\Theta\, \varphi^c_q(a) & f = a \\ \eta_\pm\, \varphi_q(a) = \eta_\pm\, \eta_- \sum_{q'} \overline{C^S_{qq'}}\, U_\Theta\, \varphi_{q'}(a) & f = a\Theta \end{cases} \tag{8-25}$$

with

$$\eta_+ = 1 \qquad \eta_- = (-1)^{2S} \tag{8-26}$$

The choice between $\eta_+$ and $\eta_-$ reflects the ambiguity in the group structure of $F_\pm$.

Now we generalize the concept of the scattering amplitude. We study simultaneously the six processes

$$i + j \to kc + lc \qquad i < j \qquad k < l \qquad i, j, k, l = 1, 2, 3, 4$$

Each of these processes is described by a functional

$$\Xi(\overline{\varphi_{kc}}, \overline{\varphi_{lc}}\, ; \varphi_i, \varphi_j)$$

as in Section 8–1. If the scattering exhibits *CPT* symmetry we have to require

$$\Xi(\overline{U_\Theta \varphi_k}, \overline{U_\Theta \varphi_l}\, ; U_\Theta \varphi_{ic}, U_\Theta \varphi_{jc}) = \Xi(\overline{\varphi_{ic}}, \overline{\varphi_{jc}}\, ; \varphi_k, \varphi_l) \tag{8-27}$$

In the Hilbert space $\mathscr{H}_i \oplus \overline{\mathscr{H}_{ic}}$ we can define a dense linear subspace $\mathscr{C}_i^\infty$ by requiring that both components $\varphi_i$ and $\overline{\varphi_{ic}}$ of $\Phi_i$ (8–23), (8–24) be in $\mathscr{C}^\infty$-spaces. On such spaces $\mathscr{C}_i^\infty$ we define a quadrilinear functional by

$$\Xi'(\Phi_1, \Phi_2, \Phi_3, \Phi_4) = \sum_{\substack{i<j \\ k<l}} (-1)^{\pi(k, l, i, j)} \Xi(\overline{\varphi_{kc}}, \overline{\varphi_{lc}}\, ; \varphi_i, \varphi_j) \tag{8-28}$$

The symbol $\pi(k, l, i, j)$ denotes the number of permutations of fermions with fermions which are necessary to bring $(3, 4, 1, 2)$ into $(k, l, i, j)$. The sign factor in (8–28) takes account of the different statistics obeyed by fermions and particles with integral spin ("bosons"). The definition (8–28) implies that $\Xi'$ vanishes on subspaces describing more than two ingoing or outgoing particles. From (8–25), (8–27), and (8–28) we get

$$\Xi'(V_\Theta \Phi_1, V_\Theta \Phi_2, V_\Theta \Phi_3, V_\Theta \Phi_4)$$

$$= \sum_{\substack{i<j \\ k<l}} (-1)^{\pi(k,l,i,j)} (\eta_\pm \eta_-)_k (\eta_\pm \eta_-)_l \, \Xi(\overline{U_\Theta \varphi_k}, \overline{U_\Theta \varphi_l} ; U_\Theta \varphi_{ic}, U_\Theta \varphi_{jc})$$

$$= \Xi'(\Phi_1, \Phi_2, \Phi_3, \Phi_4) \tag{8-29}$$

if and only if

$$(-1)^{\pi(k,l,i,j)} (\eta_\pm \eta_-)_k (\eta_\pm \eta_-)_l = (-1)^{\pi(i,j,k,l)}$$

This condition is easily seen to be satisfied only for the group $F_+ \times T_4$. We conclude that the functional $\Xi'$ is invariant under unitary transformations of the group $F_+ \times T_4$ if the functionals $\Xi$ are *CPT* symmetric with respect to the antilinear operator $U_\Theta$ (8–16), (8–27) and invariant under unitary transformations of the group $SL(2, C) \times T_4$. From now on we have to deal with this functional $\Xi'$ and its invariance group $F_+ \times T_4$.

The functional $\Xi'$ is separately continuous in all its arguments $\Phi_i \in \mathscr{C}_i^\infty$. Under permutations of the arguments it behaves as

$$\Xi'(\Phi_i, \Phi_j, \Phi_k, \Phi_l) = (-1)^{\pi(k,l,i,j)} \Xi'(\Phi_1, \Phi_2, \Phi_3, \Phi_4) \tag{8-30}$$

Treating it in the same way as the functional $\Xi$ in Section 8–1 we arrive at a representation

$$\Xi'(\Phi_1, \Phi_2, \Phi_3, \Phi_4) = \sum_q \int \prod_{j=1}^{4} [d\mu(p_j) \Phi_{q_j}(f_j)] T'_{q_1 q_2 q_3 q_4}(f_1, f_2, f_3, f_4) \tag{8-31}$$

with $d\mu(p_j)$ as defined by (8–22). The distribution $T'$ (8–31) fulfills a covariance relation like (8–4). $T'$ is also symmetric against permutations of its arguments similar to (8–30). Invariance under $F_+ \times T_4$ implies

$$T'_{q_1 q_2 q_3 q_4}(f_1, f_2, f_3, f_4) = \delta\left(\sum_i p_i\right) R'_{q_1 q_2 q_3 q_4}(f_1, f_2, f_3, f_4) \tag{8-32}$$

$$p_i = p_i^R \Lambda(f_i)$$

$$R'_{q_1 q_2 q_3 q_4}(f_1, f_2, f_3, f_4) = R'_{q_1 q_2 q_3 q_4}(f_1 f, f_2 f, f_3 f, f_4 f) \tag{8-33}$$

Note that the signs in $\Sigma_i p_i$ (8–10), (8–11) have now been absorbed in the group elements $f_i$.

## 8–3   UNITARY REPRESENTATIONS OF $F_+ \times T_4$ FOR THE SPACELIKE AND NULL ORBITS

The unitary irreducible representations of $F_+ \times T_4$ on spacelike orbits require special investigation, since their little group is a nontrivial extension of the group SU(1, 1). A spacelike orbit of SL(2, C) $\times$ $T_4$ remains an orbit for the enlarged group $F_+ \times T_4$. As reference momentum we take therefore the same vector as in Section 2  2

$$p^R = (0, 0, 0, \mu) \qquad \mu > 0$$

The corresponding little group H consists of the whole group SU(1, 1) and the element

$$\Theta' = \Theta \, \exp\left\{\frac{i}{2} \pi\sigma_2\right\} \tag{8-34}$$

with

$$\Theta'\Theta' = e_- \tag{8-35}$$

It remains for us to construct the unitary irreducible representations of this group H.

The element $\Theta'$ generates an automorphism of SU(1, 1) by

$$v \xrightarrow{\Theta'} v' = \Theta'v\Theta'^{-1} \tag{8-36}$$

which changes the parameters of the group elements as

$$v = \exp\left\{\frac{i}{2}\psi_1\sigma_3\right\}\exp\left\{\frac{1}{2}\eta\sigma_2\right\}\exp\left\{\frac{i}{2}\psi_2\sigma_3\right\}$$

$$v' = \exp\left\{-\frac{i}{2}\psi_1\sigma_3\right\}\exp\left\{\frac{1}{2}\eta\sigma_2\right\}\exp\left\{-\frac{i}{2}\psi_2\sigma_3\right\} \tag{8-37}$$

In terms of the representation functions (6–48) of SU(1, 1) the automorphism (8–36), (8–37) reads

$$D^J_{q_1q_2}(v) \rightarrow D^J_{q_1q_2}(v') = (-1)^{q_1-q_2} D^J_{-q_1, -q_2}(v) \tag{8-38}$$

In general two cases are possible. Under the automorphism (8–36) a unitary irreducible representation of SU(1, 1) goes into an equivalent representation, in which case we call such representation $\Theta'$-self-conjugate, or it goes into an inequivalent representation, in which case we call the two representations a conjugate pair. As shown in (8–38), we meet the first alternative in the case of the principal and supplementary series of SU(1, 1), and the second alternative in the case of the discrete series. The positive and negative branches of the discrete series change their positions under the automorphism.

Next we consider the cosets

$$H/SU(1, 1) = S_2$$

which form a group

$$S_2 = (E, \tau)$$
$$EE = \tau\tau = E$$
$$E\tau = \tau E = \tau$$

This group possesses two irreducible unitary representations $\chi_\pm$

$$\chi_\pm(E) = 1 \qquad \chi_\pm(\tau) = \pm 1 \tag{8-39}$$

If $D(h)$ denotes any unitary irreducible representation of H, we define a pair of such representations by [in obvious generalization of the notation (8–39)]

$$D_\pm(h) = D(h)\chi_\pm(h)$$

If this pair consists of equivalent representations, we call $D(h)\Theta'$-self-associate. If these representations are inequivalent, we say that they form an associate pair of representations of H.

According to a classic theorem in group theory all unitary representations of H and SU(1, 1) are connected in either of the two fashions. Self-associate representations of H reduce into pairs of conjugate representations of SU(1, 1) and can likewise be constructed from such pairs. The members of any associate pair of representations of H each reduce to the same self-conjugate representation of SU(1, 1). In the case of the principal and supplementary series of SU(1, 1) we have self-conjugate representations which can be extended as

$$D_{q_1 q_2}^{\sigma, J}(v) = D_{q_1 q_2}^J(v)$$
$$D_{q_1 q_2}^{\sigma, J}(\Theta') = (-1)^{q_1 - \varepsilon}\sigma\, \delta_{q_1, -q_2} \tag{8-40}$$
$$\sigma = \pm 1 \qquad \varepsilon = 0, \tfrac{1}{2} \qquad \varepsilon - q_1 \text{ integral}$$

In this fashion we get an associate pair ($\sigma = \pm 1$) of representations of H.

A conjugate pair of representations of the discrete series of SU(1, 1) can be described by the same function $D_{q_1 q_2}^J(v)$ if we let $q_1$, $q_2$ run over both branches $q_1, q_2 \geq k$ or $q_1, q_2 \leq -k$, $k = |J + \frac{1}{2}| + \frac{1}{2}$. We set

$$D_{q_1 q_2}^J(v) = D_{q_1 q_2}^J(v)$$
$$D_{q_1 q_2}^J(\Theta') = (-1)^{q_1 - k} \delta_{q_1, -q_2} \tag{8-41}$$

and have a self-associate representation of H.

In Section 8–7 we see that the representations (8–40), (8–41) of H can be used to formulate a Plancherel theorem for the group H by an almost trivial extension of the Plancherel theorem for the group SU(1, 1). This can be regarded as a proof of the theorem mentioned insofar as we restrict the assertion "all representations" to those representations involved in the respective Plancherel theorems.

A unitary irreducible representation of $F_+ \times T_4$ for a spacelike orbit can be constructed by means of covariant functions $\Phi_q(f)$ on $F_+$

$$\Phi_q(hf) = \sum_{q'} D_{qq'}^{\sigma, J}(h) \Phi_{q'}(f) \tag{8-42}$$

and a formula of the type (1–45). Almost all elements $f \in F_+$ can be decomposed uniquely as

$$f = ha(p)$$

where $a(p) \in SL(2, C)$ is any of the boosts on spacelike orbits discussed in Section 2–4, and $h$ is an element of H. In addition we require then that

$$\Phi_q(p) \equiv \Phi_q(a(p))$$

be measurable on the orbit with finite scalar product norm

$$\|\Phi\| = (\Phi, \Phi)^{1/2}$$
$$(\Phi, \Phi') = \sum_q \int d\mu(p) \overline{\Phi_q(p)} \Phi_q'(p)$$

$d\mu(p)$ was defined by (2–4).

In the case of the null orbit the little group is the homogeneous group $F_+$ itself, that is, a trivial extension of the group SL(2, C). For the unitary irreducible representations of $F_+ \times T_4$ on such an orbit, which coincide with the unitary irreducible representations of $F_+$ itself, we get from Section 3–5

$$D_{j_1 q_1 j_2 q_2}^{\sigma, \chi}(a) = D_{j_1 q_1 j_2 q_2}^{\chi}(a)$$
$$D_{j_1 q_1 j_2 q_2}^{\sigma, \chi}(\Theta) = \sigma \delta_{j_1 j_2} \delta_{q_1 q_2} \qquad \sigma = \pm 1 \tag{8-43}$$

The label $\sigma$ introduced in (8–40), (8–43) is denoted the "group theoretical signature," in brief, the "signature."

## 8–4    THE REDUCTION OF TENSOR PRODUCT REPRESENTATIONS OF $F_+ \times T_4$

We study the tensor product of two irreducible unitary representations of $F_+ \times T_4$ on timelike orbits. Representations of this kind were constructed in Section 8–2. The group $F_+ \times T_4$ is of type one, and the reduction of such tensor product representation is therefore well defined (Sections 1–3b and 1–3c). The method we use is based on the formalism of induced representations (Section 1–4) and the Plancherel theorems of the little groups.

Let $\mathscr{H}_1$, $\mathscr{H}_2$ denote two Hilbert spaces carrying the two representations $V^1$, $V^2$ (8–19) of $F_+ \times T_4$. The orbits are

$$(p_1)^2 = M_1{}^2 \qquad (p_2)^2 = M_2{}^2$$

and the spins $S_1$ and $S_2$, respectively. The spaces $\mathscr{H}_{1,2}$ are realized by covariant functions $\Phi_q^{1,2}(f)$ (8–18). We introduce the product functions

$$\Phi_{q_1 q_2}^{1 \times 2}(f_1, f_2) = \Phi_{q_1}^1(f_1)\Phi_{q_2}^2(f_2) \tag{8-44}$$

with $\Phi_{q_1}^1(f_1) \in \mathscr{C}_1^\infty$, $\Phi_{q_2}^2(f_2) \in \mathscr{C}_2^\infty$. The tensor product Hilbert space $\mathscr{H}_{1\times 2}$ is obtained as the completion of the linear space which contains all finite linear combinations of functions (8–44) with respect to the scalar product norm

$$\|\Phi^{1 \times 2}\| = (\Phi^{1 \times 2}, \Phi^{1 \times 2})^{1/2}$$

$$(\Phi^{1 \times 2}, \Phi'^{1 \times 2}) = \int d\mu_{M_1}(p_1) \int d\mu_{M_2}(p_2) \sum_{q_1 q_2} \overline{\Phi_{q_1 q_2}^{1 \times 2}(f_1, f_2)}\, \Phi_{q_1 q_2}'^{1 \times 2}(f_1, f_2) \tag{8-45}$$

where as usual

$$p_{1,2} = p_{1,2}^R \Lambda(f_{1,2}) \tag{8-46}$$

From (8–44) and (8–18) we derive the covariance constraint

$$\Phi_{q_1 q_2}^{1 \times 2}(u_1 f_1, u_2 f_2) = \sum_{q'_1 q'_2} D_{q_1 q'_1}^{S_1}(u_1)\, D_{q_2 q'_2}^{S_2}(u_2)\Phi_{q'_1 q'_2}^{1 \times 2}(f_1, f_2) \tag{8-47}$$

which carries over to all elements of the space $\mathscr{H}_{1\times 2}$. The tensor product representation $V^{1\times 2}$ is then defined in $\mathscr{H}_{1\times 2}$ by

$$V_{(f, y)}^{1 \times 2} \Phi_{q_1 q_2}^{1 \times 2}(f_1, f_2) = \exp i\{(p_1 + p_2)_\mu y^\mu\}\Phi_{q_1 q_2}^{1 \times 2}(f_1 f, f_2 f) \tag{8-48}$$

with $p_{1\times 2}$ as in (8–46).

In order to decompose the representation $V^{1,2}$ (8–48) we introduce the momentum $Q$

$$Q = p_1 + p_2$$

with the invariant "mass squared"

$$t = (Q)^2 \tag{8-49}$$

The variable $t$ ranges over the interval

$$t \leq (M_1 - M_2)^2 \tag{8-50}$$

if $f_1 f_2^{-1} \Theta \in SL(2, C)$, and over the interval

$$t \geq (M_1 + M_2)^2 \tag{8-51}$$

if $f_1 f_2^{-1} \in SL(2, C)$. In the case (8–50) we have the alternatives:

$M_1 = M_2$: $Q$ is spacelike or null;
$M_1 \neq M_2$: $Q$ is spacelike, lightlike, or timelike, but never null.

In the case (8–51) $Q$ is positive or negative timelike. Later in this section we shall see that the Hilbert space $\mathcal{H}_{1\times 2}$ can be represented as the direct integral

$$\mathcal{H}_{1\times 2} = \int_{-\infty}^{(M_1-M_2)^2 \oplus} dt J_{12}(t) \mathcal{H}_t \oplus \int_{(M_1+M_2)^2}^{\infty \oplus} dt J_{12}(t) \mathcal{H}_t \tag{8-52}$$

of Hilbert spaces $\mathcal{H}_t$, each of which carries a unitary (in general reducible) representation of $F_+ \times T_4$ on an orbit with mass squared $t$. Since the integral over the interval (8–51) can be handled the same way as the integral over the set (8–50), we concentrate our interest on the latter invariant subspace.

On the interval (8–50) we have

$$f_1 f_2^{-1} \Theta \in SL(2, C) \tag{8-53}$$

Let $Q_t^R$ denote the reference momentum on the orbit (8–49) as specified in Sections 8–2, 8–3. We can then show that $f_1$, $f_2$ submitted to (8–53) can be brought into the form

$$f_1 = u_1 b_1(t) f \qquad f_2 = u_2 b_2(t) f \tag{8-54}$$

with $f \in F_+$, $u_{1,2} \in SU(2)$, $b_1(t) = d(\eta_1)$, $b_2(t) = d(\eta_2)\Theta$, and $d(\eta)$ as in (3–22) (where we allow exceptionally for negative values $\eta_{1,2}$). The angles $\eta_1 = \eta_1(t)$ and $\eta_2 = \eta_2(t)$ are fixed by the constraint

$$p_1^R \Lambda(d(\eta_1)) - p_2^R \Lambda(d(\eta_2)) = Q_t^R \tag{8-55}$$

In fact, we have to verify only that (8–55) possesses a solution. This is done in Appendix A–3a.

From now on $\Phi^{1\times2}_{q_1q_2}(f_1, f_2)$ is any finite linear combination of functions (8–44). We insert (8–54) with $u_1 = u_2 = e$ and define

$$\Phi^{1\times2}_{q_1q_2}(t, f) = \Phi^{1\times2}_{q_1q_2}(b_1(t)f, b_2(t)f) \tag{8-56}$$

These functions (8–56) are covariant as

$$\Phi^{1\times2}_{q_1q_2}(t, u(\psi)f) = \exp\{i(q_1 + q_2)\psi\}\Phi^{1\times2}_{q_1q_2}(t, f) \tag{8-57}$$

for
$$u(\psi) = \exp\left\{\frac{i}{2}\psi\sigma_3\right\}$$

as follows from (8–47) and the shape of the elements $b_{1,2}(t)$. Under a unitary operation $V^{1\times2}_{(f,y)}$ (8–48) the functions (8–56) transform into

$$V^{1\times2}_{(f_1,y)}\Phi^{1\times2}_{q_1q_2}(t, f) = \exp i(Q_\mu y^\mu)\Phi^{1\times2}_{q_1q_2}(t, ff_1) \tag{8-58}$$

with
$$Q = Q_t^R \Lambda(f) \tag{8-59}$$

If we complete the linear space of functions (8–56) by means of the scalar product norm

$$\|\Phi^{1\times2}(t)\| = (\Phi^{1\times2}(t), \Phi^{1\times2}(t))^{1/2}$$
$$(\Phi^{1\times2}(t), \Phi'^{1\times2}(t)) = \int d\mu(f)\sum_{q_1q_2} \overline{\Phi^{1\times2}_{q_1q_2}(t, f)}\Phi'^{1\times2}_{q_1q_2}(t, f) \tag{8-60}$$

we obtain a Hilbert space $\mathscr{H}_t$ which carries the unitary representation (8–58) of $F_+ \times T_4$ on the orbit (8–49). In order to establish the direct integral decomposition (8–52) we have therefore only to compute the Jacobian $J_{12}(t)$ in

$$d\mu_{M_1}(p_1)\, d\mu_{M_2}(p_2) = J_{12}(t)\, dt\, d\mu(f) \tag{8-61}$$

The notation used in (8–61) is meant as follows: $d\mu(p)$ was defined in (8–22); $p_{1,2}$ depend on $f_{1,2}$ as in (8–46); the invariant measure on $F_+$, $d\mu(f)$, has still to be defined. Let $\xi_e(f)$ and $\xi_\Theta(f)$ denote the characteristic functions of the sets SL(2, C), respectively SL(2, C)$\Theta$, within $F_+$, namely,

$$\xi_e(f) + \xi_\Theta(f) = 1$$

$$\xi_e(f) = \begin{cases} 1 & \text{for} \quad f \in \text{SL}(2, C) \\ 0 & \text{for} \quad f \notin \text{SL}(2, C) \end{cases}$$

We set

$$d\mu(f) = \tfrac{1}{2}[\xi_e(f)\, d\mu'(f) + \xi_\Theta(f)\, d\mu'(f\Theta)] \qquad (8\text{-}62)$$

where the measure $d\mu'$ in the square bracket is the invariant measure $(1-20)$ on SL$(2, C)$. In $(8-61)$ only five parameters of $f$ are actually involved because a left factor $u(\psi)$ in $f$ can always be put into $u_1$ and $u_2$ $[(8-54)$, the covariance $(8-57)$ has to be taken into account]. In Appendix A$-3$a we find the result for the domain $(8-50)$

$$J_{12}(t) = (2\pi)^3 \Delta(t, M_1^2, M_2^2)^{1/2} \qquad (8\text{-}63)$$

The function $J_{12}(t)$ is continuous at $t = 0$ with

$$J_{12}(0) = (2\pi)^3 |M_1^2 - M_2^2|$$

The representation $V^{1\times2}$ on $\mathcal{H}_t$ $[(8-58), (8-60)]$ is, in general, reducible. In fact, the Hilbert spaces $\mathcal{H}_t$ decompose first into the direct orthogonal sum of invariant subspaces

$$\mathcal{H}_t = \sum_{q_1 q_2}^{\oplus} \mathcal{H}_{t q_1 q_2}$$

which carry pairwise equivalent unitary representations. These representations on the spaces $\mathcal{H}_{t q_1 q_2}$ are further reducible. We sketch this reduction only for the case $t < 0$. To obtain an irreducible representation we have to project a covariant component out of $\Phi_{q_1 q_2}^{1\times2}(t, f)$. This projection can be achieved by a Fourier transformation

$$\mathscr{F}_{q_1 q_2; q}(t, \sigma, J \,|\, f) = \int d\mu(h)\overline{D_{q_1+q_2, q}^{\sigma, J}(h)}\Phi_{q_1 q_2}^{1\times2}(t, hf) \qquad (8\text{-}64)$$

with the representation functions $(8-40)$ of the little group H and a similar expression obtained with the functions $(8-41)$. Here $d\mu(h)$ is the invariant measure on H normalized as

$$d\mu(h) = \tfrac{1}{2}[\xi_e(h)\, d\mu'(h) + \xi_\Theta(h)\, d\mu'(h\Theta')] \qquad (8\text{-}65)$$

with analogous notations as in $(8-62)$. In $(8-64)$ the covariance $(8-57)$ has already been considered. These Fourier coefficients $(8-64)$ have the required covariance

$$\mathscr{F}_{q_1 q_2; q}(t, \sigma, J \,|\, hf) = \sum_{q'} D_{qq'}^{\sigma, J}(h)\mathscr{F}_{q_1 q_2; qq'}(t, \sigma, J \,|\, f) \qquad (8\text{-}66)$$

They span a Hilbert space $\mathcal{H}_{t q_1 q_2 \sigma J}$ with the scalar product norm

$$\|\mathscr{F}_{q_1q_2;\,q}(t,\,\sigma,\,J)\|^2 = \int d\mu_t(Q) \sum_q |\mathscr{F}_{q_1q_2;\,q}(t,\,\sigma,\,J\,|\,f)|^2$$

$$Q = Q_t^R \Lambda(f) \tag{8-67}$$

The Hilbert spaces $\mathscr{H}_{tq_1q_2\sigma J}$ carry irreducible representations of $F_+ \times T_4$. If we split an element of $F_+$ as

$$f = ha(Q)$$

where $h \,\epsilon\, H$ and $a(Q)$ is a boost for the orbit (8–49) with $t < 0$ (Sections 8–3 and 2–4), the measure decomposes correspondingly as (Appendix A–3b)

$$d\mu(f) = J(t)\, d\mu(h)\, d\mu_t(Q)$$

$$J(t) = \frac{1}{8\pi^2\,|t|} \tag{8-68}$$

If we apply Parseval's formula for the group H (Section 8–7) we get

$$\int d\mu(f)\,|\Phi_{q_1q_2}^{1\times 2}(t,\,f)|^2$$

$$= \frac{1}{8\pi^2\,|t|} \int d\mu(h) \int d\mu_t(Q)\,|\Phi_{q_1q_2}^{1\times 2}(t,\,ha(Q))|^2 \tag{8-69}$$

$$= \frac{1}{8\pi^2\,|t|} \int d\chi(\sigma,\,J) \int d\mu_t(Q) \sum_q |\mathscr{F}_{q_1q_2;\,q}(t,\,\sigma,\,J\,|\,a(Q))|^2$$

The Plancherel measure $d\chi(\sigma,\,J)$ on H is specified in Section 8–7. Finally we read off (8–69)

$$\mathscr{H}_{tq_1q_2} = \int^{\oplus} d\chi(\sigma,\,J)(8\pi^2\,|t|)^{-1}\mathscr{H}_{tq_1q_2\sigma J} \tag{8-70}$$

## 8–5    SCHUR'S LEMMA

In Section 8–1 we mentioned that the $\Xi$ functional can be generated by a bounded operator $S-1$ which maps the Hilbert space $\mathscr{H}_{1\times 2}$ into the space $\mathscr{H}_{3\times 4}$ (8–14). For heuristic reasons we want to study in this section the consequences of an analog premise for the $\Xi'$ functional (8–28). This premise is certainly unrealistic in general, but we believe that the peculiar features of the method applied in Sections 8–7 and 8–8 can only be estimated properly after the limits of the standard procedure based on Schur's lemma have been pointed out.

We write the $\Xi'$ functional (8—28) for $\Phi_i \in \mathscr{C}_i^\infty$ in the form

$$\Xi'(\Phi_1, \Phi_2, \Phi_3, \Phi_4) = \langle \overline{C\Phi_4} \times \overline{C\Phi_2} \,|\, B|\Phi_1 \times \Phi_3 \rangle \qquad (8\text{-}71)$$

By $\Phi_1 \times \Phi_3$ we denote an element of $\Phi_{1\times 3}$ of the form (8—44). Similarly $\overline{C\Phi_4} \times \overline{C\Phi_2}$ is defined as

$$(\overline{C\Phi_4} \times \overline{C\Phi_2})_{q_4 q_2}(f_4, f_2) = \Phi_{q_4 q_2}^{\bar{4} \times \bar{2}}(f_4, f_2)$$

$$= \sum_{q'_2 q'_4} \overline{C_{q_4 q'_4}^{S_4} \Phi_{q'_4}^4(f_4 \ominus) C_{q_2 q'_2}^{S_2} \Phi_{q'_2}^2(f_2 \ominus)} \qquad (8\text{-}72)$$

The functions $\Phi_{q_1 q_3}^{1\times 3}(f_1, f_3)$ and $\Phi_{q_4 q_2}^{\bar{4}\times\bar{2}}(f_4, f_2)$ can be treated exactly as the functions $\Phi_{q_1 q_2}^{1\times 2}(f_1, f_2)$ in Section 8—4. They span Hilbert spaces $\mathscr{H}_{1\times 3}$, respectively $\mathscr{H}_{\bar{4}\times\bar{2}}$, which carry unitary representations $V^{1\times 3}$, respectively $V^{\bar{4}\times\bar{2}}$, of $F_+ \times T_4$, and so on. We have only to replace the subscripts 1, 2 by 1, 3, respectively by 4,2. Our premise on the $\Xi'$ functional amounts to the assumption that the operator $B$ defined by (8—71) is a bounded linear operator from $\mathscr{H}_{1\times 3}$ into $\mathscr{H}_{\bar{4}\times\bar{2}}$, and that the $\Xi'$ functional can consequently be extended onto these Hilbert spaces. The invariance of the $\Xi'$ functional implies then

$$V_{(f,y)}^{\bar{4}\times\bar{2}} B = B V_{(f,y)}^{1\times 3} \qquad (8\text{-}73)$$

for all elements of $F_+ \times T_4$. Schur's lemma is a bundle of assertions on a bounded operator $B$ satisfying a commutation relation with unitary representations of the type (8—73).

We consider the direct orthogonal sum

$$\mathscr{H}_s = \mathscr{H}_{1\times 3} \oplus \mathscr{H}_{\bar{4}\times\bar{2}}$$

and write each element of this Hilbert space $\mathscr{H}_s$ as a two-component vector

$$\begin{pmatrix} \Phi^{1\times 3} \\ \Phi^{\bar{4}\times\bar{2}} \end{pmatrix}$$

The space $\mathscr{H}_s$ carries the unitary representation

$$V = \begin{pmatrix} V^{1\times 3} & 0 \\ 0 & V^{\bar{4}\times\bar{2}} \end{pmatrix}$$

of $F_+ \times T_4$. With the notation

$$B^* = \begin{pmatrix} 0 & B^\dagger \\ B & 0 \end{pmatrix}$$

we can write (8–73) and its adjoint in the form

$$V_{(f,y)} B^* = B^* V_{(f,y)} \qquad (8\text{-}74)$$

In addition we consider the projection operators

$$E^{1 \times 3} = \begin{pmatrix} E & 0 \\ 0 & 0 \end{pmatrix} \qquad E^{\bar{4} \times \bar{2}} = \begin{pmatrix} 0 & 0 \\ 0 & E \end{pmatrix}$$

where $E$ denotes the unit operators in $\mathscr{H}_{1 \times 3}$ and $\mathscr{H}_{\bar{4} \times \bar{2}}$. They commute also with the representation $V$, for example,

$$V_{(f,y)} E^{1 \times 3} = E^{1 \times 3} V_{(f,y)} \qquad (8\text{-}75)$$

for all elements of $F_+ \times T_4$. As in Section 1–3b we can build a representation of the group algebra of $F_+ \times T_4$ of the unitary representation $V$ of the group. The commutation relations (8–74), (8–75) are valid also for this algebra. A standard argument in the theory of Banach algebras proves that this algebra and consequently the representation $V$ can be decomposed into a direct integral of factorial representations such that all bounded operators commuting with $V$ suffer the same decomposition. If we label equivalence classes of unitary irreducible representations of $F_+ \times T_4$ by $\alpha$, and multiples of such representations by $V(\alpha)$, we get

$$V = \int_{\alpha \in A}^{\oplus} V(\alpha) \qquad B^* = \int_{\alpha \in A}^{\oplus} B^*(\alpha)$$

$$E^{1 \times 3} = \int_{\alpha \in A}^{\oplus} E^{1 \times 3}(\alpha) \qquad \text{etc.}$$

In addition, it follows that

$$V(\alpha) E^{1 \times 3}(\alpha) = E^{1 \times 3}(\alpha) V(\alpha)$$
$$V(\alpha) B^*(\alpha) = B^*(\alpha) V(\alpha) \qquad (8\text{-}76)$$

The relations (8–76) imply that $E^{1 \times 3}(\alpha)$ reduces $V(\alpha)$, that is, with

$$V^{1 \times 3}(\alpha) = E^{1 \times 3}(\alpha) V(\alpha) E^{1 \times 3}(\alpha)$$

we have

$$V^{1 \times 3} = \int_{\alpha \in C'}^{\oplus} V^{1 \times 3}(\alpha) \qquad C' \subset A$$

and correspondingly

$$V^{\bar{4} \times \bar{2}} = \int_{\alpha \in C''}^{\oplus} V^{\bar{4} \times \bar{2}}(\alpha) \qquad C'' \subset A$$

The second commutation relation (8–76) and the definition

$$B(\alpha) = E^{\bar{4} \times \bar{2}}(\alpha) B^*(\alpha) E^{1 \times 3}(\alpha)$$

imply

$$V^{\bar{4} \times \bar{2}}(\alpha) B(\alpha) = B(\alpha) V^{1 \times 3}(\alpha) \qquad \alpha \in C' \cap C'' \qquad (8\text{-}77)$$

and

$$B(\alpha) = 0 \qquad \alpha \notin C' \cap C''$$

The function $\|B(\alpha)\|$ is obviously essentially bounded. If the bases for the representations $V^{1 \times 3}(\alpha)$, $V^{\bar{4} \times 2}(\alpha)$ are appropriately chosen, and if $n_{1 \times 3}(n_{\bar{4} \times \bar{2}})$ denotes the finite multiplicity (Section 8–4) of the respective representations, $B(\alpha)$ is a finite-dimensional matrix with $n_{1 \times 3}$ columns and $n_{\bar{4} \times \bar{2}}$ rows. Since the group $F_+ \times T_4$ is of type one, the decomposition of the representations and of $B$ is essentially unique.

In order to illustrate the relation of Schur's lemma to harmonic analysis we consider the space $\mathscr{L}^2(-\infty, \infty)$ of complex-valued functions $f(x)$ on the real line. This Hilbert space carries the right-regular representation of the group $T_1$ of translations on the real line

$$T_y f(x) = f(x + y)$$

Any element $b(x)$ of $\mathscr{L}^1(-\infty, \infty)$ defines a linear operator $B$ in $\mathscr{L}^2(-\infty, \infty)$ by the convolution integral

$$Bf(x) = \int_{-\infty}^{\infty} b(x - x') f(x') \, dx'$$

This operator $B$ is bounded by

$$\|B\| \leqq \int_{-\infty}^{+\infty} |b(x)| \, dx$$

and commutes with the right-regular representation

$$BT_y = T_y B$$

If we reduce the right-regular representation into one-dimensional unitary representations of $T_1$

$$g(\lambda) = (2\pi)^{-1/2} \int_{-\infty}^{+\infty} e^{-i\lambda x} f(x)\, dx$$

$$\mathscr{L}^2(-\infty, \infty) = \int_{-\infty}^{+\infty} {}^{\oplus} \mathscr{H}_\lambda\, d\lambda$$

$$T_y g(\lambda) = e^{i\lambda y} g(\lambda)$$

Schur's lemma tells us that $B$ is submitted to the same decomposition

$$Bg(\lambda) = B(\lambda)g(\lambda)$$

and that the function $B(\lambda)$ is essentially bounded.

On the other hand, we can present $B(\lambda)$ as the Fourier transform of $b(x)$

$$B(\lambda) = \int_{-\infty}^{+\infty} e^{-i\lambda x} b(x)\, dx$$

Schur's lemma can therefore be regarded as a means of defining Fourier transforms for functions of $\mathscr{L}^1(-\infty, \infty)$. The domain of applicability can immediately be extended beyond the space $\mathscr{L}^1(-\infty, \infty)$. We may, for example, define a bounded operator $B_\delta$ by

$$B_\delta f(x) = \int_{-\infty}^{+\infty} \delta(x - x') f(x')\, dx'$$

for functions $f(x) \in \mathscr{C}^\infty(-\infty, \infty)$ and by extension on the whole space $\mathscr{L}^2(-\infty, \infty)$ such that the unit operator results. In this case Schur's lemma allows us to identify the function $B_\delta(\lambda) = 1$ with the Fourier transform of the delta-function. Summarizing we may say that Schur's lemma is more powerful than harmonic analysis on $\mathscr{L}^1$-spaces and may also serve to decompose distribution, but on the other hand it is always limited by the premise of boundedness of the convolution operator.

Now we return to our problem proper and make use of the results of Section 8—4. Equivalence classes of irreducible representations of $F_+ \times T_4$ are labeled by the invariant mass squared $t$ (8—49) and the parameters designating the unitary irreducible representations of the little group, say $\sigma, J$ for $t < 0$. The intersection $C' \cap C''$ on which $B(\alpha)$ can be nonzero is restricted in $t$ by

$$t \leqq \min\{(M_1 - M_3)^2, (M_2 - M_4)^2\}$$
$$t \geqq \max\{(M_1 + M_3)^2, (M_2 + M_4)^2\}$$

The multiplicity of the representation $V^{1\times 3}$ is at most due to the subscripts $q_1$, $q_3$, so that we expect $B(\alpha)$ to be a matrix (say for $t < 0$)

$$B(\alpha) = B(t, \sigma, J)_{q_4 q_2, q_1 q_3}$$

From $(8-31)$, $(8-32)$ we have

$$\Xi'(\Phi_1, \Phi_2, \Phi_3, \Phi_4)$$

$$= \sum_{q_i} \int \prod_{j=1}^{4} d\mu(p_j) \overline{\Phi_{q_4 q_2}^{4\times 2}(f_4, f_2)} \Phi_{q_1 q_3}^{1\times 3}(f_1, f_3) \delta(p_1 + p_3 - p_2 - p_4)$$

$$\times \sum_{b'_2 q'_4} C_{q_2 q'_2}^{S_2} C_{q_4 q'_4}^{S_4} R_{q_1 q' {}_2 q_3 q'_4}(f_1, f_2 \Theta, f_3, f_4 \Theta) \tag{8-78}$$

with

$$p_i = p_i{}^R \Lambda(f_i)$$

We insert $(8-54)$ with $u_1 = u_2 = e$, $(8-61)$, $(8-68)$, use the notation $(8-56)$, and take into account the invariance $(8-33)$,

$$\Xi'(\Phi_1, \Phi_2, \Phi_3, \Phi_4)$$

$$= \int\limits_{t<0} dt J_{13}(t) J_{42}(t) J(t)^2$$

$$\times \sum_{q_i} \int d\mu(h) \, d\mu(h') R_{q_1 q_2 q_3 q_4}(t, hh'^{-1})$$

$$\times \int d\mu_t(Q) \overline{\Phi_{q_4 q_2}^{4\times 2}(t, h'a(Q))} \Phi_{q_1 q_3}^{1\times 3}(t, ha(Q))$$

$$+ \int\limits_{t>0} dt \dots \tag{8-79}$$

where we introduced the new notation

$$R_{q_1 q_2 q_3 q_4}(t, h) = \sum_{q'_2 q'_4} C_{q_2 q'_2}^{S_2} C_{q_4 q'_4}^{S_4} R_{q_1 q' {}_2 q_3 q'_4}(b_1(t)h, b_2(t)\Theta, b_3(t)h, b_4(t)\Theta) \tag{8-80}$$

Due to our premise that the $\Xi'$ functional be generated by a bounded operator, we can use Schur's lemma and (8–64) to rewrite the convolution integral (8–79) (the part $t < 0$ only)

$$\int_{t<0} dt J_{13}(t) J_{42}(t) J(t)^2$$

$$\times \sum_{q_i} \int d\chi(\sigma, J) B(t, \sigma, J)_{q_4 q_2, q_1 q_3}$$

$$\times \int d\mu_t(Q) \sum_q \overline{\mathscr{F}^{4 \times 2}_{q_4 q_2; q}(t, \sigma, J \mid a(Q))} \mathscr{F}^{1 \times 3}_{q_1 q_3; q}(t, \sigma, J \mid a(Q)) \quad (8\text{-}81)$$

## 8–6   SPACES OF TEST FUNCTIONS

Our main task in Sections 8–7 and 8–8 is to abandon the premise of boundedness of the convolution operator. Consequently, we have to modify the integration over $J$ in (8–81) such that it extends over a contour in the complex $J$-plane which may be different from the line $\text{Re } J = -\frac{1}{2}$, that is, the principal series. Instead of the essentially bounded function we have to deal with analytic functionals. In order to prepare such treatment of the $\Xi'$ functional, we are going to construct appropriate spaces of test functions in the present section. We restrict our investigations to $t < 0$.

We start from (8–78), insert (8–54) with $u_1 = u_2 = e$, (8–61), (8–68), and the notation (8–56). This leads us to consider the functions

$$\Psi_{q_1 q_2 q_3 q_4}(t, h)$$

$$= J_{13}(t) J_{42}(t) J(t) \int d\mu(f) \overline{\Phi^{4 \times 2}_{q_4 q_2}(t, f)} \Phi^{1 \times 3}_{q_1 q_3}(t, f)$$

$$= J_{13}(t) J_{42}(t) J(t) \sum_{q_2 q'_4} C^{S_2}_{q_2 q'_2} C^{S_4}_{q_4 q'_4}$$

$$\times \int d\mu(f) \Phi^1_{q_1}(b_1(t) h f) \Phi^2_{q'_2}(b_2(t) f \Theta) \Phi^3_{q_3}(b_3(t) h f)$$

$$\times \Phi^4_{q'_4}(b_4(t) f \Theta) \quad (8\text{-}82)$$

With (8–82) we may rewrite (8–79) as

$$\Xi'(\Phi_1, \Phi_2, \Phi_3, \Phi_4)$$

$$= \sum_{q_i} \int_{t<0} dt \int d\mu(h) R_{q_i}(t, h) \Psi_{q_i}(t, h) + \int_{t>0} dt \ldots$$

Note that the distribution $R_{q_i}(t, h)$ (8–80) behaves as

$$R_{q_i}(t, h) = \exp\{i(q_1 + q_3)\Psi\}R_{q_i}(t, u(\Psi)h)$$
$$= \exp\{i(q_2 + q_4)\Psi\}R_{q_i}(t, hu(\Psi)) \qquad (8\text{-}83)$$

whereas the functions $\Psi_{q_i}(t, h)$ (8–82) are bicovariant on H as

$$\Psi_{q_i}(t, h) = \exp\{-i(q_1 + q_3)\Psi\}\Psi_{q_i}(t, u(\Psi)h)$$
$$= \exp\{-i(q_2 + q_4)\Psi\}\Psi_{q_i}(t, hu(\Psi)) \qquad (8\text{-}84)$$

We may therefore interpret $\Psi_{q_i}(t, h)$ as a bicovariant function on H in the sense of Section 6–5, whereas $R_{q_i}(t, h)$ for fixed $t$ is later identified with the complex conjugate of a bicovariant distribution on H of the same type.

We show that the functions (8–82) are infinitely differentiable in $t$ ($t < 0$) and $h$ and possess compact support on H for every fixed $t$. This follows from our choice of the functions $\Phi_i$ as elements of $\mathscr{C}_i^{\infty}$. To prove this we extend the notion of a norm for elements of SL(2, C) (3–26) onto the group $F_+$ by

$$|a\Theta| = |a|$$

If the support of $\Phi_2(f)$ $[\Phi_4(f)]$ lies inside the domain

$$|f| \leqq N_2 \qquad [|f| \leqq N_4]$$

the integral (8–82) is necessarily restricted to the finite domain

$$|b_2(t)f| \leqq N_2 \qquad [|b_4(t)f| \leqq N_4]$$

Because of the inequality

$$|a_1| \leqq |a_1 a_2| |a_2|$$

which follows from

$$|a| = |a^{-1}| \qquad |a_1 a_2| \leqq |a_1| |a_2|$$

we have for the domain of integration

$$|f|^2 \leqq N_2{}^2 2 \cosh \eta_2 \qquad [\leqq N_4{}^2 2 \cosh \eta_4]$$

This limits the support of $\Psi_{q_i}(t, h)$ on H for fixed $t$ to

$$|h|^2 \leq |b_3(t)hf|^2 |b_3(t)|^2 |f|^2$$
$$\leq N_2^2 N_3^2 2 \cosh \eta_2 \, 2 \cosh \eta_3$$

or by three other inequalities of this kind. The infinite differentiability follows immediately from the compactness of the integration domain.

The functions $\Psi_{q_i}(t, h)$ (8–82) lie in a space $\mathscr{C}_t^\infty(H)$, and all functions (8–82) together with their limits in the sense of $\mathscr{C}^\infty$ convergence exhaust this space. Our assumption on the continuity of the $M$-functions (Section 8–1) implies that the distribution $R_{q_i}(t, h)$ is in fact a continuous function on H and on the interval $-\infty < t < 0$. Due to Fubini's theorem we may apply $R_{q_i}(t, h)$ for fixed $t$ to the test functions $\Psi_{q_i}(t, h)$ (8–82) and integrate over $t$ afterwards. In other words, for fixed $t$, $R_{q_i}(t, h)$ may be regarded as a distribution operating on the space $\mathscr{C}_t^\infty(H)$.

In the case of equal masses

$$M_1 = M_3 = M_{13} \qquad M_2 = M_4 = M_{42}$$

the limit

$$\lim_{t \to 0-} \Psi_{q_i}(t, h) = \Psi_{q_i}(0, h)$$

exists and defines a space of test functions $\mathscr{C}_0^\infty(H)$. In fact, from Appendixes A–3a and A–3b we have

$$\lim_{t \to 0-} J_{13}(t) J_{42}(t) J(t) = 32\pi^4 M_{13} M_{42}$$

whereas Appendix A–3a and (8–54) give

$$\lim_{t \to 0-} b_1(t) = \lim_{t \to 0-} b_4(t) = e$$
$$\lim_{t \to 0-} b_2(t) = \lim_{t \to 0-} b_3(t) = \Theta$$

so that

$$\Psi_{q_i}(0, h) = 32\pi^4 M_{13} M_{42} \sum_{q_2 q'_4} C_{q_2 q'_2}^{S_2} C_{q_4 q'_4}^{S_4}$$

$$\times \int d\mu(f) \Phi_{q_1}^1(hf) \Phi_{q'_2}^2(f) \Phi_{q_3}^3(hf\Theta) \Phi_{q'_4}^4(f\Theta) \qquad (8\text{-}85)$$

It is easy to verify that the supports of the functions $\Psi_{q_i}(t, h)$ are uniformly bounded on any interval $-\delta < t \leq 0$, $\delta > 0$, and that the limit (8–85) is even assumed in the sense of convergence in a $\mathscr{C}^\infty$-space, if we identify all spaces $\mathscr{C}_t^\infty(H)$.

If we replace $h$ in (8–85) formally by $u_1 h u_2$, $u_{1,2} \in SU(2)$, we get from (8–47)

$$\Psi_{q_i}(0, u_1 h u_2) = \sum_{q'_i} D^{S_1}_{q_1 q'_1}(u_1) D^{S_3}_{q_3 q'_3}(u_1)$$
$$\times D^{S_2}_{q'_2 q_2}(u_2) D^{S_4}_{q'_4 q_4}(u_2) \Psi_{q'_i}(0, h) \qquad (8\text{-}86)$$

Relation (8–86) is to be regarded as a definition which extends the function $\Psi_{q_i}(0, h)$ from H onto $F_+$ by means of covariance. If we introduce the new functions

$$\Psi_{S_{13} q_{13} S_{42} q_{42}}(f) = \sum_{q_i} (S_1 S_3 S_{13} q_{13} \mid S_1 q_1 S_3 q_3)(S_4 q_4 S_2 q_2 \mid S_4 S_2 S_{42} q_{42})$$
$$\times \Psi_{q_i}(0, f) \qquad (8\text{-}87)$$

where the brackets denote the familiar Clebsch-Gordan coefficients of SU(2), we obtain bicovariant functions on $F_+$ in the sense of Section 4–7

$$\Psi_{S_{13} q_{13} S_{42} q_{42}}(u_1 f u_2) = \sum_{q'_{13} q'_{42}} D^{S_{13}}_{q_{13} q'_{13}}(u_1) D^{S_{42}}_{q'_{42} q_{42}}(u_2) \Psi_{S_{13} q'_{13} S_{42} q'_{42}}(f) \quad (8\text{-}88)$$

By means of the definitions (8–86), (8–87) we have therefore established a one-to-one correspondence between the space $\mathscr{C}_0^\infty(H)$ and a finite number of spaces $\mathscr{C}^\infty(F_+)$ of functions on $F_+$ which are bicovariant of the type (8–88) with $S_{13}$, $S_{42}$ ranging over

$$|S_1 - S_3| \leqq S_{13} \leqq |S_1 + S_3|$$
$$|S_2 - S_4| \leqq S_{42} \leqq |S_2 + S_4|$$

It can be verified easily that the distribution $R_{q_i}(t, h)$ (8–80) is a continuous function till $t = 0$ if the masses are equal, provided the $M$-functions themselves are continuous as we assumed in Section 8–1. This implies the existence of the limit

$$\lim_{t \to 0-} \sum_{q_i} \int d\mu(h) R_{q_i}(t, h) \Psi_{q_i}(t, h) = \sum_{q_i} \int d\mu(h) R_{q_i}(0, h) \Psi_{q_i}(0, h)$$
$$= \sum_{S_{13} S_{42}} \sum_{q_{13} q_{42}} \int d\mu(h) R_{S_{13} q_{13} S_{42} q_{42}}(h)$$
$$\times \Psi_{S_{13} q_{13} S_{42} q_{42}}(h) \qquad (8\text{-}89)$$

where in analogy with (8–87) we set

$$R_{S_{13} q_{13} S_{42} q_{42}}(h) = \sum_{q_i} (S_1 q_1 S_3 q_3 \mid S_1 S_3 S_{13} q_{13})$$
$$\times (S_4 S_2 S_{42} q_{42} \mid S_4 q_4 S_2 q_2) R_{q_i}(0, h) \qquad (8\text{-}90)$$

The distribution (function) $R_{S_{13}q_{13}S_{42}q_{42}}(h)$ can also be extended onto $F_+$ by

$$R_{S_{13}q_{13}S_{42}q_{42}}(u_1 h u_2) = \sum_{q'_{13}q'_{42}} \overline{D^{S_{13}}_{q_{13}q'_{13}}(u_1)} \, \overline{D^{S_{42}}_{q'_{42}q_{42}}(u_2)} R_{S_{13}q'_{13}S_{42}q'_{42}}(h) \quad (8\text{-}91)$$

Using (8–88), (8–91) and Appendixes A–1d, A–2f we may continue in Eq. (8–89)

$$= \sum_{S_{13}S_{42}} (2S_{13} + 1)(2S_{42} + 1) \int d\mu(f) R_{S_{13}q_{13}S_{42}q_{42}}(f)$$
$$\times \Psi'_{S_{13}q_{13}S_{42}q_{42}}(f) \quad (8\text{-}92)$$

with the definition

$$\Psi'_{S_{13}q_{13}S_{42}q_{42}}(f) = 4\pi(|f|^4 - 4)^{-1/2}\Psi_{S_{13}q_{13}S_{42}q_{42}}(f) \quad (8\text{-}93)$$

For the interpretation of $R_{S_{13}q_{13}S_{42}q_{42}}(f)$ as a distribution on $F_+$ it suffices therefore to consider those subspaces of $\mathscr{C}^\infty(F_+)$ for which the left-hand side of (8–93) is infinitely differentiable at $f = e$ and $f = \Theta$.

## 8–7  REGGE POLES

The distribution $R_{q_i}(t, h)$ on H for fixed $t < 0$ is now to be submitted to Fourier transformations. We make use of the results of Chapter 6 which we generalize slightly so that they apply to the group H. If $x_{q_1 q_2}(h) \in \mathscr{C}^\infty$ is a bicovariant function on H of the type (6–60)

$$x_{q_1 q_2}(h) = \exp\{-iq_1 \psi\} x_{q_1 q_2}(u(\psi)h)$$
$$= \exp\{-iq_2 \psi\} x_{q_1 q_2}(hu(\psi))$$

we define Fourier transforms by (6–69), (8–40), and (8–65)

$$\mathscr{F}^*_{q_1 q_2}(\sigma, J) = \int \overline{x_{q_1 q_2}(h)} \, D^{\sigma; J}_{q_1 q_2}(h) \, d\mu(h)$$

$$= \tfrac{1}{4} \int_0^\infty d\eta \, \sinh \eta \{x_{q_1 q_2}(\exp \tfrac{1}{2}\eta\sigma_2) \, d^J_{q_1 q_2}(\cosh \eta)$$

$$+ \sigma(-1)^{q_2 + \varepsilon} x_{q_1 q_2}(\Theta' \exp \tfrac{1}{2}\eta\sigma_2) \, d^J_{q_1, -q_2}(\cosh \eta)\} \quad (8\text{-}94)$$

as analytic functions of $J$. In addition we introduce the analytic functions $\mathscr{F}_{q_1 q_2}(\sigma, J)$ which are obtained by analytic continuation of the complex conjugate function (8–94) off the principal series. Similarly we have Fourier transforms of the second kind (6–71)

$$\mathscr{F}^{*(+)}_{q_1q_2}(\sigma, J) = 2 \int \overline{x_{q_1q_2}(h)} E^{\sigma, J}_{q_1q_2}(h)\, d\mu(h) \tag{8-95}$$

which are also to be considered as analytic functions of $J$. An analytic function $\mathscr{F}^{(+)}_{q_1q_2}(\sigma, J)$ is again obtained by analytic continuation of the conjugate function (8–95) off the principal series. The function $E^{\sigma, J}_{q_1q_2}(h)$ is built of $E^{J}_{q_1q_2}(v)$ (6–59) in analogy with (8–40).

Inverse Fourier transformations can be exerted on these Fourier transforms by a generalization of (6–72)

$$\overline{x_{q_1q_2}(h)} = \frac{1}{2\pi i} \sum_\sigma \int_{C_\pm} dJ(2J + 1)\mathscr{F}^{*(+)}_{q_1q_2}(\sigma, J)\, D^{\sigma, J}_{q_1q_2}(h)^*$$

$$= \frac{1}{2\pi i} \sum_\sigma \int_{C_\pm} dJ(2J + 1)\mathscr{F}^{*}_{q_1q_2}(\sigma, J) E^{\sigma, J}_{q_1q_2}(h)^* \tag{8-96}$$

Here and in the sequel we use the notations $D^*$ and $E^*$ for the analytic continuations of the complex conjugate of the $D$- and $E$-functions off the principal series for either group H or F. For two functions which are both in $\mathscr{C}^\infty$ and bicovariant of the same type we have Parseval's formula in a form corresponding to (6–73)

$$\int d\mu(h)\overline{x_{q_1q_2}(h)} y_{q_1q_2}(h) = \frac{1}{2\pi i} \sum_\sigma \int_{C_\pm} dJ(2J + 1)\mathscr{F}^{*}_{q_1q_2}(\sigma, J)_x \mathscr{F}^{(+)}_{q_1q_2}(\sigma, J)_y \tag{8-97}$$

Concerning the covariance, the distribution $R_{q_i}(t, h)$ resembles the function $\overline{x_{q_1+q_3, q_2+q_4}(h)}$, whereas $\Psi_{q_i}(t, h)$ corresponds to the function $y_{q_1+q_3, q_2+q_4}(h)$. Therefore by analogy we expect the following formulas to hold

$$\int d\mu(h)R_{q_i}(t, h)\Psi_{q_i}(t, h) = \frac{1}{2\pi i} \sum_\sigma \int_{C_\sigma} dJ(2J + 1)\mathscr{T}_{q_i}(t, \sigma, J)\mathscr{F}^{(+)}_{q_i}(t, \sigma, J)_\Psi$$

$$\tag{8-98}$$

and simultaneously

$$R_{q_i}(t, h) = \frac{1}{2\pi i} \sum_\sigma \int_{C_\sigma} dJ(2J + 1)\mathscr{T}_{q_i}(t, \sigma, J)E^{\sigma, J}_{q_1+q_3, q_2+q_4}(h)^* \tag{8-99}$$

with certain contours of integration $C_\sigma$ which may depend on $\sigma$.

In fact, we can verify (8–98) and (8–99) under certain assumptions which we shall refer to as the "meromorphy hypothesis." We recall that $R_{q_i}(t, h)$ is a continuous and polynomially bounded function on H. The distribution

$$R_{q_i}^{\lambda}(t, h) = |h|^{\lambda} R_{q_i}(t, h)$$

is therefore regular if Re $\lambda$ is smaller than a certain number $\lambda_0$. Its Fourier transform $\mathscr{T}_{q_i}^{\lambda}(t, \sigma, J)$ exists as a proper function of $J$ on the principal series Re $J = -\frac{1}{2}$ and can be continued analytically into a strip around this line. We assume that this Fourier transform is meromorphic in $J$ apart from the natural branch points at the fixed positions (Section 6–4)

$$J = -M, -M+1, \ldots, -m-1 \quad \text{and} \quad J = m, m+1, \ldots, M-1$$

$$m = \min\{|q_1|, |q_2|\} \qquad M = \max\{|q_1|, |q_2|\}$$

In addition we assume that the number of poles is finite in any strip $|\operatorname{Re} J| < J_0$. In analogy with (8–96) we have the inverse formula

$$|h|^{\lambda} R_{q_i}(t, h) = \frac{1}{2\pi i} \sum_{\sigma} \int_{C_{\pm}} dJ(2J + 1) \mathscr{T}_{q_i}^{\lambda}(t, \sigma, J) E_{q_1+q_3, q_2+q_4}^{\sigma, J}(h)^*$$

which we want to continue in $\lambda$ till $\lambda = 0$. During this process of continuation the poles of $\mathscr{T}_{q_i}(t, \sigma, J)$ in $J$, which form pairs of "mirror poles" at $J_v$ and $-J_v - 1$, move towards $C_{\pm}$ from both sides. We try to avoid these poles by corresponding deformations of the contours. If we assume that no pinch between these poles and the natural branch points occurs nor that any other pair of poles pinches the contour in an unavoidable fashion (this is to say, a pinch which cannot be removed by another choice of the path in the $\lambda$-plane along which we continue), we arrive finally at the representation (8–99). $\mathscr{T}_{q_i}(t, \sigma, J)$ is also a meromorphic function whose poles we call "Regge poles."

In order to be definite, we assume that both contours $C_{\sigma}$ be obtained by deformation of $C_-$. The function $\mathscr{T}_{q_i}(t, \sigma, J)$ is assumed to fall off at least as fast as $|J|^{-3/2 - \delta}$, $\delta > 0$, for $|\operatorname{Im} J| \to \infty$. In this case we are allowed to shift the tours $C_{\sigma}$ to the left. After a finite displacement we get

$$R_{q_i}(t, h) = \sum_{\sigma} \sum_{v}^{N_{\sigma}} (2J_v^{\sigma} + 1) \left[ \operatorname{Res}_{J=J_v^{\sigma}} \mathscr{T}_{q_i}(t, \sigma, J) \right] E_{q_1+q_3, q_2+q_4}^{\sigma, J}(h)^* \big|_{J=J_v^{\sigma}}$$

$$+ \text{remainder} \tag{8-100}$$

If the pole positions $J_v^{\sigma}$ are ordered according to a falling sequence of Re $J_v^{\sigma}$, the the series (8–100) can be regarded as an asymptotic expansion of $R_{q_i}(t, h)$ in $h$ at $|h| \to \infty$. If this series has a limit when the contours are shifted to Re $J = -\infty$, the remainder is necessarily a function of $h$ which drops off faster than any power of $|h|$.

The expansion (8–100) is related to the "Regge-Mandelstam expansion" of the scattering amplitude. Since each Regge pole gives rise to a term that is proportional to a representation function of the little group H, the analogy of the connection between elementary particles and irreducible representations of the inhomogeneous SL(2, C) group suggests the picture of a "virtual particle exchange" between the particles 1, 3 on one side and the particles 4, 2 on the other. In this picture Regge poles are the reflection of physical objects whose features are independent of the properties of the particles 1, 2, 3, 4 "coupled" to them. These objects are closely related to the resonances known to exist in strong interaction particle physics. Together with elementary particles they fit into a scheme on which an analysis of phenomenological scattering data can be based. The expansion in (8–100) serves, for example, as an analysis of the asymptotic behavior of the scattering amplitude. It is only in this practical respect that the meromorphy hypothesis is of great importance. For a general mathematical framework, cuts or accumulation points of poles or isolated essential singularities would not present unsurmountable barriers.

In the same context we make the assumption of factorization

$$(2J_v^{\sigma} + 1) \operatorname*{Res}_{J=J_v^{\sigma}} \mathscr{T}_{q_i}(t, \sigma, J) = \beta^{(v)}_{q_1 q_3}(t, \sigma)\beta^{(v)}_{q_4 q_2}(t, \sigma) \tag{8-101}$$

The hypothesis (8–101) expresses the independence of the "coupling" of the Regge pole to the particles 1, 3 from the "coupling" to the particles 2, 4. A Regge pole can be characterized by the same labels as an irreducible unitary representation of $F_+ \times T_4$. If $t$ varies, the Regge pole positions are assumed to move continuously with $t$ and never to reach infinity for finite $t$. The function which gives the position of the Regge pole dependent upon $t$ is called the "Regge trajectory."

## 8–8    TOLLER POLES

For equal masses

$$M_1 = M_3 = M_{13} \qquad M_2 = M_4 = M_{42}$$

it was shown in Section 8–6 that the distribution $R_{q_i}(t, h)$ on H tends to a distribution $R_{q_i}(0, h)$ on H, which is in one-to-one correspondence with a set of distributions $R_{S_{13}q_{13}S_{42}q_{42}}(f)$ on $F_+$. If we submit these distributions on $F_+$ to a Fourier analysis on $F_+$ and afterwards apply the results of Chapter 7, we can extract additional information about the Regge poles at $t = 0$. First we have to generalize the results of Chapters 6 and 7 in a straightforward fashion so that they apply to the groups $F_+$ and H.

For a bicovariant function $x_{j_1 q_1 j_2 q_2}(f)$ on $F_+$ (4–76) which is in $\mathscr{C}^{\infty}(F_+)$

we define Fourier transforms with (8–43), (8–62)

$$\mathscr{F}^*_{j_1 j_2}(\sigma, \chi) = \int \overline{x_{j_1 q_1 j_2 q_2}(f)} \, D^{\sigma; \chi}_{j_1 q_1 j_2 q_2}(f) \, d\mu(f)$$

$$\chi = (m, \rho) \tag{8-102}$$

as analytic functions of $\rho$. In addition we introduce the analytic function $\mathscr{F}_{j_1 j_2}(\sigma, \chi)$ of $\rho$, which is obtained by analytic continuation of the complex conjugate of the function (8–102) off the real $\sigma$-axis. The Fourier transform (8–102) is independent of $q_1$ and $q_2$. According to (4–85) we have

$$\mathscr{F}^*_{j_1 j_2}(\sigma, \chi) = \sum_q K^*_{j_1 j_2 q}(\sigma, \chi) \tag{8-103}$$

$$K^*_{j_1 j_2 q}(\sigma, \chi) = [8\pi(2j_1 + 1)(2j_2 + 1)]^{-1}$$

$$\times \left\{ \int_0^\infty d\eta \, \sinh^2 \eta \overline{x_{j_1 q j_2 q}(\exp \tfrac{1}{2}\eta\sigma_3)} \, d^\chi_{j_1 j_2 q}(\eta) \right.$$

$$\left. + \sigma \int_0^\infty d\eta \, \sinh^2 \eta \overline{x_{j_1 q j_2 q}(\Theta \exp \tfrac{1}{2}\eta\sigma_3)} \, d^\chi_{j_1 j_2 q}(\eta) \right\} \tag{8-104}$$

The inverse Fourier transformation can be read off (4–86)

$$\overline{x_{j_1 q_1 j_2 q_2}(f)} = \tfrac{1}{2} \sum_\sigma \sum_{m=-j}^{j} \int_{-\infty}^{+\infty} d\rho(\rho^2 + m^2) \mathscr{F}^*_{j_1 j_2}(\sigma, \chi) \overline{D^{\sigma; \chi}_{j_1 q_1 j_2 q_2}(f)}$$

$$= \sum_\sigma \sum_{m=-j}^{j} \int_{-\infty}^{+\infty} d\rho(\rho^2 + m^2) \mathscr{F}^*_{j_1 j_2}(\sigma, \chi) \overline{E^{\sigma; \chi}_{j_1 q_1 j_2 q_2}(f)} \tag{8-105}$$

with $j = \min\{j_1, j_2\}$. An alternative form to (8–105) is obtained from (8–103) and (4–87)

$$\overline{x_{j_1 q j_2 q}(\exp \tfrac{1}{2}\eta\sigma_3)} = \sum_\sigma \sum_{m=-j}^{j} \int_{-\infty}^{+\infty} d\rho(\rho^2 + m^2) K^*_{j_1 j_2 q}(\sigma, \chi) e^{\bar{\chi}}_{j_1 j_2 q}(\eta)$$

$$\tag{8-106}$$

$$\overline{x_{j_1 q j_2 q}(\Theta \exp \tfrac{1}{2}\eta\sigma_3)} = \sum_\sigma \sigma \sum_{m=-j}^{j} \int_{-\infty}^{+\infty} d\rho(\rho^2 + m^2) K^*_{j_1 j_2 q}(\sigma, \chi) e^{\bar{\chi}}_{j_1 j_2 q}(\eta)$$

Finally we give Parseval's formula for two functions which are both in $\mathscr{C}^\infty(F_+)$

and bicovariant of the same type

$$\int \overline{x_{j_1 q_1 j_2 q_2}(f)} y_{j_1 q_1 j_2 q_2}(f) \, d\mu(f)$$

$$= \frac{1}{2} \sum_{\sigma} \sum_{m=-j}^{j} \int_{-\infty}^{+\infty} d\rho(\rho^2 + m^2) \mathcal{F}^*_{j_1 j_2}(\sigma, \chi)_x \mathcal{F}_{j_1 j_2}(\sigma, \chi)_y \quad (8\text{-}107)$$

Next we extend the expansion (7–60) onto the group H. From (8–40) and (8–43) we have

$$E^{\sigma, \chi}_{j_1 q_1 j_2 q_2}(v) = \sum_{n=0}^{\infty} (-1)^{(1/2)m - q_2} V_n^{\chi}(j_1, q_1) W_n^{\chi}(j_2, q_2) E^{\sigma', -(i/2)\rho + n}_{q_1 q_2}(v)$$

for arbitrary $\sigma$ and $\sigma'$. For an argument $h = v\Theta'$ (8–34) we obtain instead from (8–40), (8–43)

$$E^{\sigma, \chi}_{j_1 q_1 j_2 q_2}(v\Theta')$$

$$= \sigma\sigma' \sum_{n=0}^{\infty} (-1)^{(1/2)m + q_2 + j_2 - \varepsilon} V_n^{\chi}(j_1, q_1) W_n^{\chi}(j_2, -q_2) E^{\sigma', -(i/2)\rho + n}_{q_1 q_2}(v\Theta')$$

The coefficient $W_n^{\chi}(j_2, -q_2)$ can be replaced by $W_n^{\chi}(j_2, q_2)$ through

$$W_n^{\chi}(j_2, -q_2) = (-1)^{(1/2)m - j_2 + n - 2q_2} W_n^{\chi}(j_2, q_2)$$

as can be inspected from (7–31), (7–52), and (7–53). This gives

$$E^{\sigma, \chi}_{j_1 q_1 j_2 q_2}(h) = \sum_{n=0}^{\infty} (-1)^{(1/2)m - q_2} V_n^{\chi}(j_1, q_1) W_n^{\chi}(j_2, q_2) E^{\sigma_n, -(i/2)\rho + n}_{q_1 q_2}(h)$$

Because of

$$E^{\sigma, J}_{q_1 q_2}(h)^* = E^{\sigma', -J-1}_{-q_1, -q_2}(h) \qquad \sigma' = (-1)^{2\varepsilon} \sigma$$

$$E^{\sigma, \chi}_{j_1 q_1 j_2 q_2}(h)^* = (-1)^{q_1 - q_2} E^{\sigma, -\chi}_{j_1, -q_1, j_2, -q_2}(h)$$

we have also

$$E^{\sigma, \chi}_{j_1 q_1 j_2 q_2}(h)^* = \sum_{n=0}^{\infty} (-1)^{(1/2)m + q_1} V_n^{-\chi}(j_1, -q_1)$$

$$\times W_n^{-\chi}(j_2, -q_2) E^{\sigma_n, -(i/2)\rho - n - 1}_{q_1 q_2}(h)^* \quad (8\text{-}108)$$

In both cases we have the signature $\sigma_n$

$$\sigma_n = \sigma(-1)^{(1/2)m - \varepsilon + n}$$

Concerning covariance the distribution $R_{S_{13}q_{13}S_{42}q_{42}}(f)$ resembles the function $\overline{x_{S_{13}q_{13}S_{42}q_{42}}(f)}$, whereas $\Psi'_{S_{13}q_{13}S_{42}q_{42}}(f)$ can be identified with a function $y_{S_{13}q_{13}S_{42}q_{42}}(f)$ (8–92), (8–107). In analogy with (8–107) therefore we make an ansatz

$$\int d\mu(f) R_{S_{13}q_{13}S_{42}q_{42}}(f) \Psi'_{S_{13}q_{13}S_{42}q_{42}}(f)$$

or

$$= \tfrac{1}{2} \sum_\sigma \sum_{m=-j}^{j} \int_{C_\tau} d\rho (\rho^2 + m^2) \mathscr{T}_{S_{13}S_{42}}(\sigma, \chi) \mathscr{F}_{S_{13}S_{42}}(\sigma, \chi)_{\Psi'} \qquad (8\text{-}109)$$

with

$$= \tfrac{1}{2} \sum_\sigma \sum_{m=-j}^{j} \sum_q \int_{C_{\sigma,q}} d\rho (\rho^2 + m^2) \mathscr{T}_{S_{13}S_{42}q}(\sigma, \chi) K_{S_{13}S_{42}q}(\sigma, \chi)_{\Psi'} \qquad (8\text{-}110)$$

$$\mathscr{T}_{S_{13}S_{42}}(\sigma, \chi) = \sum_q \mathscr{T}_{S_{13}S_{42}q}(\sigma, \chi)$$

(8–110) gives us somewhat more freedom in the choice of the contour of integration. In both (8–109) and (8–110) the Fourier transforms $\mathscr{T}_{S_{13}S_{42}}(\sigma, \chi)$ and $\mathscr{T}_{S_{13}S_{42}q}(\sigma, \chi)$ appear as analytic functionals. Moreover, we desire a representation of the kind corresponding to (8–105) or a representation of the kind

$$R_{S_{13}q_{13}S_{42}q_{42}}(f) = \sum_\sigma \sum_{m=-j}^{j} \int_{C_\sigma} d\rho (\rho^2 + m^2) \mathscr{T}_{S_{13}S_{42}}(\sigma, \chi) E^{\sigma,\chi}_{S_{13}q_{13}S_{42}q_{42}}(f)^*$$

(8-111)

(8–106). We can indeed verify the representations (8–109)–(8–111) under similar premises [which may depend somewhat on whether we prefer the form (8–109) or (8–110), etc.] as the meromorphy hypothesis of Section 8–7. Let us concentrate our interest on the integral representation (8–111).

As in Section 8–7 we use the method of multiplying the distribution first with the power $|f|^\lambda$ and then continuing in $\lambda$ analytically till $\lambda = 0$. If Re $\lambda$ is sufficiently big negative, the Fourier transform $\mathscr{T}^\lambda_{S_{13}S_{42}}(\sigma, \chi)$ is holomorphic in the strip

$$|\text{Im } \rho| < 2 \max\{j_1, j_2\} + 1$$

The inverse Fourier transformation is performed by means of the second formula (8–105). This integral is modified such that the $\varphi$ integration extends over a contour in the lower half $\sigma$-plane which leaves all poles of the $E$-function on its left side. The sum of the contributions of these poles cancels as we know from Section 4–7. Next we continue in $\lambda$ till $\lambda = 0$ assuming that no unavoidable pinch between any poles of $\mathscr{T}^\lambda_{S_{13}S_{42}}(\sigma, \chi)$ or between a pole of this function and a pole of the $E$-function occurs. In this fashion we arrive at the contour $C_\sigma$.

Since the $E^*$-function tends to zero for $\rho \to -i\infty$, we can try to shift the contours $C_\sigma$ to $-i\infty$. After a finite displacement of the contours the poles of $\mathcal{T}_{S_{13}S_{42}}(\sigma, \chi)$, which we denote "Toller poles," give rise to a series

$$R_{S_{13}q_{13}S_{42}q_{42}}(h)$$

$$= \sum_\sigma \sum_{m=-j}^{j} \sum_{\nu=0}^{N_{\sigma,m}} ((\rho_\nu^{\sigma,m})^2 + m^2)(-2\pi i) \operatorname*{Res}_{\rho=\rho_\nu^{\sigma,m}} \mathcal{T}_{S_{13}S_{42}}(\sigma, \chi)$$

$$\times E_{S_{13}q_{13}S_{42}q_{42}}^{\sigma,\chi}(h)^*|_{\rho=\rho_\nu^{\sigma,m}} \tag{8-112}$$

which eventually converges if the contours tend to $-i\infty$. This "Toller expansion" (8–112) can be understood as an asymptotic expansion of the function $R_{S_{13}q_{13}S_{42}q_{42}}(h)$ on H at $|h| \to \infty$, if its terms are ordered according to a falling sequence of $\operatorname{Im}\rho_\nu^{\sigma,m}$. The Regge-Mandelstam expansion (8–100) of $R_{q_i}(t, h)$, which is supposed to be valid till the limit point $t = 0$, is a similar asymptotic expansion but in terms of another system of functions. We can connect both expansions by means of (8–108). As the result we obtain that each Toller pole decomposes into a family of integrally spaced Regge poles with alternating signature

$$J_n = -\frac{i}{2}\rho - 1 - n$$

$$\sigma_n = \sigma(-1)^{(1/2)m-\varepsilon+n} \qquad n = 0, 1, 2, \ldots$$

In analogy with the physical picture attributed to the Regge poles, one assumes that the residue of the Toller poles factorizes into two "coupling constants"

$$(-2\pi i)(\rho_\nu^2 + m^2)\operatorname*{Res}_{\rho=\rho_\nu} \mathcal{T}_{S_{13}S_{42}}(\sigma, \chi) = \beta_{S_{13}}^{(\nu)}(\sigma, m)\beta_{S_{42}}^{(\nu)}(\sigma, m) \tag{8-113}$$

In addition, the "quantum numbers" labeling the pole, that is, $\sigma, \rho, m$, are not to depend on the properties of the external particles, in our case on $S_{13}, M_{13}, S_{42}$, and $M_{42}$. Inserting (8–108) and (8–113) into (8–112) and comparing with (8–100), (8–101) yields

$$\beta_{q_1q_3}^{(n)}(0, \sigma_n)\beta_{q_4q_2}^{(n)}(0, \sigma_n)$$

$$= \left\{ \sum_{S_{13}} (-1)^{(1/2)m+q_1+q_3}(S_1S_3 S_{13}, q_1 + q_3 | S_1q_1S_3 q_3) \right.$$

$$\left. \times V_n^{-\chi}(S_{13}, -q_1 - q_3)\beta_{S_{13}}(\sigma, m) \right\}$$

$$\times \left\{ \sum_{S_{42}} (S_4 q_4 S_2 q_2 | S_4 S_2 S_{42}, q_2 + q_4)W_n^{-\chi}(S_{42}, -q_2 - q_4) \right.$$

$$\times \, \beta_{S_{42}}(\sigma, m) \Bigg\} \tag{8-114}$$

The factorization of the residue of the Toller pole induces the factorization of the residues of the Regge poles of the corresponding family.

## 8–9 REMARKS

Our presentation of the group theoretical approach to Regge and Toller poles follows the articles of Toller [33, 34]. The classic result of distribution theory referred to in Section 8–1 can be found in Gelfand *et al.* [12 (Volume 4, Chapter I, Theorem 5′)]. The decomposition of symmetric Banach algebras used in the derivation of Schur's lemma is treated in Naimark [25], in particular see his §41, Section 3. For the Clebsch-Gordan coefficients of SU(2) we took over Edmond's notation [8 (Chapter III)].

It should be mentioned that the group theoretical approach deals only with a rather limited aspect of the theory of Regge and Toller poles, and that a complete treatment necessitates use of analyticity properties of the $M$-functions. We believe, however, that the approach based on invariance considerations adds significantly to the physical picture attributed to the Regge and Toller poles. In an approach based on analyticity we can in particular abandon the restriction of equality of the masses which enabled us to deduce the Toller poles. Such restriction is in fact in contradiction with independence of the Regge and Toller poles of the properties of the external particles. The literature on the analytic properties of the $M$-functions in connection with Regge and Toller poles is so extensive, that we refrain from giving any detailed reference.

# Appendix

The problem of decomposing an invariant measure on a matrix group G into an invariant measure on a subgroup H times a measure on the cosets G/H appears so often that we thought it useful to accumulate the computations of these and similar Jacobians in a special appendix. Each problem is stated in shorthand notation in the heading, but written out in explicit matrix form in the text. We treat Jacobians for the group $G = SL(2, C)$ in Appendix A–1, for the group $G = SL(2, R)$ in Appendix A–2, and for the group $F_{\pm} \times T_4$ in Appendix A–3.

## A–1  JACOBIANS FOR THE GROUP SL(2, C)

In this case the variables $a_{ij}$, $\lambda$, $\mu$, $z$ are complex, and $u$ denotes a unitary unimodular matrix.

*a.*  $a = ku$

The decomposition

$$\begin{pmatrix} a_{11} & a_{12} \\ a_{21} & a_{22} \end{pmatrix} = \begin{pmatrix} \lambda^{-1} & \mu \\ 0 & \lambda \end{pmatrix} \begin{pmatrix} u_{11} & u_{12} \\ u_{21} & u_{22} \end{pmatrix}$$

leaves the phase of $\lambda$ undermined. The corresponding splitup of the measure

$$d\mu(a) = J \, d\mu_l(k) \, d\mu(u)$$

makes sense only if the function to be integrated is independent of the phase of $\lambda$. With

$$d\mu(a) = (2\pi)^{-4}|a_{22}|^{-2} \, Da_{12} \, Da_{21} \, Da_{22} \tag{1-20}$$

$$d\mu_l(k) = (2\pi)^{-3}\lambda \, d\lambda \, D\mu \qquad \lambda > 0 \tag{4-10}$$

$$d\mu(u) = (2\pi)^{-2} \, d\theta_1 \, d\theta_2 \, dt \tag{3-16}$$

$$\theta_1 = \arg u_{11} \qquad \theta_2 = \arg u_{12} \qquad t = |u_{11}|^2 \tag{3-15}$$

and

$$a_{21} = \lambda u_{21} \qquad a_{22} = \lambda u_{22}$$

$$Da_{21}\, Da_{22} = \tfrac{1}{2}\lambda^2\lambda\; d\lambda\; d\theta_1\; d\theta_2\; d|u_{22}|^2$$

$$a_{12} = \lambda^{-1}u_{12} + \mu u_{22}$$

$$Da_{12} = |u_{22}|^2\; D\mu$$

we get the Jacobian

$$J = (2\pi)\tfrac{1}{2}\lambda^2|u_{22}|^2|a_{22}|^{-2} = \pi$$

b.   $a = k\zeta$

The decomposition

$$\begin{pmatrix} a_{11} & a_{12} \\ a_{21} & a_{22} \end{pmatrix} = \begin{pmatrix} \lambda^{-1} & \mu \\ 0 & \lambda \end{pmatrix}\begin{pmatrix} 1 & 0 \\ z & 1 \end{pmatrix}$$

is unique. We compute the Jacobian in

$$d\mu(a) = J\, d\mu_l(k)\; Dz$$

From

$$d\mu(a) = (2\pi)^{-4}|a_{22}|^{-2}\; Da_{12}\, Da_{21}\, Da_{22} \qquad (1\text{-}20)$$

$$d\mu_l(k) = (2\pi)^{-4}\; D\lambda\; D\mu \qquad (4\text{-}10)$$

and

$$a_{12} = \mu \qquad a_{21} = \lambda z \qquad a_{22} = \lambda$$

follows

$$J = 1$$

c.   $a = \zeta^{-1}k\zeta$

The decomposition

$$\begin{pmatrix} a_{11} & a_{12} \\ a_{21} & a_{22} \end{pmatrix} = \begin{pmatrix} 1 & 0 \\ -z & 1 \end{pmatrix}\begin{pmatrix} \lambda^{-1} & \mu \\ 0 & \lambda \end{pmatrix}\begin{pmatrix} 1 & 0 \\ z & 1 \end{pmatrix}$$

leads to

$$a_{11} + a_{22}, = \lambda + \lambda^{-1}$$

$$a_{11} - a_{22} = -\lambda + \lambda^{-1} + 2\mu z$$

$$a_{12} = \mu$$

that is, the parameters $\lambda, \mu, z$ and

$$\lambda' = \lambda^{-1} \qquad \mu' = \mu \qquad z' = z - \mu^{-1}(\lambda - \lambda^{-1})$$

correspond to the same element of SL(2, C). In order to make the parameters $\lambda, \mu, z$ unique, we restrict $\lambda$ to $|\lambda| > 1$. With

$$d\mu(a) = J\,\theta\,(|\lambda| - 1)\,d\mu_l(k)\,Dz$$

and

$$d\mu(a) = (2\pi)^{-4}|a_{12}|^{-2}\,Da_{11}\,Da_{12}\,Da_{22} \qquad (1\text{-}20)$$

$$d\mu_l(k) = (2\pi)^{-4}\,D\lambda\,D\mu \qquad (4\text{-}10)$$

we get

$$J = |a_{12}|^{-2}\left|\frac{\partial(a_{11}, a_{22})}{\partial(\lambda, z)}\right|^2 = |\lambda|^{-2}\,|\lambda - \lambda^{-1}|^2$$

d.  $ud = ku_d$

We make the matrix $k$ in

$$\begin{pmatrix} u_{11} & u_{12} \\ u_{21} & u_{22} \end{pmatrix}\begin{pmatrix} e^{\eta/2} & 0 \\ 0 & e^{-\eta/2} \end{pmatrix} = \begin{pmatrix} \lambda^{-1} & \mu \\ 0 & \lambda \end{pmatrix}\begin{pmatrix} (u_d)_{11} & (u_d)_{12} \\ (u_d)_{21} & (u_d)_{22} \end{pmatrix}$$

unique by the postulate

$$(u_d)_{22} \geqq 0$$

The task is to compute the Jacobian in

$$d\mu_l(k) = J\,d\mu(u)\,d\eta \qquad \eta \geqq 0$$

We have

$$d\mu_l(k) = (2\pi)^{-4}\,D\lambda\,D\mu \qquad (4\text{-}10)$$

$$\theta_1 = \arg u_{11} \qquad \theta_2 = \arg u_{12} \qquad t = |u_{11}|^2 \qquad (3\text{-}15)$$

$$d\mu(u) = (2\pi)^{-2}\,d\,\theta_1\,d\,\theta_2\,dt \qquad (3\text{-}16)$$

and

$$|\lambda|^2 = e^\eta - 2t\sinh\eta \qquad \arg\lambda = -\theta_1$$

$$|\mu|^2 = 4t(1 - t)|\lambda|^{-2}\sinh^2\eta \qquad \arg\mu = \theta_2 - \pi$$

We get

$$\left| \frac{\partial(|\lambda|^2, |\mu|^2)}{\partial(t, \eta)} \right| = 4 \sinh^2 \eta$$

and finally

$$J = (2\pi)^{-2} \sinh^2 \eta$$

This result and the result of Appendix A−1a allow us to deduce that for

$$a = u_1 \, du_2 \qquad \eta \geqq 0$$

the measure decomposes into

$$d\mu(a) = \frac{1}{4\pi} \, d\mu(u_1) \, d\mu(u_2) \sinh^2 \eta \, d\eta$$

## A−2   JACOBIANS FOR THE GROUP SL(2, R)

The varibles $a_{ij}$, $\lambda, \mu, z$ are now real, and $u$ is an orthogonal unimodular matrix. In addition we make use of a Hermitian positive matrix $s(w)$ in both SL(2, R) and SU(1, 1). In the parameters $\alpha$ and $\beta$ (5−69) the measure (1−23) takes the form

$$d\mu(a) = \pi^{-2} \, \delta(|\alpha|^2 - |\beta|^2 - 1) D\alpha \, D\beta$$

a.   $a = ku$

In the decomposition

$$\begin{pmatrix} a_{11} & a_{12} \\ a_{12} & a_{22} \end{pmatrix} = \begin{pmatrix} \lambda^{-1} & \mu \\ 0 & \lambda \end{pmatrix} \begin{pmatrix} \cos \tfrac{1}{2}\psi & -\sin \tfrac{1}{2}\psi \\ \sin \tfrac{1}{2}\psi & \cos \tfrac{1}{2}\psi \end{pmatrix}$$

the sign of $\lambda$ is undetermined. The corresponding factorization of the measure

$$d\mu(a) = J \, d\mu_l(k) \, d\psi$$

makes sense therefore only if the function to be integrated does not depend on this sign. We put

$$d\mu(a) = (2\pi)^{-2}|a_{22}|^{-1} \, da_{12} \, da_{21} \, da_{22} \tag{1-23}$$

$$d\mu_l(k) = 2(2\pi)^{-2} \, d\lambda \, d\mu \qquad \lambda > 0$$

and have

$$a_{21} = \lambda \sin \tfrac{1}{2}\psi \qquad a_{22} = \lambda \cos \tfrac{1}{2}\psi$$

$$a_{12} = -\lambda^{-1} \sin \tfrac{1}{2}\psi + \mu \cos \tfrac{1}{2}\psi$$

$$da_{21}\, da_{22} = \tfrac{1}{2}\lambda\, d\lambda\, d\psi \qquad da_{12} = |\cos \tfrac{1}{2}\psi|\, d\mu$$

This yields

$$J = \frac{1}{4}$$

b.   $a = u^{-1} k u$

From

$$\begin{pmatrix} a_{11} & a_{12} \\ a_{21} & a_{22} \end{pmatrix} = \begin{pmatrix} \cos \tfrac{1}{2}\psi & +\sin \tfrac{1}{2}\psi \\ -\sin \tfrac{1}{2}\psi & \cos \tfrac{1}{2}\psi \end{pmatrix} \begin{pmatrix} \lambda^{-1} & \mu \\ 0 & \lambda \end{pmatrix} \begin{pmatrix} \cos \tfrac{1}{2}\psi & -\sin \tfrac{1}{2}\psi \\ +\sin \tfrac{1}{2}\psi & \cos \tfrac{1}{2}\psi \end{pmatrix}$$

follows

$$\mathrm{Tr}\, a = a_{11} + a_{22} = \lambda + \lambda^{-1}$$

$$\lambda = \tfrac{1}{2}[\mathrm{Tr}\, a \pm [(\mathrm{Tr}\, a)^2 - 4]^{1/2}]$$

$\lambda$ is real only for

$$|\mathrm{Tr}\, a| \geq 2$$

Because of the ambiguity with respect to $\lambda$ and $\lambda^{-1}$ and the indeterminacy of $\psi$, we restrict the measures as

$$\theta(|\mathrm{Tr}\, a| - 2)d\mu(a) = J\,\theta\,(|\lambda| - 1)d\mu_l(k)\, d\psi$$

respectively,

$$\theta(|\mathrm{Tr}\, a| - 2)d\mu(a) = J\,\theta\,(1 - |\lambda|)d\mu_l(k)\, d\psi$$

where $\psi$ is restricted to the interval $0 \leq \psi \leq 2\pi$.

We have

$$d\mu(a) = \pi^{-2}|2a_1|^{-1}\, da_0\, da_2\, da_3 \qquad (1\text{-}23)$$

$$d\mu_l(k) = (2\pi)^{-2}\, d\lambda\, d\mu$$

and

$$2a_0 = a_{11} + a_{22} = \lambda + \lambda^{-1}$$

$$2a_1 = a_{12} + a_{21} = \mu \cos \psi + (\lambda - \lambda^{-1}) \sin \psi$$

$$2a_2 = a_{12} - a_{21} = \mu$$

$$2a_3 = a_{11} - a_{22} = \mu \sin \psi - (\lambda - \lambda^{-1}) \cos \psi$$

Because of the invariance of $d\mu(a)$ and $d\psi$ with respect to multiplication with factors $u$, the Jacobian is independent of $\psi$,

$$J = \left| \frac{2}{a_1} \frac{\partial(a_0, a_2, a_3)}{\partial(\lambda, \mu, \psi)} \right|_{\psi=0} = \left| \frac{2}{a_1} \frac{\partial a_0}{\partial \lambda} \frac{\partial a_2}{\partial \mu} \frac{\partial a_3}{\partial \psi} \right|_{\psi=0}$$

$$= \left| \frac{4}{\mu} \frac{1}{2} (1 - \lambda^{-2}) \frac{1}{2} \frac{1}{2} \mu \right| = \frac{1}{2} |\lambda|^{-1} |\lambda - \lambda^{-1}|$$

c.   $a = k\zeta$

From

$$\begin{pmatrix} a_{11} & a_{12} \\ a_{21} & a_{22} \end{pmatrix} = \begin{pmatrix} \lambda^{-1} & \mu \\ 0 & \lambda \end{pmatrix} \begin{pmatrix} 1 & 0 \\ z & 1 \end{pmatrix}$$

and

$$d\mu(a) = J \, d\mu_l(k) \, dz$$

$$d\mu(a) = (2\pi)^{-2} |a_{22}|^{-1} \, da_{12} \, da_{21} \, da_{22} \tag{1-23}$$

$$d\mu_l(k) = (2\pi)^{-2} \, d\lambda \, d\mu$$

follows

$$a_{12} = \mu \qquad a_{21} = \lambda z \qquad a_{22} = \lambda$$

and

$$J = 1$$

d.   $a = us(w)$

With the standard isomorphism (5–3), (5–90), (5–91) we bring this decomposition into the form

$$\begin{pmatrix} \bar{\alpha} & -i\bar{\beta} \\ i\beta & \alpha \end{pmatrix} = \begin{pmatrix} e^{(i/2)\psi} & 0 \\ 0 & e^{-(i/2)\psi} \end{pmatrix}(1 - |w|^2)^{-1/2}\begin{pmatrix} 1 & i\bar{w} \\ -iw & 1 \end{pmatrix}$$

$w$ and $\psi$ are unique. The Jacobian in

$$d\mu(a) = J \; Dw \; d\psi$$

is independent of $\psi$. From

$$\alpha = (1 - |w|^2)^{-1/2}e^{-(i/2)\psi} \qquad \beta = -(1 - |w|^2)w\,e^{-(i/2)\psi}$$

we obtain

$$J = \pi^{-2}\left.\frac{1}{2 \operatorname{Im}\beta}\frac{\partial(\operatorname{Re}\alpha, \operatorname{Im}\alpha, \operatorname{Re}\beta)}{\partial(\operatorname{Re}w, \operatorname{Im}w, \psi)}\right|_{\psi=0}$$

$$= (2\pi)^{-2}(1 - |w|^2)^{-2}$$

e.    $a = kuk^{-1}$

From

$$\begin{pmatrix} a_{11} & a_{12} \\ a_{21} & a_{22} \end{pmatrix} = \begin{pmatrix} \lambda^{-1} & \mu \\ 0 & \lambda \end{pmatrix}\begin{pmatrix} \cos\frac{1}{2}\psi & -\sin\frac{1}{2}\psi \\ \sin\frac{1}{2}\psi & \cos\frac{1}{2}\psi \end{pmatrix}\begin{pmatrix} \lambda & -\mu \\ 0 & \lambda^{-1} \end{pmatrix}$$

follows

$$a_{11} = \cos\tfrac{1}{2}\psi + \lambda\mu \sin\tfrac{1}{2}\psi$$

$$a_{12} = -(\lambda^{-2} + \mu^2) \sin\tfrac{1}{2}\psi$$

$$a_{21} = \lambda^2 \sin\tfrac{1}{2}\psi$$

$$a_{22} = \cos\tfrac{1}{2}\psi - \lambda\mu \sin\tfrac{1}{2}\psi$$

The sign of $\lambda$ is ambiguous. The Jacobian in

$$\theta(2 - |\operatorname{Tr} a|) \, d\mu(a) = J \, \theta(\lambda) \, d\mu_l(k) \, d\psi$$

is independent of $k$ and can therefore be evaluated at $\lambda = 1$ and $\mu = 0$. We get

$$J = \left.\left|\frac{1}{a_{11}}\frac{\partial(a_{11}, a_{12}, a_{21})}{\partial(\lambda, \mu, \psi)}\right|\right|_{\substack{\lambda=1 \\ \mu=0}}$$

$$= 2 \sin^2 \tfrac{1}{2}\psi$$

*f.*   $ud = ku_d$

We make the matrix $k$ in

$$\begin{pmatrix} \cos\tfrac{1}{2}\psi & -\sin\tfrac{1}{2}\psi \\ \sin\tfrac{1}{2}\psi & \cos\tfrac{1}{2}\psi \end{pmatrix} \begin{pmatrix} e^{\eta/2} & 0 \\ 0 & e^{-\eta/2} \end{pmatrix} = \begin{pmatrix} \lambda^{-1} & \mu \\ 0 & \lambda \end{pmatrix} \begin{pmatrix} \cos\tfrac{1}{2}\psi_d & -\sin\tfrac{1}{2}\psi_d \\ \sin\tfrac{1}{2}\psi_d & \cos\tfrac{1}{2}\psi_d \end{pmatrix}$$

unique by requiring

$$\cos\tfrac{1}{2}\psi_d \geqq 0$$

We compute the Jacobian $J$ in

$$d\mu_l(k) = (2\pi)^{-2}\, d\lambda\, d\mu$$
$$= J\, d\psi\, d\eta \qquad \eta \geqq 0$$

From

$$\lambda^2 = \cosh\eta - \sinh\eta\cos\psi \qquad \lambda\mu = \sinh\eta\sin\psi$$

follows

$$J = \frac{1}{8\pi^2\lambda^2} \left| \frac{\partial(\lambda^2, \lambda\mu)}{\partial(\psi, \eta)} \right| = \frac{1}{8\pi^2}\sinh\eta$$

This result and the result of Appendix A–2a yield for

$$a = u(\psi_1)\, du(\psi_2) \qquad 0 \leq \psi_{1,2} \leq 4\pi \qquad \eta \geqq 0$$

the measure

$$d\mu(a) = (4\pi)^{-2}\, d\psi_1\, d\psi_2\, \tfrac{1}{2}\sinh\eta\, d\eta$$

## A–3   JACOBIANS FOR INHOMOGENEOUS GROUPS

Our task is to reduce the product measure on two orbits for the group $F_\pm \times T_4$ [the group $SL(2, C) \times T_4$ can be treated in the same fashion] into the measure on the orbit of the relative momentum times the invariant measure on the little group. We perform this reduction in two steps.

*a.    The Transformation of Two Momenta to Their Center of Momentum Frame*

We consider the problem

$$d\mu_{M_1}(p_1)\, d\mu_{M_2}(p_2) = J_{12}(t)\, dt\, d\mu(f)$$

where

$$p_{1,2} = p_{1,2}^R \Lambda(f_{1,2}) \qquad f_{1,2} = u_{1,2}\, b_{1,2}(t) f$$

$$u_{1,2} \in \mathrm{SU}(2) \qquad b_1(t) = d(\eta_1) \qquad b_2(t) = d(\eta_2)\Theta$$

$d(\eta)$ as in (3–22), $\eta_1(t)$ and $\eta_2(t)$ are determined by

$$p_1{}^R \Lambda(d(\eta_1)) - p_2{}^R \Lambda(d(\eta_2)) = Q_t{}^R$$

The measures $d\mu_M(p)$ and $d\mu(f)$ were defined in (8–22) and (8–62), respectively. We have to find $\eta_{1,2}$ first.

In the case $t > 0$ the equations determining $\eta_1$ and $\eta_2$ are solved by

$$\cosh \eta_1 = (2M_1\, t^{1/2})^{-1}(t + M_1{}^2 - M_2{}^2)\operatorname{sign}(M_1 - M_2)$$

$$\cosh \eta_2 = (2M_2\, t^{1/2})^{-1}(-t + M_1{}^2 - M_2{}^2)\operatorname{sign}(M_1 - M_2)$$

If $t < 0$ we have instead

$$\sinh \eta_1 = (2M_1|t|^{1/2})^{-1}(-t - M_1{}^2 + M_2{}^2)$$

$$\sinh \eta_2 = (2M_2|t|^{1/2})^{-1}(t - M_1{}^2 + M_2{}^2)$$

The point $t = 0$ is of no interest for us. The elements $b_{1,2}(t)$ are discontinuous in $t$ at $t = 0$.

In order to compute $J_{12}(t)$ we make use of the invariance of the measures on the orbits. We may set $u_1 = u_2 = e$ and restrict $f$ to a neighborhood of $f = e$,

$$f = e + \sum_{k=1}^{3} a_k \sigma_k + \text{terms of second order}$$

The Jacobian may be evaluated at $f = e$. A left factor $u(\psi)$ in $f$ is undetermined and may be put into $u_{1,2}$. We have then

$$d\mu(f) = \frac{1}{4\pi^3}\, Da_1\, Da_2\, da_3 \qquad a_3 \text{ real at } f = e$$

and the measure on the orbit

$$d\mu_M(p) = \frac{1}{2M \cosh \eta} \, dp_1 \, dp_2 \, dp_3 \quad \text{at} \quad p = p^R \Lambda(d(\eta))$$

The dependence of $p$ on $a$ up to first order is easily computed from

$$\mathbf{p} = Ma^{\dagger}(\cosh \eta e + \sinh \eta \sigma_3)a$$

This gives

$$J_{12}(t) = \frac{\pi^3}{M_1 M_2 \cosh \eta_1 \cosh \eta_2} \left| \frac{\partial((p_1)_1, (p_1)_2, (p_1)_3, (p_2)_1, (p_2)_2, (p_2)_3)}{\partial(t, a_3, \text{Re } a_1, \text{Re } a_2, \text{Im } a_1, \text{Im } a_2)} \right|_{a=e}$$

After some elementary algebra we obtain

$$J_{12}(t) = (2\pi)^3 \, \Delta(t, M_1{}^2, M_2{}^2)^{1/2}$$

for the interval $-\infty < t < (M_1 - M_2)^2$, with $\Delta(a, b, c)$ defined by

$$\Delta(a, b, c) = a^2 + b^2 + c^2 - 2(ab + ac + bc)$$

### b. The Decomposition of the Homogeneous Group into Little Group and Boosts

We consider the decomposition

$$f = ha(Q)$$

and the corresponding Jacobian

$$d\mu(f) = J(t) \, d\mu(h) \, d\mu_t(Q)$$

for $t < 0$, that is, $h \in H$. The measures $d\mu(f)$, $d\mu(h)$, and $d\mu_t(Q)$ were defined in (8–62), (8–65), and (8–82), respectively. Because of the invariance of these measures and for dimensional reasons we have

$$J(t) = c|t|^{-1}$$

with a constant number $c$.

As in Appendix A–3a we may consider neighborhoods of $f = e$, $h = e$, and

$Q = Q_t^R$ only and compute the Jacobian at this point. We have

$$f = e + \sum_{k=1}^{3} a_k \sigma_k + \text{terms of second order}$$

$$d\mu(f) = \frac{1}{8\pi^4} Da_1\, Da_2\, Da_3 \qquad \text{at} \quad f = e$$

$$h = e + \sum_{k=1,2} v_k \sigma_k + iv_3 \sigma_3 + \text{terms of second order}$$

$$d\mu(h) = \frac{1}{4\pi^2} dv_1\, dv_2\, dv_3 \qquad \text{at} \quad h = e$$

$$\mathbf{Q} = Q_0 e + Q_1 \sigma_1 + Q_2 \sigma_2 + |t|^{1/2} \sigma_3 + \text{terms of second order}$$

$$d\mu_t(Q) = \frac{1}{2\,|t|^{1/2}} dQ_0\, dQ_1\, dQ_2 \qquad \text{at} \quad Q = Q_t^R$$

In the environment of $Q = Q_t^R$ we have the boost

$$a(Q) = (|t|^{-1/2}\sigma_3\, \mathbf{Q})^{1/2}$$

$$= e + \tfrac{1}{2}|t|^{-1/2}(Q_0\, \sigma_3 + iQ_1\, \sigma_2 - iQ_2\, \sigma_1) + \text{terms of second order}$$

and therefore to first order

$$a_1 = v_1 - iQ_2 \tfrac{1}{2}|t|^{-1/2} \qquad a_2 = v_2 + iQ_1 \tfrac{1}{2}|t|^{-1/2} \qquad a^3 = iv_3 + Q_0 \tfrac{1}{2}|t|^{-1/2}$$

This implies immediately

$$c = \frac{1}{8\pi^2}$$

# References

1. Akyeampong, D. A., Boyce, J. F., and Rashid, M. A., "On the Transformation from $O_3$ to $O_{2,1}$ Bases in SL(2, C)," *Nuovo Cimento* **53**, 737 (1968).
2. Akyeampong, D. A., Boyce, J. F., and Rashid, M. A., "On the Relation between the O(4) and O(3, 1) Partial Wave Expansions," International Centre for Theoretical Physics preprint, IC/67/74 (1967).
3. Andrews, M., and Gunson, J., "Complex Angular Momentum and Many-Particle States. I. Properties of Local Representations of the Rotation Group," *J. Math. Phys.* **5**, 1391 (1964).
4. Bade, W. L., and Jehle, H., "An Introduction to Spinors," *Rev. Mod. Phys.* **25**, 714 (1953).
5. Bargmann, V., "Irreducible Unitary Representations of the Lorentz Group," *Ann. of Math.* **48**, 568 (1947).
6. Bargmann, V., "On Unitary Ray Representations of Continuous Groups," *Ann. of Math.* **59**, 1 (1954).
7. Bargmann, V., and Wigner, E. P., "Group Theoretical Discussion of Relativistic Wave Equations," *Proc. Natl. Acad. Sci. U.S.* **34**, 211 (1948).
8. Edmonds, A. R., *Angular Momentum in Quantum Mechanics*, Princeton University Press, Princeton, 1957.
9. Ehrenpreis, L., and Mautner, F. I., "Some Properties of the Fourier Transform on Semi-Simple Lie Groups I," *Ann. of Math.* **61**, 406 (1955).
10. Erdélyi, A., Magnus, W., Oberhettinger, F., and Tricomi, F. G., *Higher Transcendental Functions*, Vol. 1, McGraw-Hill, New York, 1953.
11. Gelfand, I. M., and Naimark, M. A., *Unitäre Darstellungen der klassischen Gruppen*, Akademie-Verlag, Berlin, 1957.
12. Gelfand, I. M., and Shilov, G. E., *Generalized Functions*, 5 volumes: Vol. 4 with N. Ya. Vilenkin; Vol. 5 with M. I. Graev and N. Ya. Vilenkin. Academic Press, New York, 1964, to appear, 1967, 1964, 1966, respectively.
13. Gelfand, I. M., and Yaglom, A. M., "General Relativistic-Invariant Equations and Infinite Dimensional Representations of the Lorentz Group (in Russian)," *Zh. Eksp. Teor. Fiz.* **18**, 703 (1948).
14. Gelfand, I. M., and Yaglom, A. M., "Pauli's Theorem for General Relativistic-Invariant Equations (in Russian)," *Zh. Eksp. Teor. Fiz.* **18**, 1096 (1948).

293

15. Gradshteyn, I. S., and Ryzhik, I. M., *Table of Integrals, Series, and Products,* Academic Press, New York, 1965.

16. Harish-Chandra, "Plancherel Formula for the $2 \times 2$ Real Unimodular Group," *Proc. Natl. Acad. Sci. U.S.* **38**, 337 (1952).

17. Jacob, M., and Wick, G. C., "On the General Theory of Collisions for Particles with Spin," *Ann. Phys. (N.Y.)* **7**, 404 (1959).

18. Joos, H., "Zur Darstellungstheorie der inhomogenen Lorentzgruppe als Grundlage quantenmechanischer Kinematik," *Fortschr. Phys.* **10**, 65 (1962).

19. Mackey, G. W., "Induced Representations of Locally Compact Groups I," *Ann. of Math.* **55**, 101 (1952).

20. Mackey, G. W., "Induced Representations of Locally Compact Groups II, The Frobenius Reciprocity Theorem," *Ann. of Math.* **58**, 193 (1953)

21. Mackey, G. W., *The Theory of Group Representations, Lecture Notes,* The University of Chicago, Chicago, 1955.

22. Mackey, G. W., "Infinite Dimensional Group Representations," *Bull. Am. Math. Soc.* **69**, 628 (1963).

23. Majorana, E., "Teoria relativistica di Particelle con Momento intrinseco arbitrario," *Nuovo Cimento* **9**, 335 (1932).

24. Naimark, M. A., *Linear Representations of the Lorentz Group*, Pergamon Press, Oxford, 1964.

25. Naimark, M. A., *Normed Rings*, Noordhoff, Groningen, 1964.

26. Naimark, M. A., "Decomposition of a Tensor Product of Irreducible Representations of the Proper Lorentz Group into Irreducible Representations" (3 Parts), *Trans. Amer. Math. Soc.* Series 2, **36**, 101, 137, 189 (1964).

27. Pontrjagin, L., *Topological Groups*, Princeton University Press, Princeton, 1939.

28. Rühl, W., "Complete Sets of Solutions of Linear Lorentz Covariant Field Equations with an Infinite Number of Field Components," *Comm. Math. Phys.* **6**, 312 (1967).

29. Rühl, W., "A Convolution Integral for Fourier Transforms on the Group SL(2, C)," *Comm. Math. Phys.* **10**, 199 (1968).

30. Sciarrino, A., and Toller, M., "Decomposition of Unitary Representations of the Group SL(2, C) Restricted to the Subgroup SU(1, 1)," *J. Math. Phys.* **8**, 1252 (1967).

31. Streater, R. F., and Wightman, A. S., *PCT, Spin and Statistics, and All That,* W. A. Benjamin, Inc., New York, 1964.

32. Titchmarsh, E. C., *Introduction to the Theory of Fourier Integrals,* Clarendon Press, Oxford, 1937.

33. Toller, M., "An Expansion of the Scattering Amplitude at Vanishing Four-Momentum Transfer Using the Representations of the Lorentz Group," *Nuovo Cimento* **53**, 671 (1968).

34. Toller, M., "On the Group-Theoretical Approach to Complex Angular Momentum and Signature," *Nuovo Cimento* **54**, 295 (1968).

35. Weinberg, S., "Feynman Rules for Any Spin, *Phys. Rev.* **133**, B 1318 (1964).

36. Wightman, A. S., "L'Invariance dans la Mécanique quantique relativiste," in *Relations de Dispersion et Particules élémentaires,* edited by C. De Witt, Herrmann, Paris, 1960.

37. Wigner, E. P., "On Unitary Representations of the Inhomogeneous Lorentz Group," *Ann. of Math.* **40**, 149 (1939).

38. Wigner, E. P., *Group Theory and Its Application to the Quantum Mechanics of Atomic Spectra,* Academic Press, New York, 1959.

39. Wigner, E. P., "Unitary Representations of the Inhomogeneous Lorentz Group Including Reflections," in *Group Theoretical Concepts and Methods in Elementary Particle Physics*, edited by F. Gürsey, Gordon and Breach, New York, 1964.

# Index